OILFIELDS OF THE WORLD

OILFIELDS OF THE WORLD

by E. N. Tiratsoo

Ph.D., D.I.C., B.Sc., A.R.S.M., F.G.S., F.Inst.Pet.

Consulting Geologist

First Edition, 1973

Scientific Press Ltd.
Beaconsfield, England

For S., J., N., C.

Printed in Great Britain by Creative Press (Reading) Ltd.

ISBN 0 901360 06 6

PREFACE

A BOUT 260 billion barrels* of crude oil have been produced from the world's oilfields in the period of 113 years that has elapsed since the drilling of Drake's first successful oil-well in 1859. In little more than a single century, therefore, mankind has dissipated with spendthrift lavishness much of the Earth's endowment of an irreplaceable mineral, for whose generation and accumulation many millions of years were needed.

To recover such a huge volume of liquid from depths of thousands of feet below ground level, often from locations situated in tropical forests or inhospitable deserts, has required the investment of immense amounts of capital, labour and—at least in recent years—the deployment of the most advanced technical skills. Yet, in the end, by far the largest proportion of the crude oil that has been obtained with so great an expenditure of effort has been converted into fuel of one sort or another and then destroyed by combustion. So attractive and convenient a source of flexible and easily transportable energy has oil proved to be, that it has always tended to displace the longer-established solid fuels wherever it has become available. World consumption of crude oil has in consequence roughly doubled with each decade that has elapsed since 1860, and there is as yet no sign of abatement in this remarkable rate of growth. Consumption in 1972 was 18·1 billion brl, compared with only 3·8 billion brl in 1950.

Until recently, there seemed little cause for more than superficial anxiety about the continuing availability of the supplies of crude oil which the world so lavishly destroyed each year in the course of releasing its energy content. In the past, the ratio between known oil reserves and annual consumption has generally grown in proportion to the exploration effort deployed—in a remarkable way, the greater the volume of oil that was consumed, the greater were the supplies proved to be still available. However, there is now a growing and disconcerting body of evidence to show that the end of this comfortable era is at last in sight. The new oilfields that are already needed if the huge forecast requirements of the future are to be met are proving increasingly difficult to find. On average, their discovery requires a more intense search, the use of increasingly sophisticated techniques, and the expenditure of more money per unit of production capacity than has generally been the case in the past. Furthermore, an increasing proportion of new oilfields is being discovered under the waters which cover the Continental Shelf—and progressively in deeper waters as the search moves outwards towards the slopes and rises of the continental margins. To drill a single well in an offshore area such as the North Sea may cost £2 million, while to develop a major oilfield there is likely to involve an expenditure of the order of

---

* *Approximately 36,000 million tons.*

£1000 per barrel/day of production capacity—or perhaps twenty times what it might cost in some oil-rich countries of the Middle East.

On the other hand, the continuing discussions between the international oil companies and the Middle East oil-exporting states have now made it clear that crude oil from that area, which provided one-third of world output in the early 1970's and will probably provide at least one-half by 1980, must inevitably become increasingly expensive, so that the major oil-importing developed regions—the United States, Western Europe and Japan—are likely to be faced with ever-increasing energy costs.

It is therefore perhaps timely to make an inventory of the known oilfields of the world and to attempt to arrive at a balanced picture of the actual and potential sources of oil supply. Of course, it is clearly impossible within the confines of a single volume to mention more than a small proportion of the roughly 30,000 "significant" oil accumulations that have been discovered. However, an attempt has been made to present the most important facts about the geology, geography and production history of most of the world's major oilfields, so that, with this background of the past and present reasonably firmly established, it may then be possible to consider future prospects with a clearer view.

E.N.T.

ACKNOWLEDGEMENTS

A book which attempts to describe the oilfields of nearly 70 countries must inevitably draw on the work and publications of many authors, and to all of these I am therefore greatly indebted. Wherever possible, I have attempted to give due credit by citing the sources to which I have referred. The most important and numerous of these are the *Bulletin* and other publications of the American Association of Petroleum Geologists. I am also much indebted to the Association for permission to reproduce several diagrams, which are duly acknowledged. Other organizations which have kindly given permission for the reproduction of diagrams or illustrations include The British Petroleum Co Ltd, Institute of Petroleum Information Service, Petroleum Press Service, Niedersächsisches Landesamt für Bödenforschung, and the US Geological Survey. Dr H. V. Dunnington was kind enough to supply the excellent map of the Middle East oilfields.

For statistical data I have consulted the best available published sources, and in particular the annual compilations of *World Oil,* Houston.

I must also record my thanks to Dr G. D. Hobson of the Royal School of Mines, London, for helpful comment and advice, to Miss L. M. Sandford for her invaluable assistance in all stages of the preparation of this work, and to my son John for his help with the diagrams.

E.N.T.

CONTENTS

vii

LIST OF DIAGRAMS

A typical Pennsylvanian oilfield scene in 1869, ten years after the completion of Drake's first oil-well. The derricks were of wood, and the method of drilling was by percussion tools. (Source: *Institute of Petroleum Information Service.*)

The first oil-well successfully drilled in Iran—the D'Arcy well at Chia Surkh, 1902-5.
(Source: Petroleum Information Bureau.)

CHAPTER 1

History and Uses of Petroleum

1. PETROLEUM IN HISTORY

(i) The Early Years

THE fluid mineral petroleum ("rock oil") is an extraordinary and unique substance. In its origin, it may have been related in some degree to the origin of life itself, both on the Earth and perhaps also in other parts of the universe. In its application, it has provided an enormous amount of cheap and ubiquitous energy, the most fundamental requirement of our civilization. It is, furthermore, a chemical raw material from which a huge range of products can be obtained—even including proteins for human nourishment.

The history of petroleum and its relationship to mankind goes back to the dawn of history. The early civilizations that developed in the great river valleys of Mesopotamia used the asphalt obtained from hand-dug pits as building cements, for ornamental purposes, and to caulk their boats. The legend of the Flood described in Genesis and in the epic poem of Gilgames records only one among the many inundations of the low-lying river lands of Mesopotamia, to survive which a well-caulked Ark was certainly desirable.

The Elamites, Chaldeans, Akkadians and Sumerians are all known to have mined shallow deposits of oil-derived asphalt ("pitch") in the course of their history, and in the annals of Tukulti Ninouria the town of Hit on the Euphrates is mentioned as a prolific source of this material. The ancient name of Hit was *Ihi* or *Ihidakira,* meaning "bitumen spring", and its asphalt deposits formed perhaps the first known example of a "commercial" petroleum accumulation*. The asphalt derived from Hit, which is still locally used for road-making, boat construction and as a building cement, was also exported in ancient times across the desert to Egypt, where it was used extensively in the preservation of mummies

---

* *The term "commercial" as applied to petroleum is discussed on p. 34*

B

and in ornamental work. Nile boats were caulked with it—the three-months old Moses was cradled in bulrushes "daubed with pitch". The tribute paid to Thotmes III at Karnak included a material imported from Mespotamia, which was probably asphalt from Hit.

In ancient Egypt, petroleum oils derived from seepages were used medicinally ("Syrian oil" or "green oil") and there are many recorded references to "liquid fire", which support the theory that the Egyptians originally migrated southwestwards from the Caucasus, where the flames issuing so strangely from the ground in the Baku area were later accorded a religious significance in the Zoroastrian tradition. Indeed, some form of temple was continuously in existence near the Baku "eternal fires" from ancient times right up to the beginning of this century.

Even today, the air traveller crossing the oilfield regions of the Middle East by night is vividly reminded of Old Testament allusions by the flames (as of "a leaping fiery furnace") which pierce the darkness; while the traditional fate of Sodom and Gomorrah may well be linked with the oil seepages that are still to be found around the southern borders of the Dead Sea.*

In Iran, archaeological remains in Khuzistan show that asphalt was commonly used for bonding and jewel-setting during the Sumerian epoch, i.e. about 4000 BC, while the word "naphtha" has probably been derived from the Iranian verbal root *nab*, to be moist.

After the fall of the Tower of Babel (in the construction of which "oil out of the flinty rock" is reported to have been used as a cement) and the destruction of Babylon in about 600 BC, the general use of asphalt for building purposes decreased, since the Greeks and Romans built in quarried stone.

Petroleum oils derived from seepages were used, however, in Roman times for medicinal purposes and in religious rites, while asphalt was always in demand for ship-building and repair.

After the decline of Rome, the knowledge of the properties of petroleum passed to the Arabs, who developed the first distilling process for obtaining inflammable products for military purposes. The famous "Greek Fire" included "Medean oil" from Northern Persia in its composition.† During this period, the heavier constituents of petroleum

---

* It is interesting in this context to note that in Dr. Johnson's dictionary (1755) "asphaltos" is described as follows: "A solid brittle, black, bituminous, inflammable substance, resembling pitch, and chiefly found swimming on the surface of the Lacus Asphaltites, or Dead Sea, where anciently stood the cities of Sodom and Gomorrah. It is cast up from time to time, in the nature of liquid pitch from the earth at the bottom of this sea; and, being thrown upon the water, swims like other fat bodies, and condenses gradually by the heat of the sun, and the salt that is in it . . ."

† "Greek fire" was used as a "secret weapon" by the Byzantines against the Muslims in the 7th and 8th centuries, and later by the Saracens against St Louis, and by the Knights of St John against the invading Turks at Malta. It consisted essentially of porous pots filled with paraffin and ignited by gunpowder and fuses. It is interesting to note that Herodotus (viii, 52) records a much earlier use of petroleum as a weapon of war during the siege of Athens in 480 BC, when the Persians used incendiary arrows wrapped in oil-soaked tow.

that were needed as medicaments were commonly obtained from the wrappings of disinterred Egyptian mummies.

With the Renaissance, a number of sources of shallow crude oil and asphalt were discovered in Europe, and subsequently also in the new-found Americas. Raleigh used Trinidad "pitche" for repairing his ships, and samples of various petroleum oils were brought to Europe by travellers from distant lands. Late in the 17th Century, sufficient oil was actually produced by the distillation of oil-shale (whose nature is discussed on pp. 356) to light the streets of Modena in Italy, and in 1725 Peter the Great issued various ordinances regulating the carriage of oil from Baku by boat up the Volga, showing that there was already a lively demand for petroleum at that time, principally for use as an unguent and fuel.

Rock asphalt was extensively mined in Europe from the impregnated limestone deposits at Neuchatel, Seyssel, Val de Travers and Ragusa, when the demand arose for new road-building materials after the close of the Napoleonic Wars.

Paraffin wax was first obtained on a commercial basis from shale oil at about the same time, and was subsequently extensively used in the manufacture of candles. The increasing demand for liquid fuel for lighting purposes led to the rapid development of the oil-shale industry, notably in Scotland, where it prospered and developed until recent times, when the competition provided by imported low-cost crude oils made it no longer economic.

In the New World, shallow seepages of oil and gas were plentiful, as is indicated by the occurrence of many place names such as "Burning Springs". Small quantities of crude oil were traditionally marketed for their medicinal properties, but the first truly commercial application of a petroleum product in the western world was the use of natural gas from a shallow well at Fredonia in New York State in 1820. (Many centuries earlier, in the time of the Shu Han dynasty in China—221-263 AD—there are records of natural gas from shallow wells being trans-ported locally through bamboo pipes for use as a fuel). The Fredonia gas was distributed by small-bore lead pipe to nearby consumers, including the leading local hotel, and when General Lafayette arrived at Fredonia in the middle of the night, he was welcomed by the light of gas lamps and a gas-cooked barbecue was served in his honour.

For all these early applications, only small amounts of petroleum were available, obtained from shallow pits and surface seepages, both in the New World and the Old. In August, 1859, however, the first true *oil-well* was drilled by a percussion rig in Pennsylvania. At that time, and in that area, such wells were customarily drilled to obtain brine, which was used to produce solid salt by evaporation, and traces of oil had often been noted on the surface of the water recovered. This first well, drilled by the celebrated "Colonel" Drake, produced only a small volume of oil—about 25 b/d—from a depth of 69 ft, but its success was to pre-cipitate world-wide changes. Energy had previously been provided by human and animal muscle-power and by the combustion of solid fuels (wood, peat, coal, etc.) which could only be collected with considerable effort, and had to be laboriously transported to the point where the energy was needed. Liquid or gaseous petroleum, on the other hand, could provide an "instant" source of easily-transportable energy which might, when passed through pipelines, be consumed far from the point of production. It was a much more concentrated and flexible form of fuel than any that had previously been available.

Within a year of the successful completion of Drake's well, 175 further oil-wells had been drilled in the same area. The search for petroleum spread rapidly to other parts of North America and also to other countries—Russia, Romania, Poland and the then Netherlands East Indies. A new industrial era had begun, the age of oil, in the course of which the world was to change profoundly.

(ii) Since 1859

The fact that oil was first discovered in Pennsylvania so close to its immediate markets, coupled with a favourable fiscal system which has always encouraged exploration, resulted in a very rapid rate of oil industry growth in the United States, where for many years more oil was produced than in any other country. Thus, during the first century of the industry's existence, nearly $1\frac{1}{2}$ million oil-wells were drilled in the United States, and these produced in aggregate nearly half the total volume of crude oil produced in the whole world over this period. In fact, the US proportion of annual world oil output was usually not less than 60% and sometimes more than 70% in most of the years from 1860 to 1946. (However, since then the proportion has fallen with each succeed-ing year—it was less than 20% in 1971—and although more than three-quarters of all the wells drilled anywhere in the non-Communist world are still drilled in the USA*, the average production per well is now less than in any other oil-producing country†).

Outside the USA, pre-revolutionary Russia was initially predominant, due mainly to the huge individual outputs of a relatively small number of Baku wells. Subsequently, the Russian oil industry declined, until after sustaining the ravages of two wars, its focus shifted eastwards from the old oilfields of the Caucasus to the newer oil areas of Ural-Emba and Western Siberia, and production then rapidly rose again to make the USSR once more the world's second largest oil producer. In the 1920's, Mexico, briefly, and subsequently Venezuela, in turn occupied this position. Mexican production, based on a relatively few enormously productive wells drilled along the prolific limestone reef known as the

---

* *26,224 wells out of a total of 34,083 in 1971, i.e. 76%.*   †*See Table 108, p. 363.*

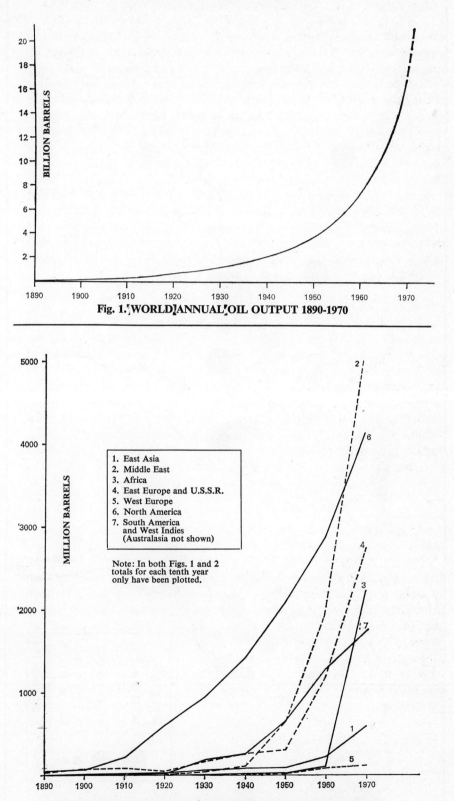

Fig. 1. WORLD ANNUAL OIL OUTPUT 1890-1970

1. East Asia
2. Middle East
3. Africa
4. East Europe and U.S.S.R.
5. West Europe
6. North America
7. South America
   and West Indies
   (Australasia not shown)

Note: In both Figs. 1 and 2
totals for each tenth year
only have been plotted.

Fig. 2. GROWTH OF OIL PRODUCTION FROM MAJOR AREAS

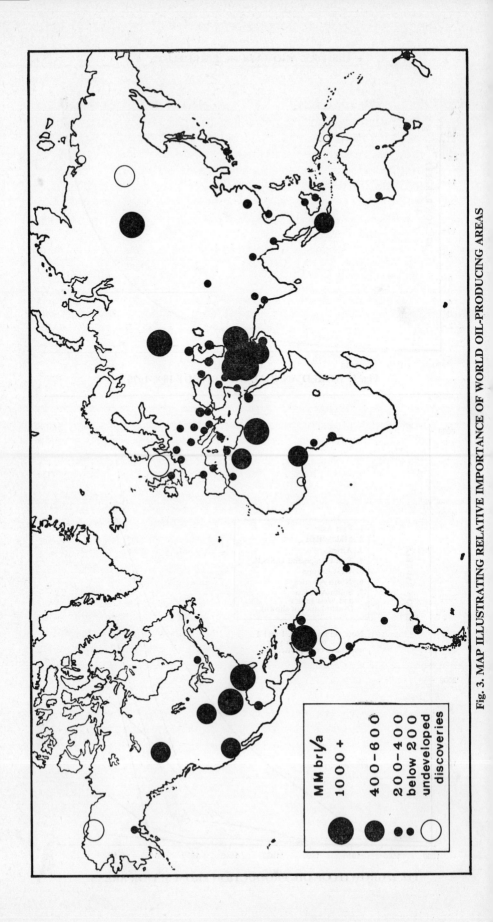

MM br√/a

1000 +

400 − 800

200 − 400

below 200

undeveloped
discoveries

Fig. 3. MAP ILLUSTRATING RELATIVE IMPORTANCE OF WORLD OIL-PRODUCING AREAS

"Golden Lane", soared for a few years but fell rapidly as these wells were exhausted and their outputs turned to salt-water; in spite of increased exploration in recent years, Mexico now ranks only 17th among the oil-producing nations of the world.

Venezuela was for many years after 1930 the world's greatest oil exporting state. Production reached a peak in 1964, much of the output being sent to the United States and western Europe. Since then, however, unfavourable tax and labour legislation has tended to discourage exploration and fresh capital investment, so that oil production has increased only very slowly over the last few years. (As a result, both Saudi Arabia and Iran moved ahead of Venezuela as oil producers during 1971.)

Several countries whose oilfields were developed early in the history of the industry still continue to be substantial producers, as for example, Romania and Germany, in Europe; Indonesia and Brunei, in Asia; Trinidad in the Caribbean; and Argentina, Brazil and Colombia in South America. However, their outputs are now small compared with those of the newer oil-producing states of the Middle East, which have shown the most rapid and impressive increases of petroleum production in recent years. This has been partly due to the geology of their oilfields, which are generally very large structures with thick and highly permeable reservoir beds, and partly also—at least in the first stages of their development—to the unitary nature of the concessions, which made very rapid production increases possible when transport facilities and markets became available.

Oil was found in Iran as long ago as 1908, in Iraq in 1927, and in Saudi Arabia in 1938; but it was not until after the end of the last war that the remarkable expansion of the Middle East oil industry really began. Today, the belt of oilfields lying between Turkey and Oman produces more than a third of the world's total oil output, and is still capable of considerable further development.

The post-war years have also seen the rapid emergence of North and West Africa as important oil-producing areas. In North Africa, large new accumulations of oil and gas have been discovered and developed in Algeria, Libya, and recently also in Egypt; while in Nigeria development has also been remarkably rapid, in spite of the dislocation caused by the Biafran war.

If world annual crude oil output is plotted against time on a scale designed to smooth-out short-term fluctuations, a remarkably consistent growth curve is obtained (Fig 1), reflecting an output which has rather more than doubled in each succeeding decade, equivalent to an average rate of growth of about 7.1 % p.a. compound throughout the history of the industry.

As crude oil and natural gas became increasingly available, they everywhere displaced the less convenient and less efficient solid fuels as

sources of energy. Thus, whereas in 1900 crude oil provided only 3.8%
and natural gas 1.5% of world primary energy* consumption, by 1970
the respective proportions had risen to 40.1% (oil) and 18.1% (gas)
(Table 1). The total petroleum-derived proportion had therefore
multiplied by a factor of nearly 11 since the start of the century.

TABLE 1

CHANGES IN THE CONSUMPTION STRUCTURE OF WORLD FUEL AND
ENERGY SOURCES DURING THE 20th CENTURY (% of Total)

	1900	1920	1940	1960	1965	1970
Coal/Lignite	94·2	86·7	74·6	52·1	43·2	36·0
Oil	3·8	9·5	17·9	31·2	36·7	40·1
Natural Gas	1·5	1·9	4·6	14·6	17·8	18·1
Hydraulic power and, from 1960 onwards, atomic power	0·5	2·0	2·9	2·1	2·2	5·8
Total	100	100	100	100	100	100

(additions vary due to rounding-off)

Sources: F. Baade and UN Statistical Yearbook "World Energy Supplies", 1962-1965.
CEC estimate for 1970 (see p. 346).

By 1971, commercial oil production was being obtained from 64
countries and totalled 17.5 B brl—roughly *125 times* the output in
1900 (Table 2). (Provisional figures for 1972 show a further overall
increase in world output of 4% to about 18.1 B brl). However, it is
interesting to note that, since man-made frontiers have little relationship
to the geological factors which control the size and productivity of individual oilfields, more than 85% of the total output comes from only 11
countries (Tables 3, 4).

TABLE 2

GROWTH OF WORLD OIL OUTPUT 1890–1972
(MM brl)

	East Asia	Austral-asia	Middle East	Africa	E. Europe & USSR	West Europe	North America	South America, W. Indies	World
1890	0·2	—	—	—	25·6	0·1	46·6	—	72·5
1900	6·0	—	—	—	69·9	0·3	64·5	0·3	141·0
1910	20·1	—	—	—	88·4	1·1	213·5	1·4	324·5
1920	29·1	—	12·2	1·0	44·8	0·6	600·2	7·1	695·0
1930	56·9	—	46·8	2·0	172·2	1·8	939·1	189·5	1408·3
1940	81·7	—	102·7	6·5	269·2	10·9	1405·8	268·8	2145·6
1950	87·7	—	640·9	16·7	307·0	19·6	2075·1	644·1	3791·1
1960	227·3	—	1923·3	105·3	1189·7	98·8	2863·6	1266·5	7674·4
1970	589·3	65·2	5154·1	2208·4	2717·8	114·7	4112·1	1729·9	16691·5
1971	607·0	114·3	5929·3	2065·0	2852·4	107·5	4116·8	1689·1	17481·4
1972†	777·8	119·8	6453·4	1968·7	2990·6	100·5	4174·8	1577·0	18102·6

† Estimated

* Defined on p. 342.

One of the walls of ancient Babylon in southern Iraq, dating from about 600 B.C., clearly showing the bituminous material which was used to cement the bricks together.

(Source: Petroleum Information Bureau.)

The temple of the fire-worshippers at Baku, photographed in 1900. Gas from the seepage zone was fed through the four corner chimneys to produce the "eternal flames".

(Source: Petroleum Information Bureau.)

"Khazzan Dubai II" en route from its onshore construction site to its offshore location 58 miles out to sea. This was the second of three 150,000-brl capacity tanks, each topped by a production platform, which have been lowered to the sea floor in this area, facilitating oil production and storage operations in situ from the Fateh oilfield in Dubai waters.

TABLE 3

PRINCIPAL OIL-PRODUCING COUNTRIES, 1971

	MM Brl	% of total
USA	3,478	19·9
USSR	2,701	15·5
Iran	1,657	9·5
Saudi Arabia	1,642	9·4
Venezuela	1,295	7·4
Kuwait	1,068	6·1
Libya	996	5·7
Iraq	622	3·6
Nigeria	556	3·2
Canada	483	2·8
Indonesia	326	1·9
11 leading countries	14,824	84·8
50 other countries	2,657	15·2
World total	17,481	100·0

TABLE 4

CRUDE OIL PRODUCTION OF MAJOR AREAS, 1971

	MM brl	% of total
Middle East	5,929·3	33·9
USA	3,478·1	19·9
USSR	2,701·0	15·5
Africa (North and West)	2,065·0	11·8
South America and Trinidad	1,689·1	9·7
Rest of the World	1,618·9	9·2
World total	17,481·4	100·0

Source of data: World Oil, October, 1972.

TABLE 5

CHANGING PROPORTIONS OF WORLD OIL OUTPUT, 1946–1971

	1946[1] %	1971[2] %
United States	63·2	19·9
USSR	5·9	15·5
Middle East	8·9	33·9
Venezuela	14·1	7·4
Rest of World	7·9	23·3
	100·0	100·0

[1] World output 2·75 B brl
[2] World output 17·5 B brl.

Table 5 illustrates the remarkable changes that have taken place since 1946 in the proportions of the world's total oil output that are obtained from the major producing areas. Over the quarter-century, the proportion of the total provided by the USA has fallen to less than a third of the 1946 figure, while the output of the Middle East countries

c

has proportionately nearly quadrupled. The shares provided by the Soviet Union and the "Rest of the World"—the latter largely due to the developments in North and West Africa—have also greatly increased, whereas the contribution of Venezuela has virtually halved.

A notable feature of the petroleum industry in these post-war years has been the remarkable acceleration that has taken place in its tempo of development. The time that elapses between an initial oil discovery and the full development of a previously unexplored area has shortened from decades to only a few years. Although the principal mechanism by which oil is discovered and produced, i.e. the rotary drilling rig, has not greatly altered in its basic design, ancillary improvements in materials and engines nowadays permit faster and safer drilling operations, rapid movement between locations, and the common attainment of depths which were considered exceptional only a few years ago. At the same time, engineering progress has reduced the time needed to construct production and export facilities, pipelines, storages, terminals and tankers.

Another notable trend which gathered momentum in the 1960's has been the shift of emphasis in petroleum exploration operations from dry land out into the shallow waters of the Continental Shelf. For this to be possible, there had to be an evolution in the design and manufacture of drilling rigs capable of operating either from floating barges or supported wholly or partly by retractable legs lowered to the ocean bottom. These rigs, which were initially used in the inland waters of Lake Maracaibo and the Caspian Sea, first operated successfully in marine conditions in the offshore areas of Louisiana and Texas. Within the last few years, their use has spread to the offshore areas of nearly every continent. In 1971, 250 offshore drilling rigs were in operation off the coasts of 75 countries, while some 7,000 platforms of various types had been built for offshore oil and gas production operations. The depths of water in which drilling can be successfully carried out have notably increased, so that whereas about 60 ft was initially considered the maximum feasible, drilling is now commonplace in waters several hundred feet deep; while techniques have been successfully pioneered which will permit drilling and production operations to take place at depths as great as 1,300 ft.

By the early 1970's, about 18% of the world's annual crude oil output was coming from offshore oilfields (Table 6), and this proportion is likely to increase every year—perhaps to as much as 30-40% of world oil output in 1980. In particular, the discovery of large oil accumulations in the northern North Sea basin is of great economic importance to the riparian states—notably Britain and Norway—who have previously been almost entirely dependent on imported oil. It is expected that some 2 MM b/d of oil will be produced from North Sea fields by 1980, rising to perhaps double this amount during the following decade.

TABLE 6

OFFSHORE OIL PRODUCTION 1970
(Outside USSR, East Europe and China)

	M b/d
Venezuela	2,761
USA (inc. Alaska)	1,577
Saudi Arabia and N. Zone	1,251
Iran	322
Nigeria	275
Abu Dhabi	269
Egypt (UAR)	257
Australia	216
Qatar	172
Brunei-Malaysia	146
Cabinda	96
Trinidad	76
Dubai	70
Mexico	37
Gabon	29
Peru	23
Other States	27
	7,604

Source: Offshore, June 20, 1971, 43.

By the early 1970's also, the development of the Arctic Basin had begun. This may well be the most important of the world's remaining petroleum provinces, with large oil and gas accumulations already discovered on the North Slope of Alaska, in the Far North of Siberia, and perhaps [also (although so far not confirmed) on the Canadian Arctic islands. The production of oil under Arctic conditions presents many problems, but it is perhaps of special significance to note that the considerations that have delayed the development of the Alaskan discoveries are not so much technical but environmental. In spite of the increasing awareness of an approaching energy shortage, agreement had not been reached by end-1972 on the route of the pipeline which it would be necessary to build to transport the oil at least part of the way to US markets. Since any pipeline across Alaska will take three years to complete once permission is given for its construction to begin, it is clear that the petroleum industry of the future is likely to be subjected to constraints resulting from environmental opposition which it has not had to contend with in the past. Some of the consequences of these emerging factors are discussed in Chapter 12.

2. PETROLEUM PRODUCTS

(i) Crude Oil

Crude oil is liquid petroleum in its natural, i.e. unrefined state. A more precise definition is that of the American Petroleum Institute which defines crude oil as "a mixture of hydrocarbons that exists in the liquid phase in natural underground reservoirs and remains liquid at atmospheric pressure after passing through surface separating facilities". /

Most crude oils are yellowish-brown to black in colour, with a characteristic greenish reflection by reflected light. They are generally lighter than water, with API gravities (defined on p.24) of $27°$–$35°$, although crude oils with gravities as low as $5°$ and as high as $61°$ have been reported.

The principal chemical elements in crude oils are carbon and hydrogen, which are combined in a series of compounds termed *hydrocarbons*. Probably more than a million varieties of hydrocarbons are present in crudes, since they can occur in a large number of isomeric forms. However, the number of hydrocarbons which actually occur in signifi- cant proportions in any particular crude are usually relatively few. Thus, for example, only 31 hydrocarbons are known to account for 75% by weight of the whole distillate fraction between 55–145°C of one Oklahoma crude.

There are three principal groups of hydrocarbons present. The most important of these comprises the *paraffins* / ("alkanes" or "saturated hydrocarbons") whose basic formula is C_nH_{2n+2}. The factor n may vary between one and 70; compounds in the range $n=1$ to $n=4$ are gases at normal temperatures and pressures; those from $n=5$ to $n=15$ are liquids, and those from $n=15$ to $n=70$ are solids.

TABLE 7

PROPORTIONS OF MOST IMPORTANT ELEMENTS PRESENT IN
NORMAL CRUDE OILS

	% wt.
Carbon	84–87
Hydrogen	11–14
Sulphur	0·6–2·0
Nitrogen	0·6–2·0
Oxygen	0·6–2·0

The second important group of hydrocarbons comprises the *cyclo- paraffins* or *naphthenes,* which are saturated ring compounds with the molecular formula C_nH_{2n}. Crudes containing large proportions of naphthenes leave a residuum of asphalt on distillation.

The third major hydrocarbon group comprises the *benzenoid* or *aromatic* hydrocarbons of formula C_nH_{2n-6} which usually comprise less than 10% by volume of most crude oils.

In addition to these (and some other, less important) compounds of carbon and hydrogen, crude oils also usually contain small proportions of sulphur, oxygen and nitrogen compounds, as well as traces of many other elements.

The sulphur compounds present may include mercaptans, thiophenes and thioethers, as well as, occasionally, free elemental sulphur; sulphur compounds being heavy are usually found mainly in the heavier crudes:

some Mexican heavy seepage oils may contain as much as 10% by weight of sulphur.

A "sweet" or "low sulphur" crude oil is defined as one which contains less than 0.5% by weight of sulphur compounds; with more sulphur than this the crude is termed "sour" or "high-sulphur" and a special refining operation is needed to remove the sulphur compounds, which would otherwise produce unacceptable corrosion and pollution effects. (Most Middle East crudes are "sour" and their import in increasing volumes to supply the needs of Western Europe, Japan, and in the near future the United States also, therefore inevitably results in atmospheric pollution problems which will become more pressing in the years ahead.)

The oxygen compounds in crude oil are mainly asphaltenes and naphthenic acids; the nitrogen compounds are homologues of pyridine or alkylated quinolines. Table 8 lists some of the other elements that have been identified in the ashes of various crude oils. Some of these, such as vanadium and nickel, are thought to have been derived from organic source materials; other elements may be related to the lithology of the reservoir rocks concerned.

TABLE 8

TRACE ELEMENTS PRESENT IN CRUDE OILS

Source of Oil	Elements Identified
Canada	Fe, Al, Ca, Mg, traces of Au and Ag
Ohio	Fe, Al, Ca, Mg, traces of Au, Ag, Cu
Mexico	Si, Fe, Al, Ti, Mg, Na, V, N, Sn, Pb, Co, Au
Argentina	Fe, Al, Ni, V, P, Cu, N, K
Japan	Si, Fe, Ca
Caucasus	Si, Fe, Ca, Al, Cu, S, P, As, traces of Ag, Au, Mn, Pb
Egypt	Fe, Ca, Ni, V
Venezuela	Fe, Ni, V
Iraq	Fe, Ni, V
Texas	Si, Fe, Al, Ti, Ca, Mg, V, Ni, Ba, Sr, Mn, Pb, Cu, and traces of Cr and Ag

Crude oils may be classified as "heavy", "medium" or "light", depending on whether their API gravities fall in the ranges 10–20°, 20–30°, or over 30°, respectively. They are also usually described by reference to the dominant hydrocarbon groups they contain as "paraffinic", "mixed base", "asphalt base" or "aromatic" (Table 9).

A characterisation factor 'K' has been devised to compare crude oils from different sources, using the average boiling point T (°F at atm. pressure) and the specific gravity S (at 60°F compared with water at 60°F):

$$K = \sqrt[3]{\frac{T+460}{S}}$$

Typical characterisation factors for different types of crude oil are: paraffinic 12.7, naphthenic 11.5, and aromatic 10.0.

The "correlation index" for crude oils established by the US Bureau of Mines[1] measures the relative amounts of paraffins, naphthenes and aromatics in each distillation fraction, and also refers to the sulphur and nitrogen contents of the oils. (The greater the proportion of aromatics or naphthenes, the higher the index.) When a number of crudes derived from geographically related oilfields exhibit similar correlation indices, there seems to be a reasonable probability of a common origin and history of accumulation.

TABLE 9

CLASSIFICATION OF CRUDE OILS

Type of Crude	Wax%	Asphalt %	% Composition of 250–300°C fraction.		
			Paraffins	Naphthenes	Aromatics
"Light paraffinic" (paraffin base)	1·5–0	0–6	46–61	22–32	12–25
"Paraffin-naphthene" (mixed base)	1·0–6	0–6	42–5	38–9	16–20
"Naphthenic" (asphalt base)	trace	0–6	15–26	61–76	8–13
"Aromatic"	0·0–5	0–20	0–8	57–78	20–35

TABLE 10

HYDROCARBON GROUPS IN TYPICAL CRUDE OILS
(Percentage proportions by weight)

Source of Crude	Naphthenes	Aromatics	Paraffins
Pennsylvania	11·7	4·9	80·5
Oklahoma	23·7	4·9	69·8
Kentucky	23·1	2·8	68·3
California	34·9	2·8	57·2
Kansas	20·3	0·4	76·6
Texas	23·1	8·7	66·0
Michigan	7·4	4·5	85·2

(ii) Natural Gas

Crude oil is always associated in the subsurface with a mixture of gases, which for convenience is termed "natural gas". The composition of the gaseous mixture is variable, and it can contain both hydrocarbons and non-hydrocarbons (Table 11)*. The hydrocarbons comprise the first few members of the paraffin series and their isomers. The number and proportion of the hydrocarbons which are present in the gaseous phase will depend on the pressure and temperature conditions prevailing. Methane, normally by far the most abundant gas associated with crude oil accumulations, is the only hydrocarbon which will always be present

[1] *Tech. Paper 610*, US Bureau of Mines, 1940.

* *For a detailed description of the various types of natural gas, the reader is referred to the author's work entitled "Natural Gas–A Study" (2nd edition, 1972), published by Scientific Press Limited.*

as a gas; the "higher" hydrocarbons from ethane C_2H_5 to heptane C_7H_{14} are usually present in proportions inverse to their boiling points (Table 12).

Natural gas occurs both in solution in crude oil in the subsurface and, where the liquid oil is saturated with gas under the prevailing conditions, as free gas which fills the pores of a section of rock overlying the oil-saturated rock layer (p. 30) and forms a "gas cap". The "gas/oil ratio" is a measure of the amount of gas held in solution in a crude oil; it is usually expressed in cubic feet of gas (at NTP) per barrel of oil, and can vary between less than 100 and several thousand cubic feet per barrel.

In some accumulations, the pressure may be high enough for the critical point of the hydrocarbon system to be exceeded, giving a vapour phase only above the dew-point curve. In such cases, a sudden release of pressure, as would result from the entry of an exploration well, may result in the process of "retrograde condensation" and the production of a very light hydrocarbon liquid termed "condensate". Another source of condensate is the normal wellhead separation process. Liquids of this type form a valuable addition to normal crude oil output, and are often used for petrochemical manufacture.

Some gas accumulations occur without associated liquid oil, and in these cases the hydrocarbon component is preponderantly methane, with usually very small proportions of higher hydrocarbons present. However, it is normally impossible to produce crude oil from a subsurface accumulation without also producing associated natural gas, since as the atmospheric pressure will always be less than that in the subsurface, dissolved gas will tend to come out of solution in the oil during production operations. This gas is therefore an unavoidable co-product of crude oil production, and used to be treated as an unwanted adjunct which had to be got rid of by burning ("flaring") at the surface or venting to atmosphere.

However, nowadays increasing efforts are being made to use natural gas constructively, both as a valuable source of fuel and as a chemical raw material. Thus, where "higher" hydrocarbons are present in substantial quantities in an associated gas, some form of vapour extraction process is generally applied to extract them, since the resultant "Liquefied Petroleum Gas"* and "Natural Gasoline"** components are of special value.

Where the proportion of extractable liquid hydrocarbon components is very small (less than 0.1 gal/Mcf of gas treated), a gas is termed "dry"; if it is 0.3 gal/Mcf or more, the gas is "wet". (These terms, of course, bear no relationship to the water content of the gases.)

The United States and Canada are by far the largest producers of

---

*Mainly butane and propane; LPG components are also obtained during the "cracking" of crude oils.   ** Iso-butane, pentane, hexane and sometimes heptane.*

The term "Natural Gas Liquids" includes both the LPG and "natural gasoline" components.

LNG (Liquefied Natural Gas) is mainly methane (sometimes with other hydrocarbon components of natural gas) which has been liquefied by a cryogenic process.

natural gas liquids—633 MM brl and 75 MM brl in 1970 respectively. About 9 MM brl was produced in Western Europe, mainly in France, with smaller volumes in Italy and Britain (about 800,000 brl from North Sea gas).

TABLE 11

OCCURRENCE OF NON-HYDROCARBON COMPONENTS OF NATURAL GASES

Gas	Typical %	Maximum recorded %
Carbon Dioxide	0–5	98
Helium	0–0·5	8
Hydrogen Sulphide*	0–5	63
Nitrogen	0–10	86
Argon	less than 0·1% of helium content	
Radon, Krypton, Xenon	traces only	

Source: E. N. Tiratsoo "Natural Gas" (2nd ed.) 1972

* A "sweet" gas is one in which there is no "detectable" odour of H_2S–i.e. one in which there is less than one part in a million of hydrogen sulphide.

TABLE 12

TYPICAL ANALYSES OF HYDROCARBONS IN NATURAL GAS

	"Wet" Gas %	"Dry" Gas %		"Wet" Gas %	"Dry" Gas %
Methane	84·6	96·0	Iso-pentane	0·4	0·14
Ethane	6·4	2·0	N-pentane	0·2	0·06
Propane	5·3	0·60	Hexanes	0·4	0·10
Iso-butane	1·2	0·18	Heptanes	0·1	0·80
N-butane	1·4	0·12			

Source: D. Heron, Paper 93, 8th Comm. Min. & Met. Congress, 1964.

The primary application for natural gas is as a domestic and industrial fuel. Because of the accuracy with which its composition can be controlled, highly efficient combustion is possible, and since the products of combustion need contain neither sulphur compounds nor carbon monoxide, it forms a clean, safe and highly flexible source of heat, burning with a clear flame when mixed with the correct proportion of air and not polluting the atmosphere with corrosive fumes.

As a fuel, natural gas has replaced the traditional coal-derived or oil-derived town gases in every country where it has become easily available and is now used in large and growing volumes for industrial, commercial and domestic heating purposes, for electricity generation, and for many "premium" applications where its special properties are particularly valued—as in glass-making and the steel industry. It is also used as a feedstock for various chemical processes, which depend on the composition of the original gas for their end-products: ammonia, acetylene and a wide range of other compounds can be produced from suitable gases.

Thus, in the United States, some 300 B cf/a of natural gas is used for petrochemical synthesis, mainly through controlled oxidation processes which produce methanol, ethanol, propanol, formaldehyde, acetaldehyde, acetone, etc., all of which can be used in a variety of further syntheses.

Ethylene is derived from the pyrolysis of ethane and propane obtained from "wet" natural gases. Propylene comes from catalytic cracker gases; butadiene from the dehydrogenation of butane and butene, which also come from catalytic cracker gases, and aromatic chemicals by the catalytic reforming of suitable crude oil fractions.

"Dry" natural gas is also used for the manufacture of carbon black, a pigment of colloidal dimensions, made by burning the gas with only a limited supply of air, and depositing the product on a cool surface. Carbon black is used in the manufacture of printing ink, polishes, varnishes and motor tyres.

The helium derived from certain natural gases, mainly those derived from some areas of the United States and Canada, is a valuable petroleum by-product. More than one Tcf* of helium gas is consumed annually in the USA, and considerable volumes are exported in liquid form.

(iii) Refined Products

It was known in the Middle Ages that inflammable liquids could be obtained by the cautious distillation of heavy crude oils derived from seepages. The basic technique used in the refining of petroleum has in fact always been the same—thermal distillation, originally on a batch basis and subsequently by large-scale, continuous operations. Nowadays, the vapours produced by distillation are condensed in baffle-plate towers which act as fractionating columns, the resulting distillates then being separated into successive "cuts" according to their boiling-point and gravity ranges. Distillation can be carried out at atmospheric pressure, under vacuum, or at various degrees of high pressure, depending on the qualities required in the end-products. Thus, high pressures and high temperatures favour the "splitting-up" or "cracking" of large hydrocarbon molecules, This is a desirable process in motor fuel (petrol or gasoline) manufacture, where high-pressure distillation is consequently used; but where cracking has to be avoided, as in the production of lubricating oils, pressures as low as possible are employed.

The chemical nature of the crude being distilled also largely affects the products obtained by thermal fractionation. Table 13 shows the typical fractions that might be obtained from a Pennsylvanian paraffin-base crude oil.

TABLE 13

PRODUCTS FROM DISTILLATION OF PARAFFIN-BASE CRUDE OIL

"Crude naphtha"	Start to 54° API gravity
"Kerosine distillate"	54°–50° API
"Crude kerosine"	50°–38° API
"Gas oil" or fuel oil	38°–35° API
"Wax distillate"	35° API to finish

* Trillion (10^{12}) cu ft.

In such an example, part of the "crude naphtha" fraction might be used to make motor fuel (after further treatment which is described below); another part might go to be used as cleaning spirit, and the balance might be used as a petrochemical feedstock, depending on prevailing requirements.

The kerosine distillate will contain constituents too volatile to use as lighting or heating oils, which will therefore be removed by re-distillation and added to the naphtha fractions. The remaining distillate, together with the crude kerosine, is then chemically treated to remove unsaturated sulphur compounds.

The "gas oil" and fuel oil fractions may be utilized as distilled, or may be blended or further distilled. Gas oil is used for the manufacture of combustible gas, as a diesel engine fuel, or it may be cracked to form motor gasoline.

TABLE 14

PROPORTIONS OF PRINCIPAL PRODUCTS OBTAINED BY REFINING TYPICAL US CRUDE OILS

	% Gasoline	% Kerosine	% Gas Oil	% Fuel Oil	% Lubricants	% Wax	% Cylinder Stock
Pennsylvania	40·0	8·0	25·0	—	7·0	2·0	16·0
Texas	—	—	42·0	15·0	39·0	—	—
W. Virginia	45·0	13·0	18·0	—	7·0	2·0	14·0
California	—	—	33·6	—	23·4	30·0	—
Oklahoma (1)	35·0	18·0	16·0	—	6·0	1·0	20·0
Oklahoma (2)	43·0	10·0	20·0	18·0	6·0	2·0	—

The "wax distillate" is a loosely-defined fraction produced by continuing the distillation of the crude under vacuum after the light gas oil has been taken off, until the residue is of a density and boiling-point suitable for use as a source of lubricants. It is then usually chilled and pressed to remove the paraffin wax, the filtered fluid being added to the light lubricant stock. Petroleum waxes are extensively used in the formulation of protective coatings in the packaging industry, in polishes, and to a lesser degree in the manufacture of candles and in the electrical industry.

The residue remaining after the evaporation of the gasoline, kerosine, gas oil, fuel oil and wax distillate, is called "cylinder stock". It can be compounded with greases to form steam engine cylinder lubricants. Alternatively, if the wax still remaining in it is removed, the resultant liquid (called "bright stock") can be compounded with lighter fractions to give motor lubricants.

The figures in Table 14 illustrate the varying quantities of the different products likely to be obtained by refining a number of typical American paraffin-base crude oils. When an asphalt-base crude is refined, the range of products obtained is different. After the crude naphtha and the kerosine, a fraction called "furnace distillate" (API gravity typically 37°

to 35°) is usually taken off, followed by gas oil between 35° API and 29° API, and then a lubricating distillate. The remainder may be used as residual fuel oil or be further distilled. Some crude oils treated like this may give a final residue of high-quality asphalt which can be used for roofing or road making; others will give a residue of "petroleum coke", used as a low-ash fuel and as a source of industrial carbon in the steel industry.

Apart from the primary objective of splitting the crude oil up into groups of different products, petroleum refining operations are also used to convert some fractions into more valuable products by processes of chemical change. The overall aim is to reduce the "pour-points"* and sulphur contents of the products, to increase the proportion of the valuable light and middle distillates, and to raise the "octane number"† of the gasoline fraction. Thus, the gasolines produced for the early low compression-ratio internal combustion engines had low octane numbers —perhaps 30, compared with the 90–97 of modern motor fuels. The improvement in quality was brought about in part by the admixture of chemical additives (e.g. tetraethyl lead) but mainly by cracking processes, which, while increasing the gasoline proportion obtained from refining a crude oil, also had the effect of improving the octane rating as a result of molecular rearrangement.

The original thermal cracking plants, using temperatures of 750°–900°F, have nowadays largely been replaced by various catalytic cracking processes, which use alumina-silica mixtures, often in "fluid-fed" (i.e. fluidized solids) plants, in which the catalyst powder is kept in a freely flowing condition by admixture with a gas.

The catalytic "reforming" of naphthas has now largely replaced catalytic cracking as the principal source of higher octane gasolines. Such "hydrocracking" operations essentially involve the catalytic cracking of naphthas in the presence of large volumes of relatively pure hydrogen obtained as a refinery by-product or by steam reforming of hydrocarbons. Naphthenes and paraffins are transformed into aromatics by this process, which is accordingly also the principal method by which pure aromatics are derived from a crude oil. Hydrocracking processes can be used as versatile methods of changing yield patterns in refineries in accordance with marketing demands. They can also improve the anti-knock quality of gasolines as anti-pollution requirements cause smaller proportions of lead additives to be used in future, and also to produce maximum yields of kerosine-type jet fuels, the demand for which is expected to at least double during the decade 1970–1980. At the

---

* Pour-point is the temperature at which some of the constituents of an oil begin to solidify on cooling, so that the oil can no longer flow freely.

† The octane number of a gasoline is a measure of its "anti-knock" quality expressed as the percentage of iso-octane in a mixture of iso-octane and heptane which matches the tendency of the given gasoline to "knock" in a standard test engine.

beginning of 1970, there were 55 hydrocracking plants operating in the world, with a total throughput capacity of 850,000 b/d; by the end of 1972 a further 22 units had been built and world hydrocracking capacity had increased to at least 1,350,000 b/d.

There are a number of proprietary variants of this basic process: all but one of these use hydrocracking catalysts in a series of fixed beds, while the other uses an ebullating (bubbling) bed.

A number of other modern refining operations should be mentioned. "Visbreaking" is the thermal cracking of a heavy viscous paraffinic residuum to make it saleable as middle distillates and fuel oil. This is carried out at temperatures of about 900°F and relatively low pressures.

"Polymerization" is a method of treating a feedstock rich in olefins derived from a catalytic cracking unit, by heating it in contact with a phosphoric acid type catalyst to produce high-octane gasoline components.

"Alkylation" has been particularly important in making aviation gasoline. It produces iso-octane by combining butylene and iso-butane in the presence of sulphuric or hydrofluoric acids. The iso-butane needed may be made from excess n-butane by "isomerization", using an aluminium chloride catalyst. (Another type of isomerization unit is used to make iso-pentane from pentane).

"Delayed coking" is a variant of thermal cracking in which the feed-stock is held at moderate pressures and high temperatures (800-900°F) for long enough for substantial quantities of coke to be formed, in which many of the undesirable contaminants of the feedstock are concentrated and thus eliminated, particularly from heavy "sour" oils. A variant is the fluid coking process which combines chemical thermal cracking with physical arrangements similar to those used in fluid catalytic cracking.

Various specialised refining techniques are used to purify and improve the qualities of lubricating oils—e.g. dewaxing and solvent treating operations. The "sweetening" processes which have been used to remove mercaptan sulphur compounds from a gasoline or kerosine usually involve mild hydrogenation. However, as a result of increased anti-pollution pressure designed to eliminate sulphur dioxide gas from the atmosphere, more severe hydrodesulphurization processes have been developed, using higher temperatures and increased amounts of hydrogen and catalyst.

Since there is a considerable variation in the physical and chemical properties of different crude oils, considerable refining skill must be deployed in matching the output of particular products with the seasonal demands and the crudes available. In general, heavy crudes give higher proportions of residual fuel oils, while lighter crudes yield higher volumes of gasolines.

Table 15 summarises the characteristics of the principal groups of products obtained from a number of typical Middle Eastern and African crude oils. In general, the North African crudes are "light" and to balance them in refinery feedstocks there has been an increased production in recent years of the heavier Middle Eastern crudes which have, however, relatively high sulphur contents. The North African oils, besides being light, are also low in sulphur, which adds to their desirability, but on the other hand they have the drawback of being "waxy"— i.e. they have a high pour-point. Blending operations are designed to match available crudes to the capacity of a particular refinery in terms of specific gravity, sulphur content, pour-point and other properties of the feedstock to be used and the products it is desired to obtain.

TABLE 15

PROPERTIES OF SELECTED EXPORT CRUDES

Yield pattern (% wt.)

	API Gravity	Sulphur %	Gasoline/ light fractions	Middle Distillates	Fuel oils	Sulphur contents of Fuel oil (% wt.)
Saudi Arabia—light	34·2	1·6	20·5	31·0	48·5	2·7
—Safaniya	27·3	2·8	16·0	23·3	60·7	4·2
Kuwait	31·3	2·5	19·4	25·3	55·3	4·0
Iran—light	34·3	1·4	22·3	30·2	47·5	2·4
—heavy	31·3	1·6	21·2	26·8	52·0	2·6
Libya—Es Sidr	42·1	0·4	32·5	32·5	35·0	0·6
—Brega	39·2	0·2	26·2	31·7	42·1	0·3
Algeria—Hassi Messaoud	44·1	0·1	35·0	36·0	29·0	0·4
—Zarzaitine	40·7	0·1	22·0	38·0	40·0	0·1
Nigeria—light	33·4	0·2	21·0	40·3	38·7	0·3
—medium	27·1	0·3	12·0	40·0	48·0	0·4

Source: Petroleum Press Service, March 1967, 96.

World requirements for the different groups of refined products change from year to year, as is illustrated by Table 16, which shows the variation in the proportionate yields of refineries in the United States and Western Europe over the decade 1961-1971.

TABLE 16

REFINERY YIELDS, 1961 AND 1971

USA	Percentage by weight 1961 %	1971 %	Western Europe	Percentage by weight 1961 %	1971 %
Gasolines	41	43	Gasolines	19	17
Kerosines ⎫ "middle Gas/Diesel ⎬distill- oils ⎭ ates"	28	28	Kerosines ⎫ "middle Gas/Diesel ⎬distill- oils" ⎭ ates"	27	31
Fuel oils	10	6	Fuel oils	39	38
Others (including refinery fuel and loss)	21	23	Others (including refinery fuel and loss)	15	14
Total	100	100	Total	100	100

Source: BP Statistical Review, 1971.

Changes also occur in the geographical balance of refinery capacity. Thus, some areas of the world—notably North America and the Communist states—have traditionally been self-sufficient in the past in their refining capacities, while Western Europe was largely dependent on the output of refineries in the Caribbean area, treating Venezuelan, Trinidad and US crudes. However, in recent years, as a result of political decisions and changing economic factors, the Western European share of world refinery capacity outside the self-sufficient areas mentioned has increased (from 20% in 1940 to almost 50% in 1970—mainly processing African and Middle East crudes), while that of the Caribbean area has shrunk proportionately (from 43% to only 12% of the equivalent refining capacity). The actual crude throughputs in 1971 are shown in Table 17.

TABLE 17
WORLD REFINERY CRUDE THROUGHPUTS, 1971

	1000 b/d	% of total
USA	11,319	23·5
Caribbean area	3,275	6·8
Western Europe	13,091	27·2
Middle East	2,373	4·9
Rest of World	18,101	37·6
	48,159	100·0

Source: BP Statistical Review, 1971.

(iv) Other Products

(a) Petrochemicals

Petrochemicals may be defined as chemicals which can be produced in one step from a petroleum feedstock; from these primary products an enormous range of further products can then be obtained, including solvents, detergents, plastics, agricultural chemicals, synthetic rubber and a variety of special products.

In the United States, the commercial production of petrochemicals began in 1930, and by 1963 petroleum-derived chemicals provided 90% of the US output of the traditional "organics". From 1950 onwards, petrochemical plants were set up in Britain, Western Europe and Japan, and in more recent years in a number of the oil-producing countries of South America, the Middle East, Asia and Africa.

Although produced from only about $2\frac{1}{2}$% of the world output of oil and gas, the value of petrochemicals is proportionately very much higher than that of any other type of oil derivative, and it is forecast that by 1985 the non-Communist world will be producing 200 MM tons/a, of which 40% will come from the USA and 44% from Western Europe. By that year, about 90% of all "organic" chemicals will probably be petroleum-based, and it is likely that this proportion will increase to 99% by the end of the century, when the world output of petrochemicals may well exceed 600 MM tons/a.

In Europe, the steam cracking of naphtha to produce olefins has always been the principal method of petrochemical synthesis, and the demand for naphtha is consequently rising at the rate of 10% p.a. and may reach 39 MM tons/a by 1975. In Britain, naphtha consumption for petrochemical production rose from 3 MM tons in 1966 to 7·3 MM tons in 1970.

In the cracking of naphtha, substantial volumes of propylene, butadienes and butenes are formed, with ethylene as the main product. From ethylene is derived polyethylene, ethylene oxide, styrene and vinyl chloride; from propylene—the most important raw material for the plastics and synthetic fibre industries—comes polypropylene, acrylonitrile, cumene, acetone and butanol; and from butadiene, synthetic rubber.

Acetylene can also be derived from petroleum naphtha (as well as from natural gas); and various aromatic compounds such as benzene, toluene, and the xylenes can be produced from suitable naphtha feedstocks.

(b) Proteins

One of the newest and most interesting applications of petroleum is as a raw material for protein synthesis to make good world deficiencies. At the present time, developing countries produce only 20% of the world's protein for 70% of the world's population. By the end of the century, by which time world population is expected to have doubled, the protein shortage is likely to be at least 20 MM tons p.a.

Now, it has been found that many micro-organisms can live and develop in hydrocarbon mixtures and can produce protein concentrates which are analogous to animal proteins and are rich in water-soluble vitamins. Paraffinic petroleum fractions treated in this way are dewaxed, which may actually be advantageous from a refining point of view. From a ton of normal paraffins, which do not need to be separated from the petroleum fraction, it is technically possible to obtain a ton of protein-vitamin concentrate containing nearly 50% protein. So, if the process could be successfully applied on a large scale, all the protein needed could be produced from only a small fraction of the world's output of crude oil.

During 1972, commercial production of high-protein yeast from petroleum was in fact started at plants in Grangemouth, Scotland (4,000 tons p.a.) and Lavéra in southern France (16–20,000 tons p.a.). Two further large-scale plants will be in production in Japan before the end of 1973, and a 100,000 tons p.a. plant will be commissioned in Sardinia in 1975. It seems likely therefore that petroleum-derived proteins will play an increasingly important part in feeding the future world—perhaps initially as an animal feed ingredient, to save imported oilseeds and fishmeal, and later as a means of fortifying conventional foodstuffs.

(c) Sulphur

"Sour" crudes and high-sulphur natural gases yield important amounts of elemental sulphur, particularly in North America, and also in France, Mexico and Saudi Arabia. Sulphur derived from crude oil and natural gas amounted to 18 MM tons in 1970, or 62% of total world sulphur in that year.

The hydrogen sulphide present in natural gases can be partly oxidized to sulphur dioxide, which is then reacted with further hydrogen sulphide to produce elemental sulphur. In modern oil refinery operations, the hydro-desulphurisation process is used to remove sulphur from petroleum distillates, usually in the form of hydrogen sulphide gas; the hydrocracking process in which heavy oils are converted to gasolines by high-pressure hydrogenation also produces hydrogen sulphide. Sulphur can be obtained from these refinery gases in the same way as from natural gas; with increased anti-pollution requirements, it is probable that the amounts of sulphur recovered from "sour" crude refining will increase, and that eventually most residual fuel oils will be desulphurised to acceptable limits.

3. UNITS OF MEASUREMENT

The literature of the petroleum industry suffers from a fundamental dichotomy in its principal units of measurement. In the United States, and wherever American influence prevails, liquid oil is measured in terms of volume, i.e. in *barrels* (of 42 US gallons). In European practice, on the other hand, including the USSR, the unit of comparison is usually one of weight, i.e. *metric tons*. In order to compare oil quantities expressed in volume with others expressed in weight, it is necessary to know the specific gravity of the crude oil in question, and this may vary over quite a wide range. (The average conversion factors for some of the most important crude oils are listed in Table 19.)

An alternative to specific gravity commonly used in commercial practice is the American Petroleum Institute degree, which is based on an arbitrary scale. This has the virtue of providing a convenient and easily-quoted unit of comparison for different crudes, whose monetary value is, of course, often largely dependent on these "gravity' differences. It is important to note, however, that *high* API gravities imply *low* specific gravities, and *vice versa*. The relationship is:

$$\text{Degrees API} = \frac{141 \cdot 5}{\substack{\text{specific gravity} \\ \text{of oil at } 60°F}} - 131 \cdot 5$$

Pure water has an API gravity of 10°, equivalent to a specific gravity of 1.00 at 60°F, while most crude oils have API gravities of 27°–35°. However, some much heavier and much lighter crudes are also known (p. 29).

Although the long ton (2,240 lbs) is still used in oil production reports from some countries, in the interests of uniformity it has been felt preferable in this work to use US barrels as the common units of oil output, and to convert as far as possible all other units into barrels. (Where no exact S.G. figure was known, an approximate conversion has been made on the basis of 7.2 barrels to the metric ton.)

This has the virtue of providing a simple method of comparing oil outputs and reserves for different fields and countries, without the need for further arithmetic. All converted figures must, however, be approximations to this degree.

Confusion has also sometimes arisen as regards the periods of time over which oil outputs are measured; as far as possible, production has been expressed here in terms of barrels *per year* (brl/a), although occasionally the output of a particular well or pool is referred to in barrels per day (b/d).

It is never easy to express large numbers in uniformly comprehensible terms; but on the basis that M denotes one thousand, it is logical to use MM for one million. For higher powers, however, there are certain complications. Since the United States is still the world's largest producer of oil and natural gas, and most statistics are derived from American sources, it has been thought reasonable that the *billion* (B) should be taken here to have its American meaning of one thousand million, i.e. 10^9, rather than its British meaning of one million million; and the Continental term "milliard" has therefore not been employed.

Similarly, the term *trillion* (T) has been used throughout with its US meaning of 1,000 billion (10^{12}) rather than its astronomical UK significance of 10^{18}.

As regards natural gas outputs, all statistics are quoted here in cubic feet, using the approximate conversion factor of 35.3 cubic feet to one cubic metre where necessary.

TABLE 18

UNITS OF VOLUME

1 cm = 35·315 cf = 219·97 Imp gal = 264·17 US gal = 6·29 US brl
1 cf = 7·48 US gal = 6·23 Imp gal = 0·178 US brl = 0·028 cm
1 US brl = 42 US gal = 34·97 Imp gal = 5·61 cf = 0·16 cm
20 M brl/d = 1·0 MM tons/a (approx.)

The *calorific values* of fuels are generally expressed in terms of British Thermal Units or kilocalories per unit of weight or volume. (The relationship is conveniently expressed : 8.9 Kcal/cm = 1 Btu/cf). As a rough rule, crude oils usually have calorific values of the order of

D

1 MM Btu/cf and typical natural gases have calorific values of 900–1,100 Btu/cf. Hence, about 6,000 cf of natural gas are about equivalent in thermal value to a barrel of crude oil.

TABLE 19

SOME CONVERSION FACTORS FOR CRUDE OILS
(*Multiply metric tons by factors below to convert to US barrels*)
N.B. There are wide variations between individual oilfields.

Abu Dhabi	7·493	Egypt	6·849	Italy	6·813	Qatar	7·709
Algeria	7·713	France	7·201	Kuwait	7·263	Romania	7·453
Argentina	6·992	Gabon	7·245	Libya	7·593	Saudi Arabia	7·370
Austria	6·974	Germany	7·223	Mexico	7·104	Trinidad	6·989
Bahrein	7·335	Indonesia	7·348	Neutral Zone	6·849	UK	7·279
Canada	7·428	Iran	7·370	Nigeria	7·361	USA	7·418
Colombia	7·054	Iraq	7·452	Pakistan	7·308	Venezuela	7·005

TABLE 20

OIL CONVERSION TABLES

Gravity degrees API	Specific gravity 60°F	Gallons/lb	brl/long ton	brl/metric ton	Cm/m ton
10	1·0000	0·1201	6·4041	6·3030	1·00123
15	0·9659	0·1243	6·6302	6·5255	1·03657
20	0·9340	0·1286	6·8569	6·7487	1·07202
25	0·9042	0·1328	7·0837	6·9718	1·10748
30	0·8762	0·1371	7·3099	7·1945	1·14285
35	0·8498	0·1413	7·5372	7·4182	1·17838
40	0·8251	0·1456	7·7632	7·6406	1·21371
45	0·8017	0·1498	7·9900	7·8638	1·24917
50	0·7796	0·1541	8·2178	8·0880	1·28478

TABLE 21

VOLUMETRIC CONVERSION FACTORS

	cm	cf	US gal	Imp gal	litres	US brl
1 cm	1·0	35·31	264·15	219·95	999·97	6·285
1 cf	0·02832	1·0	7·481	6·229	28·32	0·178
1 US gallon	0·00379	0·1337	1·0	0·8327	3·785	0·0238
1 Imp gallon	0·00453	0·160	1·201	1·0	4·546	0·0286
1 litre	0·001	0·0353	0·2641	0·2200	1·0	0·006293
1 barrel	0·1590	5·615	42·0	35·0	158·9	1·0

In comparing oil with coal, it is unfortunate that British and Continental practices differ in defining the "coal equivalent ton" (cet) in terms of which overall energy computations are often presented. The British conversion takes one ton of an average range of oils and oil products as equivalent to 1.7 tons of "average" coal; the Continental conversion takes one ton of oil as equivalent to 1.43 tons of "standard" coal. (Even more confusingly, the UN statistics use yet another relationship—1 ton oil equivalent (oe) being taken as equivalent to 1.3 ce tons; it is important therefore to check which of these factors is applicable in each case.)

TABLE 22

ABBREVIATIONS USED

cf	cubic feet
cm	cubic metres
Btu	British Thermal Units
brl	barrels
cet	coal equivalent ton
oet	oil equivalent ton
M	thousand (10^3)
MM	million (10^6)
B	billion (10^9)
T	trillion (10^{12})
b/d	barrels per day
brl/a	barrels per annum
psi(g)	pounds per square inch (gauge)

CHAPTER 2

The Nature, Occurrence and Distribution of Oilfields

1. PETROLEUM AT THE SURFACE

HYDROCARBONS are widely disseminated in small concentrations throughout the sedimentary column, and visible traces of liquid oil are not uncommon, usually occurring in the interiors of fossil casts, geodes or concretions. Petroleum is also found at the surface in larger volumes in the form of seepages ("shows") of hydrocarbon gases or liquid crude oil. These may sometimes be of local origin, but more generally are the result of the upward escape of fluids from subsurface accumulations of petroleum.

High-pressure hydrocarbon gases which reach the surface by travelling along faults or fissures may produce the characteristic phenomena of "sedimentary vulcanism". The gas often carries wet mud up with it, to be spattered around the surface vent until a mound is built up with a crater at the summit. If the release of gas is intermittent, the resultant "mud volcano" tends to palpitate, and eruptions in shallow seas may result in the formation of new islands—as has happened in recent years in the Caspian, the Caribbean, and off the Makran coast.

In Trinidad, Venezuela and the Baku area of the USSR, "mud volcanoes" grow along fairly regular lines, which probably coincide with the strikes of faults in the subsurface. The periodicity of the Caspian "mud volcanoes" has also been shown to be related to the gravitational influence of the Moon[1]. In Burma, a basin (or "paung") filled with agitated soft mud is often produced rather than the more common mound or cone. The critical factor which determines whether a mound or basin will be formed is the viscosity of the mud-sand mixture that is

---

[1] Y. Tamrazyan, *Nature*, Dec. 15, 1972, 407.

being ejected. In Central Australia, the 7-mile wide crater of Gosses Bluff, originally believed to have been the result of meteoric impact, is now thought[2] to be the vent of a single enormous eruptive "mud volcano".

Oil may sometimes be held in emulsion in the erupting mud and its presence may therefore not be superficially apparent; on the other hand, the admixture of asphaltic oil may result in the production of "tar cones", as at Mene Grande in Venezuela.

When oil seepages occur at the surface, they usually form pools of heavy, viscous liquids. In some petroleum-producing areas seepages are common, and sometimes also the oil impregnates the surface rocks; in these cases, a large proportion of liquid oil may make the rock relatively plastic, or alternatively the effects of oxidation and polymerization may convert a heavy oil into a semi-solid end-product (*see below*).

Seepages of light, fluid oil are also known, although infrequent; examples from Borneo and Eastern Venezuela are said to be light enough to be directly usable as fuels in internal combustion engines. Such oils may be the product of "wet" gas condensation, or the result of filtration through clays which have removed the heavier constituents of normal crudes. Seepages of light oil cannot travel far from their points of efflux, since the oil is very soon precipitated in streams or rivers, particularly those containing suspended clays. Paraffinic oils in particular are rapidly dispersed by surface waters, leaving very little trace of their presence.

Oil seepages in marine waters are curiously rare, perhaps because any crude oil released in this way will tend to be polymerized and precipitated as a result of contact with clay particles held in suspension in the water. In offshore areas of California, for example, where sea-floor seepages are likely to occur due to the relatively thin "cover" overlying many subsurface oil accumulations, there is often difficulty in determining whether the semi-solid asphaltic material washed onto the beaches is in fact the result of crude oil seepage or of contamination from tanker washings. However, recent investigations have identified a number of offshore oil seepages, the most notable of which lie between Huntington Beach and Point Conception, particularly in the Coal Oil Point area.

Surface seepages were the only sources of petroleum for many centuries, and up until quite recent times shallow, hand-dug pits provided local supplies of oil in some parts of the world, such as Burma and China. At a few places—Pechelbronn in Alsace, Grosni in the USSR and Wietze in Germany—commercial volumes of oil were obtained by drainage from shafts driven into shallow, oil-saturated sandstone beds outcropping at the surface.

In the early history of exploration for petroleum, wells were often

---

[2] T. Ranneft, *Bull. AAPG,* March 1970, 54(3), 417.

drilled in the immediate neighbourhood of oil and gas seepages, since these were correctly considered as providing evidence that petroleum had accumulated in the vicinity. In areas such as Pennsylvania, where there were many seepages from shallow pools, this technique was quite successful. However, although a seepage is usually an indication of an accumulation not far away, the oil may have reached the surface by migration laterally along a fault or bedding plane, so that a well drilled near the "show" can still be far removed in the vertical plane from the underlying accumulation. Furthermore, the existence of a seepage must indicate the loss of a portion of the subsurface oil and gas; the larger the seepage and the longer it has been active, the less "live" oil will remain and the greater will be the chance that its quality will have been affected by the action of meteoric waters and bacteria. Surface seepages therefore have to be viewed with caution as indicators of subsurface oil accumulations.

Liquid petroleum exposed to the effects of atmospheric oxidation and bacterial attack becomes *inspissated*—i.e. the lighter components evaporate and the remainder polymerize and oxidise to form a variety of viscous, tarry, semi-solid or solid residual products, whose mode of occurrence and nature varies with their tectonic history and the chemistry of the crude oil from which they were derived.

Such residual petroleum deposits include the "native bitumens" or "vein asphalts", which sometimes fill vertical fissures, often many feet wide, running through several thousand feet of strata; the mineral waxes, which are derived from paraffinic crudes; the "tar sands" (described on page 353), which may extend over areas of hundreds of square miles; and the horizontal "lakes" of asphalt which occur in different parts of the world—most notably in Trinidad and Venezuela—and which are believed to represent the last stages in the oxidation of asphaltic crude oils released in large volumes as a consequence of the destruction of subsurface reservoirs.

2. THE NATURE OF OILFIELDS

In the subsurface, accumulations of gaseous and liquid hydrocarbons, associated nearly always with salt-water, occur filling the pore spaces of "reservoir rocks" (generally sandstones and carbonates), and are confined within the limits of some form of geologic "trap".

The most typical form of subsurface accumulation is the anticlinal trap which consists of an arch or "upfold" of porous rock, bounded above, below and laterally by impervious beds. Within such a trap there are usually three distinct layers of rock distinguishable by the fluids they contain: gas-saturated rock near the apex or culmination of the anticline—(the "gas cap"), followed by a layer of rock which contains oil, overlying in turn the lowermost layer of salt-water saturated rock.

The fluids occupy the intergranular pore spaces of the rock body, and also any cavities or fissures that may be present.

Although "oil", "gas" and "water" are the terms traditionally used, there are in fact many different hydrocarbons present, and generally also several non-hydrocarbon gases (Table 11, page 16), so that the relationship between the many possible systems and their phases is extremely complex. Furthermore, the separation between the layers is often not clear-cut, since there are generally zones in which gas merges into oil and oil is interspersed with water.

It is important to note that the layers of rock in such an accumulation which are apparently filled with hydrocarbons, in fact also contain a considerable proportion of salt-water. This forms a film around the individual rock particles and fills the intercapillary passages of the rock. The interstitial water is retained by adhesive and surface tension forces, and is therefore normally "fixed", in contrast to the other oilfield waters described below. Interstitial water may occupy up to 50% of the pore volume of a reservoir, so its presence is obviously of great importance in petroleum production operations. Besides reducing the volume of pore space available for oil and gas in the reservoir, it will restrict the flow of non-aqueous fluids through the rock and consequently prevent more than a fraction of the petroleum being recovered. Furthermore, some of the salt content of the interstitial water may be precipitated as the reservoir pressure falls, with the result that the crude oil then contains an unacceptable proportion of sodium chloride crystals, which can only be removed by electrolytic processing.

The salt-water which normally saturates the rock stratum underlying an oil accumulation is called "bottom-water" if it is encountered by drilling *through* the oil layer, or "edge-water" if it is met by drilling laterally *beyond* the oil-water contact. These are "free" waters, in the sense that they are free to move within their aquifers according to the prevailing hydraulic gradients, in contrast to the "fixed" interstitial water.

Oilfield waters are generally called "brines" because, as with sea-water, they contain sodium chloride as the dominant dissolved solid. There are usually also carbonates and bicarbonates of sodium, potassium, calcium and magnesium present, but none of the sulphates that characteristically occur in meteoric waters.

In general, also, oilfield waters contain much higher contents of dissolved salts than the average sea-water (up to 30% by weight compared with about 3·7%), and so the disposal of the brine which is often produced with the oil in older fields sometimes involves considerable problems. (In the USA, an average of three barrels of brine are produced with each barrel of oil. In one state—Kansas—more than 5 MM b/d of brine is produced, which if reduced to solids by evaporation would be equivalent to 65,000 tons of salt.)

It seems likely that oilfield waters represent the concentrated saline fluids produced by compaction effects during the primary migration process (p. 42); the salt-water of the ancient seas originally retained in the sediments was altered by the loss of a considerable proportion of its H_2O content as the result of adsorption on to clay mineral surfaces, so that the compaction currents which were eventually expressed were much more concentrated than the original sea-water, and in addition, had lost their sulphates due to bacterial action.

The amount of gas held in solution in a crude oil varies with the physical characteristics of the gas and of the oil, and with the temperature and pressure conditions in the reservoir. Where free gas occurs as a "gas cap" filling the rock layer overlying the oil, the reservoir is termed "saturated", since there must be sufficient gas present to saturate the liquid oil phase if excess gas is able to appear as a separate "gas cap". On the other hand, a reservoir in which all the gas present is dissolved in the oil is termed "undersaturated", and in this case there may only be two layers present—oil and water. However, when the prevailing physical conditions are changed in the production process, the gas will come out of solution at the "bubble point"; this is often also termed the "saturation pressure", since the reservoir temperature is usually constant, and it can vary from the reservoir pressure down to atmospheric.

"Associated gas" is defined as the free natural gas in contact with crude oil in a reservoir in the form of a "gas cap", so that the production of the gas must be related to the production of the oil. "Dissolved gas" is the natural gas held in solution in a crude oil in a reservoir. (Except in the USA, the term "associated gas" is generally understood to include both "gas cap" and "dissolved" gas components.)

The term "reservoir" is used to describe the parts of a geologic trap which are filled with hydrocarbons and are under a single pressure system. An accumulation can comprise a single reservoir, or a number of reservoirs which may be separated in either the horizontal or vertical planes. A single accumulation is termed a "pool", and an "oilfield" may contain several pools related to a common geological feature.

In some parts of the world, a number of oilfields occur at local culminations or permeability zones within an extensive reservoir bed, and these may form an almost continuous "oil belt" or "trend" (e.g. the Smackover "trend" in the United States). Again, a number of accumulations may be found in a *common geological environment*—for example, associated with carbonate "reefs", large-scale faulting, salt dome movements, etc. over a relatively wide geographic area. In such cases, the regions involved are conveniently termed "petroleum provinces".

The area which an oilfield covers on the ground can vary in extent between a fraction of a square mile and many square miles. Thus, the *Santa Fé Springs* field in the Los Angeles basin covers only 1,515 acres, whereas *Pembina* in Canada covers 755,000 acres. Of course, area of reservoir is only one of several factors which control the size of an oilfield when measured in terms of its oil content (i.e. amount of "oil-in-place"*). The thickness and lithological characteristics of the reservoir beds are also most important. Thus, the largest oilfields in terms of their original oil contents are generally those with the thickest and most uniform development of reservoir rocks. For example, the *Burgan* field in Kuwait, one of the largest in the world, owes its immense volume of oil mainly to the great thickness of more than 1,100 ft of nearly uniform reservoir sandstones which it contains.

In the standard classification of oilfields by size adopted by the American Association of Petroleum Geologists, a "recoverable volume" of oil of at least 1 MM brl is the qualification for the smallest "commercial" field (Group E). Accumulations of this size have sometimes been described in American practice as "profitable", but the term "significant" is nowadays preferred to "profitable", since the latter obviously involves other than volumetric considerations.

A "major" discovery has generally been defined in the USA as one with at least 50 MM brl of recoverable oil reserves; in recent years, this criterion has been reduced to 25 MM brl, and it is noteworthy that only one in about 220 wells drilled in the United States each year nowadays discovers an oilfield of these relatively modest dimensions. At the other end of the scale, the classification of "giant" in American practice has been used to describe fields with at least 100 MM brl of recoverable oil,† of which there are about 310 currently known in the United States.

In other parts of the world, however, this definition of a "giant" is unsatisfactory, since some accumulations may be much larger than US "giants", while still being less economically attractive to develop, due to the great distances which separate them from their potential markets. The phrase "typical Middle East oilfield" is therefore sometimes used to signify an accumulation ten times as large, i.e. containing at least 1,000 MM (1 B) brl of recoverable oil, which would presumably be worth developing in most circumstances.

Recent compilations[3] of the world's most important oilfields have used a recoverable oil content (using present techniques) of half this amount, i.e. 500 MM brl as the convenient international standard for a "giant" field. On this basis, it is instructive to note that while there are only 187 "giant" oilfields in the whole world—i.e. less than 0·7% of all the approximately 30,000 "significant" oilfields that have been discovered are "giants"—nearly three-quarters of all the oil that has ever been produced has come from these fields; and, furthermore, nearly

---

* *The relationship between oil-in-place and recoverable reserves is discussed in Chapter 12.*
† *or one trillion (10^{12}) cu ft of gas.*
[3] AAPG "Geology of Giant Petroleum Fields" ed. M. Halbouty, 1970, p. 502 *et seq.*

82 % of the original recoverable reserves of these "giants" still remained to be produced as recently as mid-1970.

Size or volumetric content of oil is only one among a number of factors which determine whether a particular oil accumulation is "commercial" or "economic". Such factors include the depth, extent and thickness of the reservoir bed, the degree to which it is saturated with hydrocarbons, the ease with which it can be drilled and developed, the reservoir energy available, the chemical nature and marketability of the crude and the distance of the accumulation from points of consumption.

A "commercial" oilfield can therefore really only be defined as one whose development *in the existing circumstances* would be worth the investment of the necessary capital and effort. Obviously, an oil accumulation which is discovered in or near densely populated and industrialised areas could be relatively small both in its reserves and productivity and yet be profitable, whereas another accumulation situated in a remote region where the factors of accessibility, distance, local demand, and perhaps political stability were considerably less favourable would need to be proportionately larger and more prolific to be worth developing. Similarly, offshore fields must be large enough to offset the extra costs and risks that their development entails compared with onshore fields.

3. THE ORIGIN OF PETROLEUM

Early theories about the origin of petroleum were generally related to one or other of a variety of laboratory syntheses of the simpler hydrocarbons by "inorganic" (or more correctly "abiogenic") chemical reactions. Although in recent years there has been some revival of interest in more sophisticated versions of such theories, it seems improbable that petroleum could have been formed as a deep-seated magmatic product, in view of its relatively shallow habitat, its content of heat-sensitive compounds, and the fact that it nearly always occurs in sedimentary rocks. Furthermore, no known abiogenic reactions could produce the complex mixture of hydrocarbons normally present in crude oils.

On the other hand, methane gas is an important constituent of the atmospheres of the outer planets of the solar system and of their satellites, while the tails of comets are largely composed of solid methane. Furthermore, traces of asphaltic hydrocarbons are claimed to have been detected in certain meteorite samples. It is reasonable to conclude, therefore, that *some* terrestrial hydrocarbons may have had an abiogenic origin, related to the cosmic history of the Earth, while most crude oil components were derived from biogenic materials, as is evidenced by the optical activity of petroleum distillates, and the occurrence of branched-chain alkanes and nitrogenous porphyrins in crude oils—probably derived from the chlorophyll or haemoglobin molecules of living organisms.

To what degree the biogenic source materials were of vegetable as opposed to animal derivation is problematic. Methane is so commonly produced from decaying humic material that it is tempting to believe that oil was formed by an analogous bacterial process under different conditions. Furthermore, a close geographic relationship often exists between petroleum and coal deposits in some parts of the world. However, no hydrocarbon other than methane appears to be formed by normal vegetable decay, and a stratigraphic distinction can always be observed between coal-bearing and oil-bearing strata.

Although petroleum hydrocarbons have been found in reservoir rocks as old as the Cambrian and even the Pre-Cambrian, land vegetation did not exist in quantity until much later. The first coals were formed in Devonian times, so that the ancient oils that occur in Ordovician, Silurian and Cambrian beds could scarcely have been generated from the decay of terrestrial vegetable matter. On the other hand, the general evidence for a preponderating marine animal source for petroleum is quite persuasive. Thus, many crude oil accumulations have been found closely associated with beds containing the remains of marine organisms—fish, molluscs, lamellibranchs, foraminiferae, etc., so it seems likely that these and similar sea creatures may have provided much of the basic material from which the oil was formed. There is no reason to assume that any particular marine organism was the exclusive source, and many types may have contributed; part at least of the chemical variations noted in different crude oils may in fact be related to local differences in the composition of the source organisms.

Great concentrations of animal life have always existed in the "paralic" seas bordering the Earth's continental margins, where the upwelling currents provide the highest concentrations of phytoplankton to support life. For example, in California a single Eocene shell bed is known which is more than 4,000 ft thick. Only a small fraction of the total marine animal material that has existed in the world's seas could quantitatively have produced all the petroleum that has been found to date. Thus, an investigation[4] of the Tertiary oilfield zone of the Los Angeles Basin showed that only about 0·08% of the organic matter that originally reached the sediments in that area was ever converted into producible oil.

Marine vegetable materials—algae, seaweeds etc.,—may also have provided a proportion of the biogenic source material, since they would have been likely to exist in large amounts in the near-shore environment in which, as described below, the conversion of living organisms into elementary petroleum probably took place.

[4] K. Emery, *AAPG*, "Habitat of Oil", 1958, 955.

All living matter, both plant and animal, is largely made up of proteins, carbohydrates, lipids and pigments. Of these, the lipid fatty acids would have provided the most likely materials for hydrocarbon conversion, since a simple process of bacterial decarboxylation could have removed their oxygen components, leaving paraffinic hydrocarbons.

The first stage of this conversion process was probably completed soon after the death of the organisms concerned, since fatty acids are to be found in only small traces in both ancient and modern sediments, although they must have provided 10–20% of the original organic detritus*. Under anaerobic conditions, the bacterial removal of oxygen from fatty acids probably proceeds rapidly.

Supporting evidence for an anaerobic environment includes the presence in crude oils of certain compounds which would have been destroyed in the presence of oxygen, and the frequent occurrence of hydrogen sulphide in natural gases, presumably formed by the parallel bacterial reduction of dissolved sea-water sulphates.

Now, sea-water may become anaerobic either by reason of its depth or because of non-circulation. Since the lithology of the sediments in which petroleum is found indicates relatively shallow-water depositional conditions, the lack of oxygen must therefore have been due to stagnation.

Favourable environments would therefore have been semi-enclosed seas and gulfs in which concentrations of marine organic life could have flourished in a limited layer of well-aerated surface water, where abundant phytoplankton supported a dense concentration of fauna and flora. Below this surface layer, the water must have been increasingly depleted in oxygen, with high concentrations of hydrogen sulphide at depth. In these circumstances, there would be little benthonic life to consume the falling pelagic organic débris, which would not be oxidised and destroyed as normally, but instead buried and preserved for bacterial diagenesis.

The sedimentary evidence for the existence of anaerobic conditions is in general the presence of sulphides (notably pyrites) and black or dark-coloured deposits. The formation of pyritic black shales has often occurred in the geologic past, but it is not certain whether an entirely landlocked basin was necessary to produce them or whether stagnant areas in the open sea sufficed. Some barrier to free water circulation was certainly required: this could have been either a physical obstacle or a density barrier resulting from variation in sea-water salinity, perhaps as the result of periodic desiccation.

Lime-secreting creatures could not live in strongly anaerobic conditions, so that where limestones were formed there must have been both

---

* Small amounts of hydrocarbons also occur in modern sediments, but these are different in both their nature and range from those in crude oils.

a stagnant bottom layer of water containing the reducing bacteria, and an upper oxygenated surface zone in which organisms with calcareous shells could thrive. Limestones containing hydrocarbons might then be formed.

If the sea-floor subsided, and sedimentation was relatively rapid, a thick mass of sediments containing the products of the bacterial reduction of fatty acids would begin to accumulate. The form which these precursors of petroleum (sometimes called "protopetroleum") would have taken is uncertain. Perhaps they combined with other organic compounds to form the ubiquitous "kerogen", which makes up about 90% of the organic material contained in clays and shales. This is a dispersed, dark-coloured material, insoluble in acids and organic solvents, which is made up of bacterially altered vegetable and animal detritus. In its chemistry, "kerogen" is variable, and the name describes a range of residual materials rather than a specific compound. The molecular weight is about 3,000, and the basic structure consists of stacked sheets of aromatic rings in which atoms of nitrogen, sulphur and oxygen also occur. At the edges of the sheets, various organic compounds are attached, including normal paraffin chains. Thermal "cracking" can detach these chains from the main "kerogen" molecule to produce stable paraffin compounds.

Now, crude oils contain a number of heat-sensitive compounds whose presence indicates that the oils could not have been subjected to temperatures of, at the most 300°F—perhaps less—in the course of their histories. Crude oil could therefore not have been produced, as was once thought, by the high-temperature distillation of a "kerogenous" clay or shale. It is possible, however, that the much gentler heating of such a potential source rock, carried on over lengthy periods of geologic time (the process termed "eometamorphism"[5]) could have "cracked" the "kerogen" molecules and thus released their paraffin chains. Further lengthy heating, assisted by the catalytic effect of the contiguous clay minerals, could then perhaps have produced soluble "bitumen" compounds, and then the various saturated and unsaturated hydrocarbons, the asphaltenes and other components that are characteristic of crude oils[6].

Increases of temperature sufficient to initiate eometamorphic "cracking" processes could have been produced by the blanketing effect of the sedimentary overburden accumulating within a subsiding basin. The lowermost layers of clays would be raised to a temperature of 150°F, which is believed to be the minimum necessary, when buried to a depth of a few thousand feet—the effective depth in each case being related to the local geothermal gradient. (This is a measure of the

[5] K. Landes, *Bull. AAPG*, 51, (6) 828, 1967.
[6] B. Tissot *et al.*, *Bull. AAPG*, 55, (12) 2177, 1971.

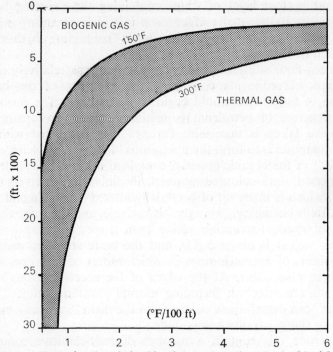

Fig. 4. Diagram illustrating the relationship of thermal gradient, depth of burial and zones of liquid oil generation and destruction.

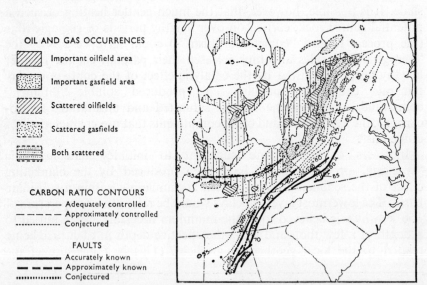

OIL AND GAS OCCURRENCES

Important oilfield area

Important gasfield area

Scattered oilfields

Scattered gasfields

Both scattered

CARBON RATIO CONTOURS

———— Adequately controlled

– – – – Approximately controlled

·············· Conjectured

FAULTS

———— Accurately known

– – – – Approximately known

·············· Conjectured

Fig. 5. Carbon Ratios in the Appalachian Geosyncline. The contours indicate lines of equal heat metamorphism (see also p. 225)

intensity of outward heat flow from the Earth—the result of radio-activity and tectonism; it produces a temperature increase of about 2°F/100 ft, on average, but this can be much higher near orogenic regions.)

Once started, the evolution of hydrocarbons from a kerogenous source rock could be expected to develop exponentially with deeper burial and further temperature increases[7]. At about 300°F, however, any liquid hydrocarbons produced would be converted into gas, and only gas would be formed above this temperature.* (The equivalent depth might be 15–25,000 ft, and there certainly seems evidence to show that com-paratively little liquid oil occurs below such depths.

In mobile areas of high thermal gradient, liquid oil would be formed at shallower depths than in stable areas of low thermal gradient; but on the other hand the *range* of the depths between which oil could be generated would be correspondingly narrower.

Modern techniques of measuring the maximum temperatures sustained by a potential source rock in the course of its history are of value in determining its "maturity"—i.e. the extent to which it is likely to have produced hydrocarbons, and whether these were in the liquid or gaseous phase. Such assessments of source rock viability are of obvious importance in estimating the oil prospects of an area.

Techniques used include the measurement of the degree of darkening of fossil pollen grains, which is diagnostic of the intensity of "cooking" a sediment has undergone; and the assessment of palaeotemperatures by electron spin resonance spectrometry (ESR)[8]. (Of course, palaeotemp-eratures often markedly differ from modern temperatures, since the bed concerned may have been raised or lowered in the course of its history, while some or most of the overburden may have been removed by erosion.)

Long-sustained heating, combined with clay mineral catalysis, would also have been likely to affect the chemical compositions of crude oils after they had once been formed in this way.

There is in fact some evidence of a general increase in the "paraffinicity" of crude oils with the age of the reservoirs in which they are found. That is to say that the proportion of hydrogen increases, there are fewer unsaturated compounds, and the crudes tend to grow lighter with age. Thus, by plotting the average densities of some 7,600 American crude oils against the ages of their reservoir rocks[9], a fairly uniform decrease of density has been shown to occur relative to age—from Pliocene oils (21° API) to Ordovician oils (40° API), although

[7] G. Philippi, *Geochim. Cosmochim. Acta,* 29, 1021, 1965.

* Large volumes of natural gas are, of course also found in "cool" reservoirs. Possibly this may be gas of a different provenance or the result of long-distance migration.

[8] W. Pusey, *Petroleum Times,* Jan. 12, 1973, 21.

[9] G. Hopkins, *Journ. Pet. Technol.,* 2, 7, 1950.

there are notable exceptions (e.g. the Eocene and Permian crudes of Texas and nearly all the Cambrian oils).

In some areas also, notably the Gulf Coast of the United States, there is a progressive evolution of crude oil characteristics with age— from younger, heavier, darker, more aromatic crudes to older, lighter, paler, more paraffinic varieties[10].

In general, younger oils tend to have a smaller gasoline fraction and to be relatively free from paraffin wax. To some degree also, gas/oil ratios are often higher in the oils from older formations, perhaps simply because these are usually deeper beds and have thus been subjected to higher temperatures. However, there are many exceptions to these general relationships.

4. MIGRATION PROCESSES

The process by which hydrocarbons leave the fine-grained clays and shales in which they are believed to have been formed and enter the coarse-grained "reservoir rocks" in which they accumulate is termed "primary migration". The subsequent movement, concentration and segregation of gas and oil within the reservoir rocks in some form of "geologic trap" is "secondary migration".

The essential properties of a reservoir rock are its relatively high porosity and permeability. The *effective porosity,* which depends essentially upon the way the particles which constitute the rock have been deposited and consolidated, is the ratio of all the *interconnected* pores to the overall external volume of a unit of rock (conveniently expressed as a percentage).

While the volume of fluid contained in a unit of rock is controlled by its porosity, the *permeability* of the rock is the measure of its capacity to transmit its fluid content. Primary permeability depends on the number and the dimensions of the interconnected super-capillary sized pore spaces in the matrix of a rock unit. A solid of no porosity would be impermeable to a fluid, but one of high porosity might also be impermeable; on the other hand, a rock could have a very high secondary permeability as a result of fractures, but very little porosity. There is no simple quantitative relationship between permeability and porosity but only the qualitative relationship that a rock is permeable by reason of its effective porosity.

By far the most important types of reservoir rocks are sandstones and carbonate rocks. Thus, one survey[11] showed that about 59% of the production from a number of major oilfields came from "sandstones" and 40% from "carbonates"—these terms being loosely used to include,

---

[10] D. Barton, in AAPG, "Problems of Petroleum Geology," 109, 1934.
[11] G. Knebel and G. Rodriguez-Eraso, *Bull. AAPG,* 4, 554, 1956.

on the one hand, orthoquartzites, greywackes, arkoses and conglomerates, and on the other, precipitated carbonates, muddy limestones, "reef" limestones, and many varieties of dolomite. Other rocks occasionally provide petroleum reservoirs, but these are relatively rare; they include diatomites, cherts, locally fractured shales, and even crystalline rocks.

The porosity and permeability of reservoir rocks may be *primary* or *secondary* in origin. Primary reservoir properties depend essentially on the size, shape, grading and packing of the constituent rock particles, and the degree and manner of their initial consolidation. Secondary post-depositional factors may either enhance or diminish these properties; they will generally be improved by the effects of partial solution and recrystallization, by weathering after temporary exposure, or by the development of fractures and fissures as a consequence of tectonic movements. They will be adversely affected by deposition from percolating fluids, and by increases of pressure or temperature due to overburden or shearing.

Secondary porosity and permeability in carbonate rocks may be the result of the chemical action of meteoric waters containing dissolved CO_2; sub-atmospheric weathering following uplift and exposure has produced good secondary reservoir properties in many limestone reservoirs. Calcite is much more soluble than dolomite, so that when a "mixed" carbonate rock has been subjected to atmospheric weathering, "cavernous" secondary porosity may result.

Fractures in limestones and dolomites are also important sources of secondary porosity and permeability. They may result from lateral folding movements or from changes in vertical pressure. Many prolific oilfields occur in carbonate reservoir rocks of relatively poor primary permeability but high secondary (fracture) permeability. Primary carbonate rock porosity is generally much less important in North American non-reef carbonate reservoirs than is fracture porosity, since in many potential reservoir "lime-sands", calcitic cementation has taken place so that only secondary fracturing can then produce commercial reservoirs. In the Middle East, on the other hand, primary carbonate porosity is very important—e.g. the Kuwait Minagish oolites and the Arab Formation bioclastic "grainstones". However, in Iran, the Asmari Limestone, with low primary permeability, nevertheless has excellent reservoir properties as a result of its extensive fissure system.

The first explanations put forward for the concentration and stratification of the gas, oil and salt-water found in subsurface reservoirs were on a hydrostatic, i.e. gravitational, basis. Since oil is heavier than gas and lighter than water, it seemed to follow that the density differences between the three fluids accounted for their separation. This explanation ignored the special factors involved in the movement of liquids through rocks.

Rock pores typically form a complex network of interconnected

E

passages surrounding the individual rock particles, and many of these passages are of capillary or even sub-capillary proportions. The great majority of sedimentary rocks have been laid down in an aqueous environment, so that their pore passages are likely to be water-wet.* Accordingly, it would be impossible for droplets of oil to move along the narrow, tortuous passages except under pressures sufficiently great to overcome the considerable surface tension resistance they would encounter; gravitational or buoyancy forces would be quite insufficient, as would capillarity effects.

Some other source of energy was clearly necessary. It seems likely that this was provided by the increasing pressure of the sediments themselves. Thus, the lowermost layers of a mass of accumulating sediments would experience a steadily increasing geostatic pressure, together with a temperature increase related to the depth of burial and the local geo-thermal gradient. The clay beds present would suffer progressive compaction, since clay minerals such as montmorillonite have a characteristic atomic lattice structure wherein a large proportion of water is normally loosely combined with the clay, from which it can subsequently be removed in stages by heat and pressure, converting the montmorillonite into illite. Clay beds may therefore be reduced by compaction and dehydration to only a fraction of their original volumes.

The primary migration of the hydrocarbons being generated within a "kerogenous" clay source bed would begin when the depth of burial, with the resultant pressure and temperature increases, had become sufficient to initiate the first stages of clay dehydration. This would be at a depth rather greater than that at which the oil generation process had begun. The corresponding rise in temperature would make the nascent hydrocarbons mobile enough to leave the source beds in which they would otherwise tend to be retained by capillarity and adhesion, and to move away in solution, suspension, or emulsion within the currents of water being expressed from the compacting lattice layers; they would make up only a very small proportion of these fluids—of the order of a few hundred parts per million.

As compaction progressed and additional quanta of water were released from the lattice layers of the clay minerals, further phases of hydrocarbon migration would be initiated, and the process might be continued over a period of time equivalent to the deposition of several thousand feet of sediments.

---

* *Occasional examples of oil-wet reservoir rocks have been reported, but they are rare (the Oklahoma City and Bradford, Pennsylvania, fields are thought to be two such cases). In such instances, it is not clear how the original water has been removed from the rocks—possibly it was a consequence of exposure to the atmosphere at some time in their history, or through the action of a surface-active agent in the invading crude oils. In the relatively rare cases where a sandstone is of aeolian rather than marine origin (as is found in China), there would perhaps be an opportunity for the oil entering in a mixed migration current to be the initial wetting fluid, while any gas might pass on to accumulate in a different water-wet reservoir.*

Clearly, a limiting factor must have been the induration of the sedimentary mass, since unless extensive fault systems or permeable channels were available, the compaction currents would clearly not be able to move through the pore spaces of lithified strata. The general direction taken by these moving fluids would presumably have been upwards and outwards, from the regions of higher to those of lower geostatic pressures. However, in a mass of loose, compacting sediments the migrating fluids would move along complex paths, so that lateral movements between source rocks and reservoir beds could occur, and also, in certain circumstances, downward migration from a higher source bed to an underlying reservoir.

There has been considerable debate over the question of the depths at which the primary migration process becomes operative. It is now generally believed that while it may start when the depth of burial is only 2,000–3,000 ft, "flush" migration in fact takes place at considerably greater depths of perhaps 5,000–8,000 ft. In some parts of the world, notably the Ventura and Los Angeles Basins of California, there is some evidence of primary migration at even greater depths of 11,000 ft and below[12].

It is also not at all certain over what lateral distances the primary migration process can operate. On the one hand, many examples are known of chemically distinct crude oils occupying individual reservoirs which are separated by only small thicknesses of non-permeable rocks. Such occurrences, as in the coastal fields of Peru and California, support the view that migration normally takes place over small distances only, and must be completed before lithification. On the other hand, it seems necessary to accept the occurrence of long-distance migration to account for the huge volumes of oil and gas that have been concentrated in many of the world's major hydrocarbon accumulations.

It is possible that in favourable circumstances extensive "carriers" in the form of permeable beds, unconformities, or even bedding planes, may have served as conduits along which migrating hydrocarbons were able to move for scores or even hundreds of miles, even though the surrounding strata were largely indurated and impermeable to fluid movements.

It is unfortunately impossible to identify such "carriers" in practice, since hydrocarbons moving in much larger volumes of migrating water could not be expected to leave distinctive traces, and even if such traces remained they would long ago have been destroyed by bacterial action.

The efficiency with which the primary migration process operated in removing hydrocarbons from their source rocks is also uncertain. Probably there was a critical relationship between the rates of oil generation and migration in different rocks. If a clay bed were to be

---

[12] G. Philippi, *loc. cit.*, 1965.

finally compacted before all its organic matter which could be converted into petroleum was in fact so altered, the primary migration operation would obviously cease. In fact, the study of ancient non-reservoir sediments indicates that at its best, the process of primary migration probably only removed a small proportion of the hydrocarbons which had been formed, much of which remains still disseminated in the source beds, as bituminous material or insoluble "kerogen".

Finally, it should be noted that oil source rocks presumably include fine-grained carbonate muds as well as organic clays, and calcilulites have usually not compacted to anything like the same degree as clays, as is evidenced by the lack of crushing of delicate fossils. Fine-grained, freshly deposited calcium carbonate is usually mainly aragonite which is later converted to calcite, with a consequent small decrease of porosity, but this process is certainly not sufficient to act as an expulsive process for contained fluids, except on a very small scale. Perhaps, therefore, the lack of an adequate expulsion mechanism accounts for the fact that many carbonate source rocks are also reservoir rocks, no primary migration having occurred. Where adequate permeability is available, secondary migration due to buoyancy separation would presumably operate to separate the gas, oil and water layers. It has been suggested that gas formation will be favoured in these circumstances, as the alteration process by which the original organic material is converted into petroleum will have longer to operate than when the oil moves away in the primary migration process.

5. THE ACCUMULATION OF PETROLEUM

Let us consider the situation that is likely to prevail in the compacting sedimentary column when not only clays are present but also porous sands, perhaps loosely cemented into sandstones. The water expressed from buried clays by compaction forces will tend to move outwards and upwards in the directions of lesser geostatic pressure, until a permeable bed is encountered. Within such a bed, the pressure will in any case be somewhat less than in the surrounding clays, because of its resistance to compaction. If the first permeable bed entered by the migrating fluids comprises a relatively small volume of rock, for example a sandstone lens, oil will be trapped at the upper surface between the sandstone and the overlying clays and will commence to accumulate at this interface. From whatever direction the mixed currents may have entered the lens, they will tend to leave it at the highest point, i.e. the point of least pressure, and the oil will therefore begin to accumulate at this level and will then spread along the sand/clay interface and downwards through the body of the sandstone until eventually, if the oil supply in the fluid still entering is sufficient, the lens will become completely oil-saturated. (However, even when apparently oil-saturated, the sandstone will still contain a certain amount of water retained around its constituent grains in "pendular" contacts, and this, together with the thin layers of water adsorbed by individual grains, and the water held in pores with outlets too narrow to allow oil or gas entry, will make up the "interstitial" water content of the rock mentioned on p. 31).

So far, the fluid leaving the sand at the point of minimum pressure will have been nearly all water, but after the sandstone has become oil-saturated, the liquid flowing out will contain about the same proportion of oil as that flowing in. The next sand body lying in the path of the migrating fluids will then act as a new focus of accumulation, and so on.

The efficacy of a permeability barrier or a coarse-fine interface will be relative rather than absolute, depending upon a number of variable factors. Under hydrostatic conditions, the upward buoyancy of the oil will tend to drive it into the capillary openings of any rock surface lying in its path, so that the greater the oil column, the greater will be the capillary displacement pressure that results and hence the finer will be the capillaries which the oil can enter. In addition, changes of temperature, pressure, oil viscosity or hydraulic flow may cause oil to penetrate a permeability barrier which was once effective.

Whenever a relatively large, "blanket" type bed of permeable sandstone, rather than a small lens of rock, is penetrated by hydrocarbon-bearing compaction currents, the mixed fluids will tend to segregate in the course of their subsequent movements within the bed. The gas and oil components will travel faster updip than the associated water, and will continue to move along the steepest gradients until a barrier to further progress is encountered. The minimum angle of dip needed to permit this segregational movement to occur would depend largely on the density and chemical nature of the oil and the permeability of the rock. It would also be influenced by the direction of flow and velocity of any hydraulic currents that might already be flowing within the bed. The minimum angle of dip required for secondary migration to occur in normal circumstances is thought[13] to be of the order of half a degree, i.e. a rise of about 50 ft/mile.

Now, if at some point within the permeable bed there was a zone of low potential energy where the movement of the compaction currents was arrested by structural, stratigraphic or hydrodynamic factors, an accumulation of petroleum would begin to form within the resultant "geologic trap".

Of course, such a trap must be correctly located both in space and time if it is to interfere effectively with the migration processes. Most traps are in fact *composite* or *combination* traps, inasmuch as more than one trapping factor was nearly always present; certainly, every structural

---

[13] W. Wilson, Rept. of Intn. Geol. Congress, VI, E, 25, 1950.

must be to some degree a stratigraphic trap, insofar as the superior
osity and permeability of the reservoir bed relative to the surrounding
ata is always necessary for any accumulation of hydrocarbons to be
formed. Similarly, few stratigraphic traps are devoid of structural
features. Hence, the feature which appears to have been most important
in producing a particular accumulation must be selected as diagnostic.

(A) STRUCTURAL TRAPS

(i) Anticlines

By far the most common varieties of structural traps are the
anticlines, in which at least 80% of the world's known oil and gas
accumulation have been found—largely because of the relative ease
with which they can be located by geological and geophysical techniques.

Anticlines typically take the form of upfolds of rock which are oval in
horizontal plan. Symmetrical anticlines have dips which are roughly
equal on both sides of their axial planes, while asymmetrical anticlines
have axial planes dipping towards a flank. An extreme case of asymmetry is the "overturned" or "recumbent" fold; when this has actually
been displaced by lateral pressure along a thrust plane over a stable
surface, a "nappe" is formed, as in the Alps.

A truncated or *"bald-headed"* anticline is one which is intersected by
an unconformity, so that the apex has been removed, leaving potential
reservoir beds present only on the flanks.

Although most anticlines have been produced by lateral pressure,
depositional or *compaction anticlines* are also known, resulting from
the "draping" and subsequent compaction by gravity of beds above a
pre-existing "high", e.g. a granite ridge, a bioherm, a fault-scarp, or a
"basement" feature. Thus, in the "buried-ridge" oilfields of Kansas and
the Texas Panhandle, the local sediments were laid down over the
pre-existing granite uplifts of the Nemaha Ridge and Amarillo Mountains, and subsequently compacted and further folded.

Regional uplifts or arches several hundred miles long are known in
the USA and USSR, but hydrocarbon accumulations are generally
limited to relatively small productive areas along the crestal culminations
of such major folds. Most oil-producing anticlines are only a few miles
long and wide, although they can vary considerably in their size, dip,
pitch and closure. In general, strong folds will have associated faults and
therefore potentially fractured reservoirs, although some accumulations
are known which have been preserved by enveloping clays within overthrust structures. At the other extreme, some anticlines in "Shelf" areas
(p. 58) have flank dips of less than a degree and yet are major oilfields.
When anticlines do contain hydrocarbons they may be filled to the
limits of their "spill-full" contours, or on the other hand they may
contain considerably smaller volumes of oil or gas than their dimensions and relief could accommodate, and are referred to as "starved".

The *closure* of an anticline is the vertical distance between its highest point and the plane of the lowest closed structural contour—the "spilling plane", which is the level at which hydrocarbons could escape if the trap were "spill-full" to its maximum capacity. The *"closed volume"* is the volume between the highest point and the "spilling plane".

When anticlines do contain hydrocarbons they may be filled to the limits of their "spill-full" contours, or on the other hand they may contain considerably smaller volumes of oil or gas than their dimensions and relief could accommodate. A useful indication of the hydrocarbon content of a structure is given by the height of the *oil and gas column*— i.e. the vertical distance between the oil/water contact plane and the highest point at which oil or gas is found in the trap.

When the longitudinal axis of an anticline is inclined, the anticline is said to *plunge* in the direction in which the axis becomes lower. Plunge of the axis in opposite directions from a high point creates a dome*, culmination or "high"; plunge from opposite directions towards a low point gives a "saddle". An anticlinal feature superimposed on a regional homocline, its axis lying in the direction of the regional dip, will lack closure and produce an open anticlinal "nose" when its axis dips throughout in the direction of the dip of the homocline. An effective trap can then result only if the "open" side is closed by some additional structural or stratigraphic feature, e.g. a fault or a permeability barrier.

(II) Fault Traps

A rock fracture which results in the relative displacement of the strata on either side of the break is called a fault. Faults are generally due to shearing forces related to earth movements and may vary in size from very small local dislocations to major breaks more than 100 miles long and involving many miles of displacement.

From the point of view of petroleum geology, the most important basis of classification of faults is the relationship between the directions of movement of the blocks of strata intersected by the fault. If the block above the fault-plane has gone down relative to the block below the fault-plane, resulting in a local tensional lengthening of the crust, the fault is called "normal"; but if it has moved relatively upwards, the fault is termed "reverse". A shallow-angle reverse fault, where the average dip of the fault plane is less than 45° is called a "thrust fault".

A structural depression bordered by roughly parallel normal faults with their downthrow inwards is termed a "graben"; while the faults concerned here are usually normal, they may also be reverse. "Half-grabens" are said to result from faults which have their downthrown sides in the same direction. A long, narrow "graben" which has sunk as a

---

* *This term is usually restricted to folds that are less than twice as long as they are wide. Domes are typically produced in overlying beds by the upward movement of an underlying intrusion—e.g. a salt mass.*

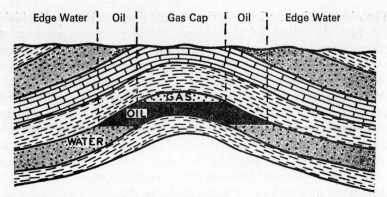

Fig. 6. Idealized cross-section of an anticlinal trap showing gas, oil and water layers.

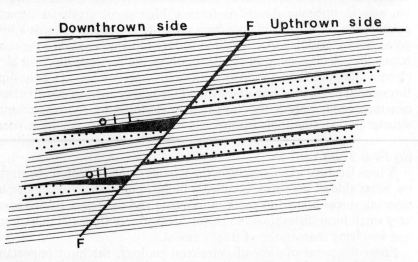

Fig. 7. A fault trap, with oil in porous reservoir beds sealed against the fault-plane.

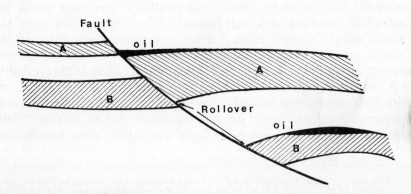

Fig. 8. Typical "growth fault" and "rollover" traps as found in US Gulf Coast fields.

result of crustal tension while being covered with sediments forms a "rift valley" if it is of regional dimensions and is bordered by normal faults.

To some degree, there is a general relationship between faulting and folding. Thus, while some folds are not faulted at all, it is more common for faults to develop in regions of maximum folding stress—i.e. near the zones of sharpest curvature at the culmination of anticlines. These faults may be sufficiently important to have allowed the partial or complete escape of hydrocarbons if the fracturing occurred during the later phases of folding after initial accumulation had taken place. On the other hand, faults can provide physical barriers to oil movement by setting an impermeable bed against a reservoir formation. The trapping capacity of a fault is in fact directly related to the differences in displacement pressures between the reservoir and boundary rocks which it brings into contact[14]. Faults can also themselves provide a barrier to updip migration against which secondary hydrocarbon migration and accumulation can build up. This sealing capacity of a fault will be related to the content of impervious "gouge" material (usually clay) which is present between the fault walls.

In general, faults can produce oil accumulations when they act in combination with other structural and stratigraphic factors; thus, the intersections of a fault plane with a structural "nose" can form a sealed trap, provided that an impermeable surface is brought into contact with the reservoir formation, while fissuring and fracturing associated with faulting may induce secondary porosity and permeability in compact rocks and hence favour oil accumulation within them. Faults also often break up reservoirs into non-communicating sections and are thus important features in oilfield development.

Where accumulations are directly produced by faulting, they are nearly always found on the upper side of the fault, since in these cases oil and gas will tend to flow towards the fault plane. However, on the US Gulf Coast, "growth faults" are important structural factors. These are normal faults which show a substantial increase in "throw" with depth, and across which there is a marked thickening of equivalent beds between the upthrown and downthrown blocks. The resultant effect on the downthrown block is termed "rollover" or "reverse drag", and traps may result on the downthrown side.

(iii) Other Structural Traps

Any mechanism which arrests the flow of hydrocarbons migrating updip within a permeable bed will tend to form an accumulation. If impermeable beds lie above and below the channel of migration, the hydrocarbons in the compaction currents will accumulate against the face of any sealing medium that intersects their path; this could be a salt mass (as described below), an igneous intrusion, a plug of asphalt

---

[14] D. Smith, *Bull. AAPG*, 50, Feb. 1966, 363.

or heavy oil resulting from atmospheric evaporation and oxidation. Traps of this type are sometimes called *sealed monoclines (homoclines)*.*

Salt domes are pillar-like masses of salt, roughly circular in section and often several miles in diameter, which occur in "halokinetic" areas in different parts of the world. They are called "piercement" domes when they penetrate the surface, making topographic hills, or they may be "non-piercement" domes, found only by subsurface drilling. In Germany and the US Gulf Coast, many oilfields are associated with exposed and buried salt domes. Salt dome oilfields are also known in Russia, Romania, Iran and Mexico. (There are also, of course, many salt domes in the vicinity of which no trace of oil has been found, both in these countries and elsewhere.)

Hydrocarbon accumulations typically occur around the tops and sides of these salt masses, in their "cap rocks", which are generally uplifted limestones, or in "super-cap" or "flank" sandstones. The salt is of sedimentary origin and probably represents the last stages of the drying-up of ancient seas. In its composition it may include variable proportions of gypsum and anhydrite. It is believed that as a result of increasing vertical pressure, the salt was squeezed from deep-seated evaporite layers, moving upwards in plastic (or "diapiric") flow through faults and zones of weakness in the overlying sediments. The relationship of salt domes to oil is thus essentially structural, and hydrocarbon accumulations occur in their vicinity when the traps formed by the passage of the salt masses through the sediments are closed and also "timely" with regard to the flow of local compaction currents. Thus, many oilfields are associated with the salt domes of the US Gulf Coast, since there the sediments penetrated are soft clays and sands, so that arched and dipping structures have resulted; whereas in the Middle East the salt has had to fracture relatively rigid limestone beds, with the result that the potential structural traps have largely been destroyed.

(B) STRATIGRAPHIC TRAPS

A stratigraphic trap may be defined as one which is bounded on one or more sides by zones of *low permeability*. Few such traps are in fact completely devoid of structural elements, and most are really "combination" traps, since they show evidence of structural deformation in addition to the wedging-out or lensing of a permeable reservoir bed. In fact, the unconformity or "truncation" traps owe much of their trapping capacity to structural factors, but are nevertheless considered to be "stratigraphic" in the sense that the confining factor is essentially *a change in permeability*. They can be classified according to whether the truncation has been erosional or depositional.

1. *Primary Stratigraphic Traps*

A primary stratigraphic trap results from the formation of a zone

---

* *The terms "monocline" and "homocline" are virtually interchangeable.*

of high permeability during the deposition or diagenesis of a sedimentary sequence which includes source rocks.

Where these traps have a pronounced three-dimensional aspect, and are bounded by an air or water surface at the time of formation (e.g. the reefs and sandbars in groups (a) and (b) below), they have been termed "palaeogeomorphic" traps, or "buried landscapes"[15], in contradistinction to traps formed by lateral facies changes—group (c).

(a) Carbonate Reefs

The fossil carbonate reefs which have been found by drilling in certain parts of the world—Mexico, Alberta, the Ural area and Libya, among others—form a group of often very prolific stratigraphic traps.

"Bioherm" is the term used to describe a thick organic carbonate deposit of this type when it is of limited area and produces a local surface elevation, while "biostrome" describes a relatively uniform reef laid down over a wider area, generally in bedded form.

Ancient reefs have been found ranging in ages from Pre-Cambrian to Tertiary, but with their principal development in Palaeozoic and Mesozoic times.

The largest examples can be very large indeed—extending laterally for several hundred miles and with a thickness of hundreds of feet. (Thus, for example, the Permian Guadalupe Capitan Reef of West Texas and southeastern New Mexico is more than 400 miles long, with a maximum thickness of 1,200 ft.)

Reefs may be classified as "barrier reefs"—typical bedded biostromes built up by sedentary organisms such as crinoids and corals; "patch" reefs—small, thick and relatively isolated and unbedded limestone or dolomite lenses; and atoll or "pinnacle" reefs, which are small in cross-sectional area but of characteristic high relief (e.g. *Golden Spike* in Alberta). An oil production rate of 260,000 b/d which has been recorded from a single well in the Golden Lane of Mexico, exemplifies the enormous potential productivity of reef reservoirs, due to the volume of extremely porous and permeable rock they contain. Thus, a thickness of 959 ft of oil "pay" was found in one 1,200 ft thick section of the *Idris A* (renamed *Intisar*) bioherm in Libya, illustrating the massive nature of the oil reservoir that can exist in these carbonate bodies; this oval bioherm (one of several discovered in a small area) is only about $2\frac{1}{2}$ by 3 miles in its greatest dimensions, but contained some 2 B brl of oil-in-place before production began.

(b) "Marine Bar" Sandstone Lenses

Another type of primary stratigraphic trap is the "marine bar" sandstone body enclosed in a sequence of impermeable clays or shales. The

---

[15] R. Martin, *Bull. AAPG*, 50 (10), 1966, 2279.

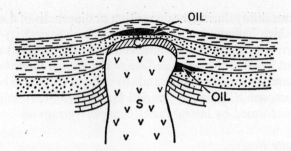

Fig. 9. Typical salt-dome mass (S) intruding pre-existing sediments, producing anticlinal and flank-sealed traps for later accumulation of hydrocarbons. The cap-rock (C) is shown.

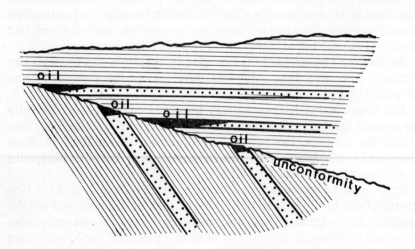

Fig. 10. Trapping of oil in permeable beds lying above and below an unconformity.

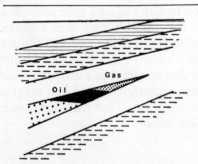

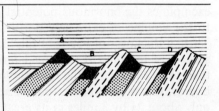

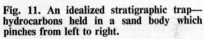

Fig. 11. An idealized stratigraphic trap—hydrocarbons held in a sand body which pinches from left to right.

Fig. 12. Examples of palaeogeomorphic traps—in crestal (A), valley (B), escarpment (C), and dip-slope (D) positions below a mature erosion surface. (After G. Rittenhouse.)

lenticular sandstone and the surrounding shale may be sharply differentiated, or they may interdigitate and grade into each other. Fossil sand-bars, made up of coarse, well-sorted sands, usually elongated in section, and of lenticular cross-section with convex side uppermost, form many reservoirs. In particular, the scouring of the sea floor by fast currents heavily laden with sediment may have resulted in the deposition of coarse sand bodies in the hollows thus formed, lying roughly perpendicular to ancient coastlines. These long and often sinuous "channel" or "valley fill" sands are also called "cheniers" or "shoestrings" in the USA. In Kansas, they are believed to be the result of the filling of previously scoured channels in the underlying Pennsylvanian Cherokee shales. Some of the sand bodies have been traced back to the shoreline sands which run roughly at right angles to them, showing that the streams emptied their contents into open water along the ancient continental shores. Petroleum traps of this type also occur in similar deposits in the Caucasus, Gabon, Brazil, Oklahoma and East Venezuela. In recent years, the discovery of buried submarine canyons off the edge of the Continental Shelf holds out possibilities for hydrocarbon traps in similar fossil valleys (now lying both onshore and offshore) which may have been partly filled by turbidity currents during Pleistocene glacier meltings so that coarse sandstones and silts are interbedded with the usual clays. (An example is the Paleocene Meganos buried channel in the central part of the Sacramento Valley, California.)

(c) Facies Change Traps

The sediments eroded from a land mass and deposited in a neighbouring sea will generally diminish in coarseness as the depth of water increases, and variations in sea depth caused by local uplift or subsidence will result in a variation in the coarseness of the sediments deposited. Thus, a permeable sandstone bed which is deposited roughly parallel to the ancient shoreline, may grade off laterally into less permeable beds. If subsequent tilting occurs, so that impervious strata deposited further out to sea now lie updip of this sandstone mass, permeability barriers will then surround it on every side. Such shoreline "changes of facies" from relatively coarse sands to impermeable clays and shales can produce important traps, whose essential feature is the local variation of facies and permeability in beds of the same age: sandstone may grade into shale, or permeable limestone into impermeable limestone or clay.

Facies-change traps of this type will only be completely closed on all sides if the permeability phase-out zone is curved, with its concave side down-dip, or if it is combined with a structural feature such as a plunging anticline or "nose". Hydrocarbon accumulations will then be confined by the updip limits of permeability—i.e. the "pinch-outs" of the permeable beds.

Major hydrocarbon accumulations have been found in many simple traps of this type in which sandstone reservoir beds are limited by facies changes combined with minor structural closures. Perhaps the most important examples lie in the Bolivar Coastal Fields area of Venezuela.

Other sandstone fields which are examples of this type of trapping are *Burbank* and *Hugoton* in Kansas, *San Ardo* in California, *Pembina* in Canada, and the Pliocene fields of *Ferrandina* and *Pisticci* in Italy.

In compact carbonate rocks, porosity may be induced by dolomitization, and the updip limits of dolomitization will therefore usually coincide with the limits of hydrocarbon accumulation. An excellent example is the huge *Lima-Indiana* field of Indiana and Ohio, in which the accumulation of hydrocarbons is controlled by local dolomitization (which here is thought to be primary in its origin). Another example is the *Carthage* field in Texas, where a porous dolomitized limestone "pinches-out" updip into an impervious limestone.

2. Secondary (Unconformity) Traps

These traps are the result of the presence of an *unconformity*—i.e. a break in the sequence of local geologic deposition, marked by an erosion surface, above and below which the beds are of different ages. The intervals of time marked by unconformities can vary greatly; sometimes the beds above and below the erosion surface are nearly parallel ("disconformity"); more usually they lie at an angle ("angular unconformity").

Traps may be formed both below and above the plane of unconformity, which may thus provide either the upper or lower confining limits of a trap. Sometimes (e.g. *Prudhoe Bay* in Alaska) both types of trap are found in the same accumulation.

(a) Traps below the unconformity

As a result of post-depositional processes which have affected the upper surface of the underlying bed, greatly increased porosity and permeability may be produced in it by water circulation, weathering, recrystallization, etc. When this surface is then covered by impermeable beds laid down by an advancing ("overlapping" or "transgressing") sea, an excellent lithologic trap may be formed below the unconformity.

The transgression by advancing seas over a previously exposed and truncated (i.e. "bald-headed") uplift has produced many important hydrocarbon traps, where impermeable beds were laid down across the exposed updip edges of potential reservoir formations. Many examples of this type of trap are known in the United States, of which two of the most famous are the *Kraft-Prusa* and *Oklahoma City* fields (pp. 231/4), where the principal reservoirs lie in the Cambro-Ordovician Arbuckle dolomite below the pre-Pennsylvanian unconformity. Other examples are *El Dorado* (Kansas) and *Yates* (Texas).

(b) Traps above the unconformity

When permeable beds meet an underlying unconformity they are said to "buttress out" or "wedge out" against it, and traps are formed in these cases above the plane of unconformity, which is then the underlying seal, provided suitable overlying cover is also available.

This type of trap is often formed where permeable sandstones are unconformably laid down by *offlap*—i.e. marine regression—against an impermeable uplift, and subsequently sealed by the deposition of overlying shales. A good example is the Texas Panhandle accumulation, in which arkosic "granite wash" reservoir sandstones wedge out against the buried Amarillo uplift.

(c) Hydrodynamic Traps

In some anticlinal oilfields and gasfields (particularly in the Rocky Mountains province) it has been observed that the oil-water or gas-water contact planes are not horizontal but have a demonstrable dip across the structure. To explain these occurrences, it has been suggested[16] that under certain circumstances, the down-dip hydrodynamic flow of formation waters may act as a trapping agency, neutralising the up-dip movement of hydrocarbons.

In most cases, a petroleum trap must be *closed* in both vertical and horizontal planes by the confining factors if a viable accumulation is to result; hydrodynamic traps appear to be exceptional in this respect.

Where hydrodynamic forces are active, local hydrocarbon accumulations may lie asymmetrically within structural traps or alternatively they may form pools in open structures such as "noses" and "terraces", or on the flanks of homoclines or even of synclines, as appears to be the case with the gasfields in the *San Juan* Basin of New Mexico. Furthermore, the effect of a hydrodynamic pressure gradient in a structural trap would be to displace the oil accumulation down the flank of the structure in the direction of the flow of water; and this may be the reason for the curious distribution of some of the pools in the Rocky Mountain oilfields.

Barren Traps

A trap in any potential reservoir bed which contains salt-water but no hydrocarbons is termed "barren". The absence of hydrocarbons may be due to one or more of the following causes: (i) inadequate source rock or reservoir rock, (ii) inadequacy of the trap itself, (iii) dispersal or destruction of the accumulated hydrocarbons. Since both source rocks and reservoir rocks may be expected to be relatively plentiful in a sedimentary sequence, the other factors are the more likely explanations of failure.

---

[16] see for example M. K. Hubbert, *Bull. AAPG*, 37, 1954, 1953; and 50, 2504, 1966.

Probably the most critical feature in the formation of a trap is its *timeliness* relative to the inception of the local secondary migration processes. If the trap is structural, it must clearly have been formed before the migrating fluids reached it; and if the closure *at that time* was insufficient to retain the hydrocarbons which entered, then an accumulation will not have been formed, even if subsequently the relief of the fold was increased. Thus, for example, the *Kelsey* dome in Texas is barren in the Woodbine sandstone although it is a more pronounced surface structure than the *Van* structure only a few miles away; the generally accepted explanation is that it did not come into existence as a fold until the Eocene, by which time the *Van* oil had already accumulated.

With stratigraphic traps resulting from post-diagenetic increases in porosity due to weathering or other factors, the period in which accumulation could take place must be considerably later in time than where the reservoir rock is a sand-bar or carbonate reef formed penecontemporaneously with the source rock. A striking example of late accumulation is the *Florence* field of Colorado, where the oil could clearly not have accumulated until after fissuring had occurred, and this could only take place after the lithification of the Pierre shales.

Trapping conditions may also have been inadequate in that although the trap itself was timely, i.e. folding or faulting had occurred at the "right" time relative to primary migration, secondary migration still did not produce an oil accumulation within it. The oil may perhaps have remained thinly diffused along the upper surface of the carrier bed, unable to move upwards through the coarse-fine interface but also not moving updip to coalesce and concentrate into an accumulation. This may have been due to the insufficient local velocity of the water currents, or in some cases to the retentive forces of oil-wet minerals. The traces of heavy oil which are commonly found along the upper surfaces of otherwise water-wet reservoir beds may in fact be evidence of this inadequacy of trapping. If such beds are subsequently exposed at the surface, due to erosion or uplift, their water content will eventually evaporate and leave tarry oil deposits to form a potentially deceptive type of "show".

The dispersal or destruction of a hydrocarbon accumulation once formed may have occurred as the consequence of tectonic activity—intense folding, faulting, igneous activity, strong metamorphism—tilting which allowed the trap to open, or erosion which exposed the reservoir to surface waters.

In some cases, traces of the original oil may remain in the form of asphaltic or bituminous residues, inspissated oil or wax deposits along the fault-planes, or there may be indications, such as blackened fossil pollen grains, of the incipient metamorphism of the reservoir rock.

Many examples of the remains of ancient oilfields are known in different parts of the world where the sedimentary cover has been inadequate to protect against faulting and erosion; in such cases, only heavy oil or solid asphalts may remain. However, other reservoirs may have been completely eroded away in the course of geologic time, leaving no traces. Their hydrocarbon contents have presumably been oxidised and dispersed, or possibly have been re-accumulated in some later migration-trapping cycle.

It is generally not essential for a reservoir bed actually to be uncovered to the atmosphere by erosion for its oil content to be lost. The destructive activities of bacteria transported downward by meteoric water circulation have probably transformed or destroyed many oil accumulations in the past. It seems likely therefore that besides providing the increase of temperature needed to generate a crude oil, a reasonably thick sedimentary cover would be needed to protect any accumulation, both from the ravages of water-borne bacteria and also from the fracturing resulting from tectonic movements. In favourable circumstances, the timely deposition of a relatively thin but very impermeable "cap rock" could also provide an effective seal—particularly if the protective bed was sufficiently plastic to bend rather than fracture when subjected to tectonic forces (many Middle East oilfields have been preserved in this way by seals composed mainly of evaporite beds).

Where an unconformity occurs which is stratigraphically not far above a buried fold, it is always possible that the oil which had once accumulated in the structure has been subsequently lost by a combination of upward leakage and downward water circulation, even though the reservoir bed itself was never exposed. Subsequent resubmergence and the deposition of younger sediments may thus disguise the reason for the dispersal of the original accumulation.

Traps which contain only non-associated gas are also difficult to explain. The absence of oil in such cases may be due to a variety of factors—changes in source materials and conditions of genesis, differential migration, metamorphism, bacterial transformation, etc. In general, the increases in temperature associated with depth of burial and high thermal gradients certainly seem to favour the genesis of gas rather than of liquid oil (p. 39), although this relationship is not everywhere apparent.

6. THE DISTRIBUTION OF PETROLEUM

(i) The Framework of the Earth

The continents comprise about 30% of the Earth's present surface area. They are made up of varying thicknesses of sedimentary rocks—with an average of about 3 km and a maximum of perhaps 15 km—underlain by about 20 km of metamorphic gneisses and granites, which in some areas outcrop at the surface. Below this layer and differentiated from it by a change in seismic velocity at the "Moho"* is a deeper basaltic layer thought to be composed of magmatic rocks (eclogites or peridotites), which nowhere reaches the surface. Geophysical data show that the total thickness of the continental crust varies between about 25 km and 75 km.

In contrast, it has been deduced from seismic data that the oceans are underlain by a relatively uniform basaltic crust, only about 12 km thick on average, above which lies a thin sedimentary cover of mainly abyssal-type sediments.

The continents are all made up of a relatively few geotectonic components: the "Shields", platforms, block-faulted areas, intracratonic and marginal basins, mountain belts and volcanic plateaus. Different continents have different proportions of these basic elements.

(a) The Stable Areas

About a quarter of the present land area of the Earth consists of very ancient, highly metamorphosed Pre-Cambrian rocks, which have seldom if ever been covered by the sea. These areas have been termed "Shields" or "Cratons" and form the stable nuclei of the continents.

At various times, the "Shields" sustained gentle tilting and submersion of their edges, so that shallow seas encroached and deposited sediments around the central nuclei; the thickness of these sediments does not usually exceed three miles. Much of the Earth's present land surface is made up of the resultant "Platform" or "Shelf" areas, which form gently-dipping, low-relief regions of sub-continental extent—e.g. Western Siberia and the Great Plains of North America.

As a result of deep-seated crustal processes and large-scale, vertical ("epeirogenic") movements, the platform areas were often subjected to regional warping effects, and in some cases were broken up into smaller blocks.

Some of these blocks have remained stable since Pre-Cambrian times, while others have subsided and been covered by younger strata. The blocks are often bordered by belts of relatively gentle "paratectonic" folds, produced by the deflection of regional stresses around them. Examples of stable fault-blocks bordered by folds of this type are the Colorado Plateau of the United States, the Moroccan Meseta, the high plateau of Algeria, and the Pannonian "block" of eastern Europe.

The broad "swells" resulting from epeirogenic movements formed the regional arches or "anteclises" (anticlinal structures often many miles long, as in Siberia)—and the sags the intracratonic basins (or "synclises") mentioned below. More localised vertical ("taphrogenic") movements produced the block-fault mountains and rift valleys. Taphrogenic

---

* The Mohorovicic seismic discontinuity.

movements also affected the underlying crystalline "basement" rocks in some parts of the world, producing deep-seated structures which are sometimes reflected in the overlying strata.

"Shelf" areas are customarily divided into stable and less stable categories. The unstable shelves form the transition or "hinge" zones between the stable and mobile areas of the Earth's crust and are characterised by having been subjected to cyclical, oscillatory rhythms of sedimentation. Although the well-aerated shallow-sea deposits laid down on them would be unlikely to include source rocks, the many widespread, gently-dipping sandstone and limestone beds would form excellent reservoir rocks. If, as seems probable, extensive migration of hydrocarbons subsequently took place out of the mobile basinal areas bordering the "hinge" zones, major accumulations of petroleum could be formed wherever suitable trapping conditions (mainly "graben" or "half-graben" structures) had been produced.

One analysis has shown that "giant" hydrocarbon accumulations are more likely to be preserved in these tectonically stable environments than outside them. Thus, 83% of 221 defined "giants" has been found in platform, semi-platform or platform margin areas[17]. Another statistical analysis has shown[18] that 50% of the oil content of 236 "major" oilfields outside the USSR lay in the "hinge" areas, compared with $13\frac{1}{2}\%$ in the "Shelf" zones. The latter regions were found to contain more numerous, but less prolific fields than the "hinge" areas.

(b) The Mobile Areas

Around the margins of the stable areas, tectonic forces which were perhaps related to the movements of "lithospheric blocks"*, at different times produced a series of downwarpings or depressions in the crust of the Earth. Within these *sedimentary basins,* whose floors continued to subside over lengthy periods of time, great thicknesses of sediments accumulated.

It seems likely that both source and reservoir rocks would be included in this infilling, so that in due course hydrocarbons would have been generated in the central areas of the basins and, with increasing geostatic pressure and rising temperature, would have migrated towards the peripheries, where both structural and stratigraphic traps would have

---

[17] M. Halbouty, *Bull. AAPG,* 55 (2) 341, Feb. 1971.

[18] G. Knebel and G. Rodriguez-Eraso, *loc. cit.* 1956.

* *According to the theories of "plate tectonics", the Earth's surface—both continents an oceans—is made up of a few rigid, aseismic "lithospheric blocks", whose boundaries are marked by zones of seismic and volcanic activity. It is believed that these "plates" can move as a result of temperature-induced convection effects between the lithosphere and the underlying asthenosphere. When two plates are propelled against each other, the leading edge of one may be crumpled to form an "island arc" or a mountain chain, while the leading edge of the other is thrust downwards into the Earth's mantle, with the evolution of intense heat and resultant vulcanism. The gap is then filled with new upper mantle material welling up along the trailing edge.*

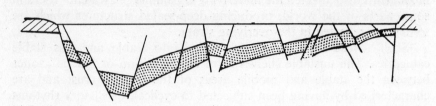

Fig. 13. "Graben" and "half-graben" fault structures exemplified in a cross-section of the Rhine Valley.

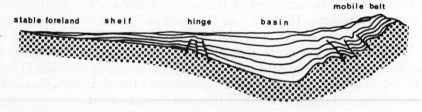

Fig. 14. Idealized cross-section of an asymmetric sedimentary basin lying between a stable "foreland" and a "mobile belt".

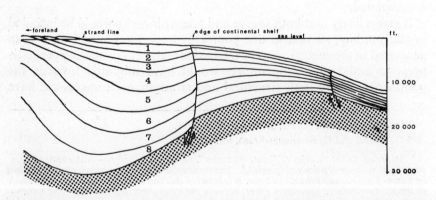

Fig. 15. Cross-section of a typical "open" basin as found on the US Gulf Coast. There appears to be no borderland, and the foreland is far inland. The age of the beds range from Quaternary (1) through Pleistocene to Cretaceous (2 to 7), then Jurassic and older (8). (After Moore).

been likely to occur. Of course, every sedimentary basin, and every point within it, would not have been equally favourable for petroleum generation. Too thin a sedimentary section would be insufficient to produce much oil; too thick a section might result in high temperatures and the conversion of the oil to gas. In addition, much must have depended on the history of sedimentary filling. Thus, those basins in which partial closing or "silling" occurred, as in Lake Maracaibo or the Caspian Sea, would probably have generated greater volumes of oil than others which remained relatively open throughout their histories. There seems to be a tendency for older reservoirs to occur on the outer flanks, and younger oil to be found basinwards, presumably because these areas represented successive zones of optimum oil generation, i.e. of critical depth of burial. (Examples of this can be noted in the US Gulf Coast, where the oil is found ranging in age from Jurassic to Pleistocene in parallel belts, or in the Mesopotamian basin, where Jurassic oil on the "foreland" shelf is followed in turn by Cretaceous and Tertiary accumulations towards the deep basin.)

Sedimentary basins have been classified in different ways, according to their shapes, histories or tectonic origins. However, mention can only be made here of three principal types of basins which are of special interest in petroleum geology.

The most important group from this point of view are the *"asymmetric basins"* which lie between the mobile mountain chains and the stable areas. In cross-section, these basins have their steeper flanks on the mountain sides, and in many cases their gentler flanks seem to have moved with time towards the stable platforms by a tectonic process which is not yet understood. In consequence, some of the basin sediments became absorbed into the mountain "borderland", while new deep basin troughs were formed by the gravity sagging of part of the stable "Shelf" or "foreland".

"Dynamic" basins of this type have always been favourable areas for the accumulation of petroleum, since they contain varied, shallow-water sediments which are similar in their nature to those which have gone to make up the neighbouring orogenic mountain chains, but which are unmetamorphosed and less deformed. The great thickness of clays and shales deposited in the deeper, central areas of the basins provide potential source rocks, while the sandstones and limestones lying on the shallower flanks make excellent reservoir rocks. Traps result from the many unconformities, or the folds produced by gravity slumping and lateral compression. Periodic desiccation results in the deposition of evaporite beds which make excellent seals; all the basic requirements for the generation and accumulation of petroleum are therefore likely to be present, and asymmetric basins of this nature in fact include some of the world's most important oilfields—as, for example, those mentioned below.

In East Venezuela, the Maturin Basin lies between Tertiary mountain chains on the north and the stable Guyana "Shield" on the south. It is asymmetric, with a steeper northern flank, and its principal axes of deposition have migrated southwards between Eocene and Pleistocene times, probably because of isostatic adjustments.

In Western Canada, the asymmetric Alberta Basin is bordered to the west by the mobile belt of the Rocky Mountains, while on the east and northeast the relatively flat and undisturbed sediments of the Interior Plains platform thin towards the Pre-Cambrian "Shield" which is exposed in the northeast corner of Alberta.

In the Middle East, the Mesopotamian Basin lies between the tight folds of the Zagros Mountains and the gently dipping platform area which borders the Arabian "Shield".

A second important group of basins comprises the *"open" basins* which occur on the margins of the continents—particularly on the flanks of the Atlantic, and to some extent the Indian Oceans. These basins have probably resulted from continental "rifting" processes; there are no "borderlands" discernible, while the "forelands" may be far inland. Basins of this type are usually filled with a wedge of sediments that is thickest along the present-day coastline but thins both inland and out to sea. Extensive lateral and step-faults are common on the seaward side of such basins, which are often associated with salt dome uplifts. From the point of view of petroleum geology, the Gulf Coast area of the United States is the most important example of an "open" basin; other smaller productive basins of this type have been discovered in recent years on the west coast of Africa and along both the east and west coasts of South America.

The *intracratonic basins* differ from either of the previous groups inasmuch as they are underlain by "continental" type crust, whereas the others are underlain by "intermediate" type crust—i.e. crust which is intermediate in its nature between "continental" and "oceanic" types. In this context, it must be remembered that, perhaps as a result of "plate" movements, some continental areas are apparently separating, while others are being drawn together. Hence, the present-day margins of the continents do not everywhere coincide with the limits of the oceanic and continental types of crust, so that cratonic basins may extend offshore (e.g. the North Sea basin), while intermediate-crust basins may occur in areas which are now cratonic (e.g. the South Caspian basin).

(The oceanic basins which lie beyond the continental margins have not, as yet, been shown to hold commercial petroleum accumulations, although some oil indications have been found.)

Most intracratonic basins are associated with Palaeozoic fold trends. They include the circular or elliptical depressions which lie within stable areas, such as the Michigan and East Interior Coal Basins of North America; the complex, multi-cycle "composite" basins of the Rocky

Mountains, West Texas and the Ural-Volga area, where the formation of the structural basins were not synchronous with their infilling; and rift, "graben" or "half-graben" depressions on the edges of many stable areas, as for example the Balcones fault zone of the United States, the Sirte Basin of Libya, or the Suez "graben" of Egypt.

(c) Oilfields and Orogenies

In addition to these groups of sedimentary basins, all of which are roughly circular or oval in plan, another type of downwarping developed at various times. This took the form of a long and relatively narrow trench or "trough", sometimes extending for many hundreds of miles. Such troughs were not always continuous, but could be made up of a series of separate but contiguous linear depressions, which, in the course of time, became filled with great thicknesses of typical shallow-water sediments, perhaps several miles deep. At various times in the course of the Earth's history, these masses of sediments were subjected to processes of "orogenesis"—violent folding, and uplift, accompanied by volcanic activity*.

As a result, the uplifted and contorted sediments sometimes spilled outwards over the stable areas lying on the flanks of the troughs, subsequently to become the roots of mountain chains, often hundreds of miles long. Being composed of material derived from the weathering of the upper crust (with an average specific gravity of 2·7), they were isostatically uplifted whenever the downward drag weakened, since the specific gravity of the surrounding deeper crustal rocks was probably 3·0–3·4.

There have been three principal phases of orogeny during post-Cambrian times, although there is an increasing body of evidence for many other periods of lesser activity. In the Caledonian (Silurian-Devonian), Hercynian or Variscan (Carboniferous-Permian), and Alpine (Tertiary) periods, "geosynclines" were formed, filled with great thicknesses of rapidly accumulated sediments, compressed laterally into folds and isostatically uplifted into mountain belts in the cores of which there was great igneous activity and magmatic penetration. Many of these mountain belts were subsequently largely obliterated either by the foundering of their roots, by epeirogenesis or by erosion.

Remains of Caledonian mountain belts occur in parts of Britain and Scandinavia, while in North America, the Appalachian mountains are a complex of two orogenic systems corresponding roughly to the Caledonian and Hercynian system of Europe. A Hercynian belt still survives in Europe as the Ural Mountains, and remnants of a southern

---

* The term "geosyncline" has been used to describe these thick piles of sediments accumulating in a subsiding trough and affected by orogenesis—although considerable debate has centred round attempts to classify and distinguish between different varieties of geosyncline.

belt are found in southwest Ireland and the Harz and Bohemian Mountains of Central Europe.

Most of this southern Hercynian belt foundered to form the Tertiary Tethys geosyncline, which extended for many hundreds of miles between a northern landmass of Eurasia and a southern landmass of Gondwanaland in late Mesozoic times. In the late Mesozoic and early Tertiary, a complex series of orogenic mountain-building movements began which eventually produced the mobile mountain chains extending across the world from the Pyrenees via the Alps, Carpathians, Caucasus, Taurus and Himalaya ranges to the mountains of Indonesia and New Zealand. About the same time, another narrow geosynclinal belt was forming along the western coasts of the Americas and around the Pacific Coast of Asia. This in due course produced the group of Tertiary mountain ranges which now circle the Pacific, including the Rocky and Andes mountain systems.

The youngest of these mobile belts are predictably areas where earthquake shocks are common. In fact, a plot of the epicentres of major earthquakes shows close coincidence with the Mediterranean and circum-Pacific orogenic belts. Volcanic activity is also typical of the central, structurally weakest zones of the orogenic mountain chains, so that the preservation of any hydrocarbon accumulations in these belts would be unlikely. However, their influence on petroleum accumulation processes has been important, since the lateral compression forces which formed them also produced less violent folding and many structural traps within the neighbouring sedimentary basins.

Oilfields associated with the Caledonian and Hercynian orogenies are understandably relatively infrequent, due to the time such accumulations have been "at risk". The fields along the western front of the Ural Mountains in the USSR provide some important examples, however, while in the United States, a number of major Palaeozoic oil accumulations occur along the outer rim of the Appalachian Mountains.

The structures produced by the "Alpine" orogeny on the other hand, provide many examples of petroleum accumulations in anticlinal traps. Thus, in Europe, oil pools of varying sizes occur in the foothills of the northern Alpine fold zone, from the Pyrenees to southern Germany, Austria, Transylvania and the Carpathian fields of Romania. Thereafter, the northern fold line crosses the floor of the Black Sea, and emerges to form the structures of the Caucasus oilfields, continuing into the Apsheron Peninsula. It then strikes across the Caspian Sea into Central Asia by way of Cheleken and Turkmenia to Ferghana, with a number of associated oilfields, and to the northern rim of the Elburz-Hindukush ranges. The Kwen Lun and neighbouring ranges north of Tibet continue across China to join the circum-Pacific girdle south of Japan.

The southern branch of the Alpine orogeny is associated with oil accumulations in the foothills of the Atlas Mountains, the Apennines of Italy, and the Dinaric Alps of Yugoslavia; it then continues in the great arc of the Zagros Mountains, in the foothills of which are many prolific Middle East accumulations. From Arabia, the same trend may strike through Baluchistan, south of Bandar Abbas, or alternatively may link up with the Carlsbad ridge off Socotra.

In Pakistan and India, the Alpine fold line stretches from Karachi to the Salt Range and then along the southern Himalayas into Assam and Burma. Oilfields occur in the north Punjab, in Assam, and in the Arakan Yoma and Pegu Yoma hills of Burma. The same folding movement extends to the Barisan Mountains of Sumatra and then passes through Java, Madoera and the Soenda Islands. Another branch turns north through Borneo to Sakhalin, Kamchatka and Japan. Oil is found in varying quantities along both these belts.

On the other side of the Pacific Ocean, many important oilfields lie along the Rocky Mountains of North America, from Alberta through Montana and Wyoming into Colorado. A younger, late Tertiary belt lying farther west is associated with the oilfields of California and Mexico.

In South America, the trend of the Rockies is continued in the Andes, and this range with its subsidiary mountain belts provides the dominating structural feature. The Brazilian "Shield" has acted as a north-eastern "foreland", and oil is found all the way along the Andean foothills, as far south as Tierra del Fuego, and in the foothill structures of the Cordillera ranges of Colombia, Venezuela and Trinidad.

(ii) Distribution in Time

Commercial accumulations of petroleum have been found in sediments of nearly every age. However, it seems that all ages were not equally favourable for the generation and concentration of oil and gas, presumably due to variations in the richness of the marine raw materials available and the efficiency of the conversion, accumulation and sealing processes.

A comparison of the past production plus the unproduced oil reserves of some of the larger US oilfields with the geologic ages of their principal reservoirs has indicated[19] that in the United States the Ordovician, Cretaceous and Tertiary were the most favourable periods for oil accumulation; while relatively the least oil is found in the very old (Cambrian and Pre-Cambrian), the very young (Pleistocene), and the mainly terrestrial (Triassic) beds. However, the discovery of Triassic accumulations on the North Slope of Alaska shows that this formation is far from devoid of petroleum prospects.

In other parts of the world, the relationship of hydrocarbon distribution to rock age is different, although it is generally true to say that

---

[19] G. Hopkins, *J. Petrol. Tech.*, 2, 6, 1950.

most hydrocarbons have been found in Tertiary and Mesozoic reservoirs[20].

It was calculated, in 1962[21], that as much as 83% of all the world's oil and gas produced up to that year, and 87% of the remaining reserves, lay in Tertiary and Mesozoic strata; the upper half of the Tertiary alone accounted for more than 30% of all the oil ever discovered.

Confining the analysis to "giant" (500 MM brl) fields only, it was estimated[22] that 24% of their ultimate (produced and proven) oil and gas reserves lie in Tertiary beds, 63% in Mesozoic beds and 13% in Palaeozoic beds. If the calculation excludes the Middle East Gulf region, the results are notably different—with 40% in the Tertiary, 39% in the Mesozoic and 21% in the Palaeozoic.

The predominance of younger reservoirs is to be expected since accumulations in these rocks will have been "at risk" from tectonic disturbances or erosion processes for the shortest times. The USSR seems to be an exception to this general rule; according to one analysis[23], more than half (52·1%) of proved reserves lie in Palaeozoic rocks. (This is now probably an underestimate, in view of the recent discoveries in Siberia.)

The very ancient oil accumulations have had to survive for so long that it is remarkable that any of them still exist at the present time. However, many important pre-Silurian (mainly Ordovician) oilfields lie in a belt across North America through Kansas, Oklahoma, Texas and New Mexico, with accumulations occurring in the Arbuckle, Ellenburger and Simpson formations. The Trenton Limestone of Ohio, Indiana, and Michigan has also produced large volumes of oil and gas. Cambrian oil is less common, but nevertheless several billion barrels of oil and equivalent amounts of gas have been produced from Cambrian reservoirs in the United States, Siberia and North Africa.

Until very recently, no trace of fossil organic life, and consequently no petroleum, had been detected in the great thicknesses of Pre-Cambrian rocks, which represent a time interval of some 2,500 million years. The Pre-Cambrian was therefore considered to be of no interest from the point of view of petroleum geology. However, the discovery of "wet" gas in Australian Pre-Cambrian rocks has caused some revision of thought. Thus, a well drilled in 1963 in the Amadeus Basin of Central Australia produced a strong flow of methane with some propane from a Pre-Cambrian dolomitic limestone. Similar gas shows have also been reported from the Siberian "Shield". It is therefore possible that hydrocarbons will be found in other sections of relatively unmetamorphosed Pre-Cambrian strata. They may be abiogenic in origin; or

[20] G. Knebel and G. Rodriguez-Eraso, *loc. cit.* 1956.
[21] F. Gardner, *Oil Gas J.*, 60, 42, 15 Oct. 1962, 265.
[22] M. Halbouty, AAPG "Geology of Giant Petroleum Fields", 1970, 502.
[23] Mirchink, *et al.*, *Neft. Khoz.*, 1959, 4.

alternatively, it is possible that the biogenic source organisms had not yet learned to use the calcium carbonate in the ancient seas to form hard skeletal structures, which would explain the absence of associated fossils. Pre-Cambrian hydrocarbon accumulations will in any case be very rare, needing to have survived the destructive effects of so many orogenies since their formation.

On the other hand, the very youngest strata may not have been buried deep enough to allow petroleum to form and migrate within them, and this may account for the relative scarcity of accumulations in beds which are less than about one million years old. Pleistocene oil occurs in only a very few areas, (Louisiana, California, Eastern Venezuela) and there is generally some doubt as to whether it is not in fact a migrated older crude.

Conclusion

The preceding necessarily brief review of the conditions governing the formation and preservation of oilfields shows that at least four essential conditions must be fulfilled. First, a source rock must be available which is of the requisite "maturity", the product of appropriate depth of burial and geothermal gradient; secondly, a permeable reservoir rock must have been connected to the source rock via a channel suitable for fluid migration and must have been in the correct stratigraphic relationship with it; thirdly, an effective closed trap must have been formed in which the oil could accumulate; and fourthly, the filled trap must have been preserved intact for long periods of geologic time.

Furthermore, it is essential that each of these prerequisites should have existed in the correct relationship in *time* as well as in space. Thus, a suitable reservoir rock must have been available and in contiguity with the source rock by the time the latter expelled the bulk of its hydro-carbons in the process of primary migration; then, the trap must have been formed in time to receive and retain the fluids moving in secondary migration; and finally, the sealing medium must have been effectively in place over the filled trap in time to prevent its destruction by erosion or tectonism.

The factor of "timeliness" is therefore the fifth essential requirement for an oil accumulation to exist today; the fact that accumulations are relatively rare—and large accumulations rarer still—probably indicates that this requirement has only infrequently been satisfied.

Where is petroleum most likely to occur? Since it is predominantly a product of the sediments, it is evident that most oil will occur in areas where there has been thick sedimentation.

Considered on a world scale, the two major sediment-filled crustal downwarp areas are those surrounding the Gulf of Mexico and the Middle East Gulf. These regions form two opposing "oil poles", containing as they do, in the west the oilfields which stretch in a great arc from Louisiana to Trinidad, and in the east the huge accumulations of the Middle East. Probably 80% or more of the world's oil has been found within these eastern and western "oil pole" regions, in which all the conditions necessary for the formation and preservation of petroleum have clearly been satisfied relatively frequently.

Several other major sediment-filled depressions are also known which contain many oilfields—the newest discovered being the Arctic Basin—but these two areas are by far the richest.

Apart from these generalisations, there are great variations between the concentrations of oil that have been found in different sedimentary areas and in different parts of the same area. In some basins where there are large traps, for example, relatively few of these may be filled with hydrocarbons, due presumably to the failure of one or other of the necessary factors; in other basins, there may be many, but only relatively small oil and gas accumulations; while in still others, there are but a few very large "giant" fields. Many complex factors acting over long periods of time have served to bring about these variations; the influences of some of these factors, such as the local geothermal gradients have only been appreciated relatively recently, so that their effects have in many cases not been completely assessed.

In particular, the factors controlling the occurrence of "giant" accumulations are not entirely understood. Obviously, very large traps, plentiful source material, efficient maturation, expulsion and preservation mechanisms must have been needed. Perhaps, in addition, there is some fundamental tectonic relationship which is not yet properly understood. Thus, it has been proposed[24] that the shear patterns produced by deep-seated wrench-faults (i.e. faults where the dominant movement of one block relative to the other is horizontal and the fault-plane is nearly vertical) have influenced the linear distribution of petroleum traps in some parts of the world—California, Alaska, Colombia-Venezuela and the Gulf of Guinea. But these and other *ex post facto* explanations for the present-day distribution of "giant" fields are still not enough entirely to explain their apparently random geographical distribution, which is probably related to the complex history of movement of the underlying lithospheric "plates"[25].

---

[24] J. Moody, Bull. AAPG, 57 (3) March 1973, 449.
[25] D. Tarling, *Nature*, 243, June 1 1973, 277.

CHAPTER 3

Western Europe and the North Sea

THE European continent has been much affected by Caledonian, Hercynian and Alpine folding movements. The stable pre-Cambrian "Shield" of Scandinavia is surrounded by a shelf area of broad sedimentary basins, broken up into smaller tectonic units by later orogenies. In the Mediterranean region, there is a belt of relatively narrow "Tethyan" geosynclines following the line of the Pyrenees, Alps and Apennines, and extending via the arc of the Carpathians deep into Eastern Europe.

Table 24 lists the major sedimentary basins of Western Europe. Nearly all the oil that has been found in these basins comes from Upper Tertiary or Mesozoic (Lower Cretaceous-Jurassic) strata, except for the British accumulations in Carboniferous beds. The oilfields are relatively small, although locally important, and Western Europe in fact produced only about 3% of its rapidly growing oil requirement in 1971.

The total demand for oil in that year was about 15 MM b/d, and by 1980 consumption is expected to rise to more than 23 MM b/d (about 8,600 MM brl/a).

The balance of 97% of Europe's oil consumption has been traditionally made up by imports, mainly from the Middle East and Africa. (In 1971, for example, oil provided 62.1% of the energy needs of Western Europe, and 82.4% of its oil requirements came from the Middle East.) Production from the newly-discovered oil province in the North Sea will eventually substantially reduce this dependence on imports; but a huge investment of capital and technology will first be needed (p. 99).

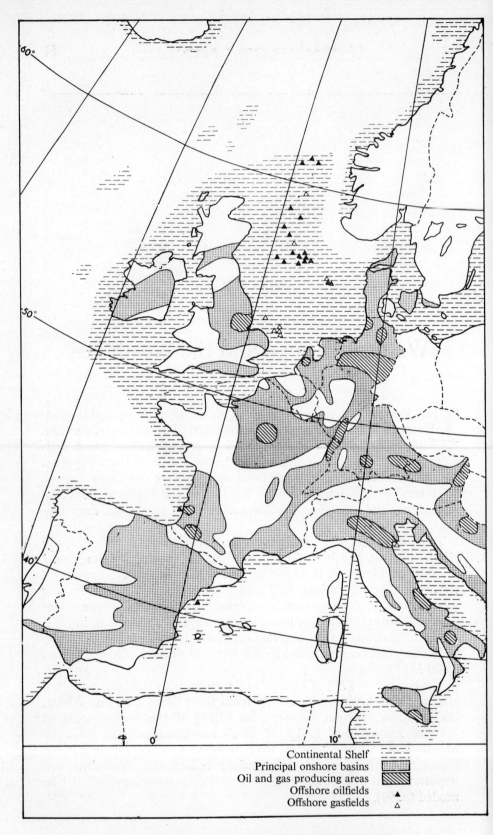

Fig. 16. SEDIMENTARY BASINS OF WESTERN EUROPE

(based on O.E.C.D. Report "The Search for and Exploitation of Crude Oil and Natural Gas in the European Area of the O.E.C.D.")

TABLE 23

WESTERN EUROPE—ENERGY CONSUMPTION

	Energy consumption MM tons c.e.	Market shares % Solid fuels	Liquid fuels	Natural gas	Hydro/ Nuclear elec.
1950	562	88	12	—	—
1958	771	71	25	1	3
1969	1246	37	53	6	4
1975*	1640	23	61	11	5

Source: ECE 1971.   *Estimated

TABLE 24

WESTERN EUROPE—ONSHORE BASINS

Country	Total Area sq km	Main Sedimentary Basins	Basin Areas sq km
W. Germany	245,289	North-West Province	60,000
		Rhine Trough	7,500
		"Molasse" Basin	34,000
Austria	83,851	Calcareous Alps	15,000
		"Flysch" Zone	4,500
		Tertiary "Molasse" Basin	8,600
		Vienna Basin	3,900
		Graz Basin	7,300
France	551,441	Paris Basin	130,000
		Aquitaine Basin	66,000
		Alsace	4,250
		Jura-Bresse-Savoie	30,000
		S.E. France	43,000
Italy	301,081	Po Plain Basin	50,000
		Apennine-Adriatic Basin	10,000
		Tertiary "Molasse" Basin	8,000
		Apennine "Foreland" Zone	20,000
		Sicilian Basins	17,000
		Other Basins	37,000
Netherlands	33,734	E. Permian Basin	8,000
		E. Mesozoic Basin	8,000
		W. Mesozoic Basin	10,000
UK	243,999	S.E. England Mesozoic Basin	10,000
		N.E. England Permian Basin	6,000
		S. Scotland Carboniferous Basin	4,500
		Lancashire Basin	5,700
		Midlands Basin	16,000

Source: The Search for and Exploitation of Crude Oil and Natural Gas in the European Area of the O.E.C.D., 1962.

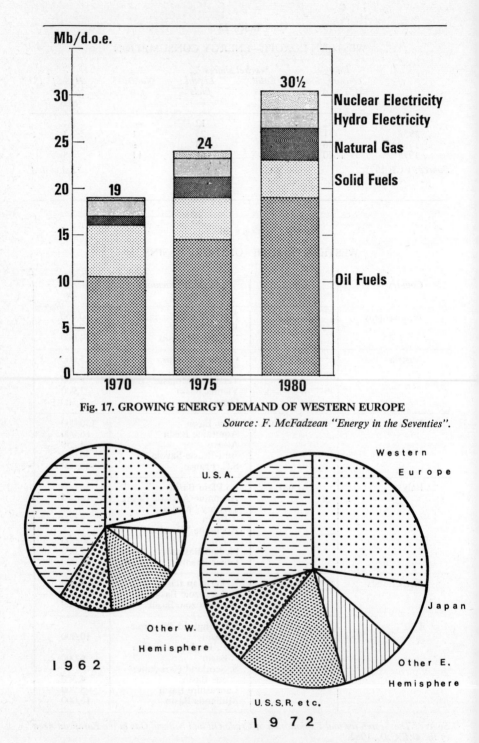

Fig. 17. GROWING ENERGY DEMAND OF WESTERN EUROPE

Source: F. McFadzean "Energy in the Seventies".

Fig. 18. WESTERN EUROPE'S INCREASING SHARE OF WORLD ENERGY
CONSUMPTION

Source: BP Statistical Review.

TABLE 25

WESTERN EUROPE—OIL PRODUCTION, 1880–1970
(thousands of brl)

Year	Austria	France	W. Germany	Holland	Italy & Sicily	Spain	U.K.	Total
1880	—	—	29	—	2	—	—	31
1890	—	—	109	—	3	—	—	112
1900	—	—	314	—	12	—	—	326
1910	—	—	1,032	—	51	—	—	1,083
1920	—	356	246	—	35	—	3	640
1930	—	523	1,182	—	59	—	1	1,765
1940	2,808	496	7,371	—	85	—	124	10,884
1950	10,200	909	8,107	—	63	—	340	19,619
1960	16,874	14,233	40,076	13,378	13,613	—	649	98,823
1970	18,994	17,225	53,356	13,112	10,259	1,153	642	114,741

TABLE 26

WESTERN EUROPE—OIL PRODUCTION AND ONSHORE RESERVES, 1971

	Production MM brl	Reserves MM brl
Austria	17·08	186·5
France	13·64	98·9
Germany	54·17	560·0
Holland	11·72	261·2
Italy	9·41	271·1
Spain	0·91	12·0
UK	0·64	18·0
	107·57	1407·7

Source: World Oil, Aug. 15, 1972.

TABLE 27

WESTERN EUROPE—ESTIMATED OFFSHORE RESERVES, 1971

	MM brl
Denmark	250
Norway	2,000
Spain	150
UK	3,000
	5,400
Total offshore and onshore reserves:	6,808 MM brl

AUSTRIA

The oil produced since 1934 within the frontiers of present-day Austria has all come from two sedimentary basins—the Vienna Basin in the east and the "Molasse" Basin in the west of the country. Production before the 1939–45 war was confined to a few small fields in the Vienna Basin, but after the war several larger oil and gas accumulations were discovered in this area, and in 1962 hydrocarbons were also found in the "Molasse" Basin.

H

The Inner-Alpine Vienna Basin, which accounts for about 90% of the hydrocarbon output, covers an area of about 2,300 sq miles in Austria, and extends into Czechoslovakia. It is a trough of major subsidence[1] which contains a thickness of some 15,000 ft of Later Tertiary sediments, including marine Tortonian, brackish-water Sarmatian and freshwater Pannonian rocks, varying in facies from gravels and coarse sands on the edges of the basin to finer-grade sands and clays in the centre[2].

The basin is bounded by a series of faults, which are particularly important on the northwest. Here a series of "roll-over" fold structures form hydrocarbon traps sealed against the major Leopoldsdorf-Steinberg fault. Most of the resulting small oilfields have faulted reservoirs of Sarmatian and Tortonian sandstones. They include (from southwest to northeast) *Pirawarthe, Gaiselberg, Zistersdorf, Gosting, Neusiedel* and *Van-Sickle-Plattwald*. *Mühlberg*, further to the northeast, has large reserves of non-associated gas in Pannonian and Sarmatian sands, in addition to some Tortonian oil.

Structural uplifts in the "basement" have also produced several oil and gas accumulations. Thus, the *Matzen-Auersthal* field is the largest oilfield in Austria—consisting of a faulted, flattish anticline with oil in Tortonian and Helvetian (Miocene) sandstones and non-associated gas in several Pannonian, Sarmatian and Helvetian reservoirs. Although this field is slowly declining, it still produces nearly half the total Austrian oil output. In the neighbouring *Schonkirchen-Tief* field, the productive zones are mainly the deeper Triassic "Hauptdolomit"[3], the first Mesozoic oil reservoir found in the Vienna Basin. *Tallesbrunn*, east of Matzen, has Sarmatian and Helvetian gas, while *Aderklaa* has some Tortonian oil and much larger gas reserves in deeper Cretaceous, Jurassic and Triassic beds.

The most important Austrian gasfield is *Zwerndorf-Baumgarten*, discovered in 1952, which has Tortonian and Triassic reservoirs; it is a flat, faulted anticline which lies partly across the frontier with Czechoslovakia.

In the Vienna Basin, the thick layers of marly clay between the Miocene sands have provided effective cover and possibly have also acted as source rocks.

The asymmetrical "Molasse"* Basin occupies about 5,100 sq miles of Upper Austria, and contains a moderate thickness of sediments which vary in age from Oligo-Miocene in the southwest to Mio-Pliocene in the northeast. Natural gas was first discovered in 1960 in Oligo-Miocene

---

[1] R. Janoschek, "Habitat of Oil", *AAPG*, 1958, 1134.

[2] L. Happel, *Sci.Petr.*, 6, 1, 1953, 34.

[3] H. Hawle *et al.*, 7 WPC Mexico, 1967, PD(3) 6.

*The terms "flysch" and "molasse" describe the geosynclinal sediments which are typical of the Alpine pre-orogenic and post-orogenic phases respectively.

sands[4] at *Wildendürnbach,* an asymmetrical anticline with an east-west axis. *Lindach* is another gasfield in this area with a lens-shaped sandstone reservoir of similar age.

The *Voitsdorf* oilfield, discovered in 1963, has two reservoirs; an upper Paleocene sandstone and a deeper Upper Cretaceous bed. The structure is an upthrown fault block sealed by overlying Oligocene shales, and output is about 1·5 MM brl/a.

Several small Eocene-Paleocene fields have been discovered since 1956 (*Puchkirchen, Ried, Kohleck,* etc.) which are all structurally related to upthrown fault blocks.

Production and Reserves

Oil production, which reached a peak of 20 MM brl in 1965, fell to 17·1 MM brl in 1971. Most of the output comes from the Vienna area and Lower Austria, mainly from one field—*Matzen-Auersthal.*

Austrian reserves were estimated[5] to be about 210 MM brl of recoverable oil in 1969; by end-1971 they were about 186·5 MM brl, so that an increasing dependence on imports is probable.

TABLE 28

AUSTRIA—OIL PRODUCTION, 1971

	MM brl
Vienna Basin	
Matzen-Auersthal	8·59
Schonkirchen-Tief	3·61
Zistersdorf-Gaiselberg	0·60
Other fields	1·97
"Molasse" Basin	
Voitsdorf	1·28
Other fields	1·03
Total	17·08

FRANCE

Most of the oil and gas discovered in France has been found in two large sedimentary basins—the Aquitaine Basin, lying between the Pyrenees and the Massif Central, and the Mesozoic Paris Basin.

The Aquitaine Basin

The Aquitaine Basin is a broad geosynclinal area of about 30,000 sq miles. Its northern and north-eastern edges form a shallow platform corresponding to the buried slopes of the Armorican massif which is

---

[4] F. Brix *et al.*, 6 WPC Frankfurt 1963, 1, 3.
[5] R. Janoschek & K. Götzinger, "Exploration for Petroleum in Europe & N. Africa", IP, 1969, 179.

exposed in Brittany and the Massif Central, while in the south there is a trough corresponding to the Cretaceous "flysch" and Tertiary "fore-deep" of the Pyrenees. The main structural features of the basin are controlled by the general northwest-southeast trend of the Armorican system, and by the roughly east-west framework of the Pyrenees. All the important hydrocarbon concentrations in the Aquitaine Basin are related to the Cretaceous/Jurassic unconformity.

The first accumulation discovered within this basin was natural gas and condensate at *Saint Marcet* in 1939. The main gas reservoir here is an Upper Cretaceous breccia unconformably overlying a diapiric core of older Triassic, Jurassic and Cretaceous rocks, probably tectonically related to deep-seated Triassic salt movements. There is also some Middle Jurassic gas. Only the neighbouring *Charlas* anticline has been found to be gas-bearing among a number of similar structures drilled in this area; the reservoir here is again of Upper Cretaceous age. Further westwards in the basin, the *Lacq* field was discovered in 1949, near Pau. The first discovery was of shallow oil, but a deeper test (in 1951) proved a large reservoir of very sulphurous high-pressure gas, with associated condensate. The structure of this field is a much-faulted anticline with gentle dips. The oil reservoir is a highly permeable Senonian limestone; the gas reservoir is partly a Neocomian-Valanginian dolomitic limestone and partly similar Upper Jurassic Purbeckian-Portlandian beds. A smaller anticlinal gasfield has been discovered at *Meillon,* a few miles southeast of Lacq, with which it is now being developed.

The first oilfield to be discovered in the Aquitaine Basin was found at *Parentis,* southeast of Bordeaux, in 1954. This field, with its small associated neighbouring structures (*Mothes, Lucates, Lugos, Mimizan, Cazaux,* etc.) produces at the rate of some $7\frac{1}{2}$ MM brl/a (more than half the total French oil output) and has estimated recoverable reserves of about 190 MM brl.

The main structure here is an anticline which was discovered by seismic surveys. The axis runs almost directly east-west, and the fold is slightly asymmetrical, with a steeper south flank. The producing horizon is a partly dolomitized Lower Cretaceous limestone, which is probably equivalent to the deep Neocomian gas reservoir at Lacq.

The small *Cazaux* field in this area is interesting in that the oil horizon here is of a different age and facies, being an Albian sandstone.

An offshore extension of *Parentis* has also been discovered (*Antares*), and exploration is being carried out off the coasts of the neighbouring Gironde and Landes Départements.[6]

Paris Basin
The Paris Basin covers an area of 50,000 sq miles in northern France,

---

[6] J. Cabrit, *World Oil,* March 1969, 71.

probably extending also to southeastern England. It contains a large volume of Mesozoic rocks on a metamorphic or Pre-Cambrian "basement". Unstable tectonic conditions have resulted in the production of a number of local uplifts and folds, and the traps in this area are mainly small anticlines. The reservoir formations are usually of Dogger oolitic limestones or Neocomian sandstones, with occasional deeper Triassic production.

The first oilfields to be discovered in this basin were *Coulommes* and *Châteaurenard* in 1958. The latter field is interesting in that there the Neocomian sands dip and wedge westwards, while the trap is sealed to the eastward by a fault zone. Other small oil accumulations have since been found at *Chailly-Chartrettes, St. Martin-de-Bossenay, St. Firmin des Bois, Chuelles* and *Villemer*.

Southeastern France

For many years before the discovery of hydrocarbons in the Aquitaine and Paris Basins, oil had been produced in Alsace at *Pechelbronn* in the Rhine Valley, where the oil sands were extracted by mining. (This is one of the few areas of the world where oil sand mining has been economic.)

The Pechelbronn Basin is a part of the elongated Rhine "graben", which forms a faulted depression about 180 miles long between the Palaeozoic uplifts of the Vosges Mountains in France and the Schwartzwald in Germany. A complex system of faults dissects the strata, the main faults paralleling the Rhine Valley, while cross-faults divide the area into a number of individual blocks which dip gently into the centre of the depression.

Oil has been found mainly in Oligocene rocks of Lattorfian (Sannoisian) age, in a number of sand lenses, not usually more than a few feet thick and of limited extent, which occur in a great local thickness (more than 5,000ft) of Oligocene marls and limestones. The "basement" rocks are Jurassic and Triassic beds, with the Cretaceous completely absent. Some Mesozoic oil has also been obtained.

Wells are drilled in the normal manner to exhaust the sand lenses, and when no further production can be obtained in this way, the practice has been for a level to be driven into the sand from the nearest mine shaft. The oil is then collected as it oozes from the walls and ceiling of the level into a trench in the floor, whence it flows by gravity to a central sump, and thence is pumped to the surface. An alternative method of working is to direct the level or cross-cut so as to intersect the cap rock just above the sand, and then to make a trench in the floor deep enough to drain the sand. Oil can then be drawn from the pits by suction.

In the Pechelbronn mines, it has been found that the average proportion of the oil-in-place that can be extracted by drilling is about 17%, and by mining 43%, leaving a residue of 40% unrecoverable.

The annual oil production was once nearly half a million barrels, but the field is now almost exhausted.

A number of small local oil accumulations have been found in faulted homoclinal blocks south of Pechelbronn in the Rhine Valley (*Schirrheim, Soufflenheim, Eschau, Staffelden, Reiningue,* etc.). Reservoirs in most cases are Oligocene sands, but at *Eschau* and *Staffelden* the reservoir formation is a Jurassic (Bathonian) limestone.[7] (The production from the German oilfields in the same basin is discussed on p. 85.) Four of these small fields were abandoned during 1968, so that only *Eschau* was still producing at the end of that year.

Other Areas

There is a small oil pool at *Gabian*, Languedoc, in the sub-Pyrenean zone, where oil occurs in a sand of Middle Triassic age in an anticline associated with local thrusting; and an isolated Oligocene accumulation was found in 1951 at *Gallician* in the Rhône Delta. A number of oil-impregnated asphalts are known in the Savoy "Molasse" basin and the Oligocene strata of the Limogne Valley, and there is one small Triassic gasfield at *Lons-le-Saunier* (*Valempoulières*) in the western Jura.

The total French oil production reached a peak of 21·49 MM brl in 1966, but has decreased yearly to only 13·62 MM brl in 1971.

TABLE 29

FRANCE—PRINCIPAL OILFIELDS AND PRODUCTION, 1971

	MM brl
Cazaux	2·76
Lavergne-la-Teste	0·51
Parentis	7·44
Chailly	0·30
Coulommes	0·48
Other fields	2·13
Total	13·62

GERMANY

Germany is by far the largest crude oil producer in Western Europe, with an output of 54·2 MM brl in 1971, which, however, provided less than 20% of the country's oil requirements.

A little seepage oil was produced from shallow pits in the Hanover area in the middle of the last century, and soon after 1870 crude oil was obtained by mining and draining the impregnated sandstones of Pechelbronn in Alsace, then in German occupation, and of Wietze. The first of the salt dome accumulations in the Hanover Basin was found in 1904, and there was also some small early production from Bavaria.

---

[7] W. Bruderer and M. Louis, "Habitat of Oil", *AAPG*, 1958, 1123.

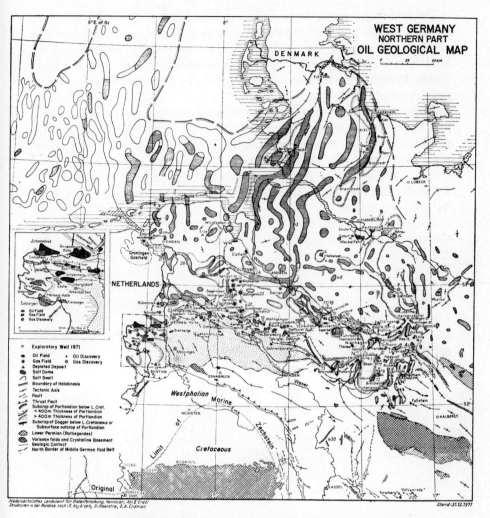

Fig. 19. NORTHWESTERN GERMANY—OILFIELDS AND GASFIELDS

Recent exploratory wells indicated by number on map:

1. Hohenhorn-West 2	11. Wendesse 2	21. Cloppenburg Z-2
2. Neuwerk 1 and 1A	12. Oberg T-1	22. Ahlhorn Z-1
3. Basdahl 3	13. Barsinghausen 3	23. Horstedt Z-1
4. Volkensen-Nord 1	14. Heessel 1	24. Hude Z-1
5. Scheessel Z-1	15. Ostenholz T-1 and 1A	25. Wardenburg Z-1
6. Schneverdingen Z-3	16. Hodenhagen 2	26. Leer Z-2
7. Schmarbeck Z-1	17. Hollige 1	27. Jemgum Z-3
8. Ebstrof Z-1	18. Hollige 1A	28. Lahn Z-1
9. Hofer-Habighorst 1	19. Drentwede Z-1	29. Apeldorn Z-3
10. Abbenser Muhle 1 and 1A	20. Rechterfeld Z-2	30. Solling-Devon 1

Source: Niedersachsisches Landesamt fur Bodenforschung, Hannover.

The first major gasfield was discovered in 1938 near the Dutch frontier, and this was followed by oil discoveries in the Emsland area from 1942 onwards. After the 1939–45 war, many new fields were found in Schleswig-Holstein and in the area between the Weser and the Ems rivers. Production from the "Molasse" Trough of Bavaria was also developed in the 1950's.

There are now some 86 known petroleum accumulations in Germany, and about 4,000 wells, with relatively small individual outputs. The production of oil reached a peak in 1965 and has since then been slowly declining. The principal producing reservoirs are sandstones of Lower Cretaceous (Valanginian) and Upper Jurassic (Dogger) ages, although there is also some Middle and Lower Jurassic and Upper Triassic production.

Proved and probable oil reserves (of which 95% lie in Northwest Germany) amounted to 560 MM brl at the end of 1971.

In contrast to the declining crude oil output, the production of natural gas has increased remarkably during the last few years, due principally to Emsland developments and new discoveries made between the Weser and Ems rivers. Large volumes of non-associated gas occur in Lower Triassic (Bunter) sandstones, Upper Permian Zechstein "Haupt"- and "Platt"- dolomites, and Upper Carboniferous sandstones, as well as, to a lesser degree, in Tertiary and Jurassic strata.

TABLE 30

WEST GERMANY—OIL PRODUCTION, 1971

	MM brl	% of total
North of Elbe River	5·74	11·2
Between Elbe and Weser Rivers	16·15	30·0
Between Weser and Ems Rivers	13·61	24·8
West of Ems River	14·44	26·6
Upper Rhine Valley	1·40	2·4
Alpine "foreland"	2·83	5·0
	54·17	100·0

The petroleum accumulations so far found in Germany can be considered in three groups, of which the first is by far the largest both in area and oil production: these are Northwest Germany, the Alpine Foreland and the Rhine graben.

Northwest Germany

The Zechstein Basin began to form in the middle of Upper Permian time, stretching from England to Russia, with an embayment into the northwest of Germany. It was bounded to the north and northeast by the Fenno-Scandian "Shield", and to the south and southwest by the Rhenish Massif and the Harz Mountains. Its sediments were mainly dolomites, limestones, shales and marls, which later, as the sea became desiccated and landlocked, gave way to great thicknesses of gypsum and salt; these beds have provided the material for most of the numerous salt domes of the area. After brackish-water conditions set in at the beginning of the Triassic, the basin was affected by different degrees of depression. In the south, the salt is still at shallow depths and can be mined; in the north, sinking was continuous down to the Pleistocene, with the resultant formation of a great mass of Mesozoic and Tertiary sediments overlying the salt.

The northwestern oil zone can be divided tectonically into a marginal zone of folding in the south and southwest, and a zone of salt dome structures in the interior part of the basin. In the fold zone, gentle structures occur, with salt intercalated between other sediments. (*Bentheim,* the first German gasfield to be discovered in 1933, is an example of this type of structure.)

In the inner part of the basin, there are at least 200 salt domes, most of which have cores of Zechstein salt and have undergone their main process of uplift in the Upper Cretaceous, although every degree of salt injection is found.

The date of the salt movement is not certain, but Mesozoic salt movement seems to have been of importance in controlling ridge and basin distribution. The alignment of these elongated salt "ribbons" with "basement" trends seems to indicate that the salt movement was induced by tectonism and is therefore related to sub-salt structure.

Most of the crude oil found in this area occurs in Cretaceous and Jurassic reservoirs, whereas the most important gas reservoirs are of Permian age, although gas accumulations also occur in Jurassic, Triassic and Carboniferous beds.

The question of the age and origin of the hydrocarbons has given rise to much debate. Originally, the Wealden beds of the Lower Cretaceous were thought to be the source beds, but the discovery of oil in mid-Zechstein dolomites below the salt at Volkenroda in Thuringia indicates a more probable Permian source. Mid-Zechstein oil has also been found in the Fallstein anticline, north of the Harz Mountains. It is possible that the Stinkschiefer bed which is the stratigraphic equivalent of the Zechstein "Hauptdolomit" may have been the source of that oil, since it is frequently bituminous and contains extensive fish remains. The Posidonierschiefer of the Lias is also bituminous, with fish remains. Oil may well have migrated into the upper reservoir sands by way of faults resulting from the upward movement of the salt.

The petroleum accumulations of Northwest Germany can be conveniently considered in four groups, lying respectively in the areas north of the Elbe River, between the Elbe and the Weser, between the Weser and the Ems, and west of (and in the estuary of) the Ems.

(i) North of the Elbe

The fields in this area are all associated with the flanks or crests of salt structures, many of which show evidence of two different ages of salt —Lower Permian Rotliegendes in the cores of the domes and Zechstein on the flanks. Deep troughs of Jurassic sediments occur between the salt masses, and the source rocks are believed to be the bituminous Upper Lias shales and clays.

The first accumulation to be discovered (in 1937) was the complex *Heide* diapir in the west of Schleswig, from which commercial oil was originally obtained from Permian limestone, although several years later (in 1951) much larger production was found in Wealden sandstones. A number of other salt dome flank fields have been found in eastern Holstein, e.g. *Boostedt, Plon, Bramstedt, Kiel, Plon-Ost, Preetz*—all of which have Middle Jurassic (Dogger) reservoirs. The most northerly field is *Schwedeneck*, which is an unconformity trap near the edge of a salt structure, with a probable offshore extension.

The *Reitbrook* field in the south, near Hamburg, was discovered in 1937 and during the 1939–45 war this was the largest oil-producing field in Germany. The output in 1940 was $2\frac{1}{2}$ MM brl, but it has now declined to only about 260,000 brl/a. The structure is the faulted, crestal zone of a large circular Upper Permian salt dome which has pierced several thousand feet of Cretaceous and Tertiary beds. The reservoir is composed of intensely fissured and fractured limestones of Maestrichtian (Upper Cretaceous) age, sealed below a core of impermeable Tertiary clays. There is also some oil in Lower Eocene sands and some gas in shallow Lower Oligocene and Upper Eocene sands, which have been produced sporadically from as long ago as 1911 at Neuengamme on the crest of the structure. Deeper drilling on the flanks of the dome after 1960 led to the discovery of another oil reservoir in Dogger sandstones extending into part of the adjoining Mackelfeld-Hohenhorn syncline. This extension is called the *Reitbrook West* field, and the adjoining *Mackelfeld-Süd* and *Sinstorf* fields—actually just south of the Elbe—are geologically similar accumulations.

Plon-Ost, Reitbrook and *Preetz* are now the largest oil producers north of the Elbe; but the whole area accounts for only 11 % of German oil output.

(ii) Between the Elbe and the Weser

This was the first area in which commercial oil production was established. It still accounts for 30 % of the national output, more than any other area, due to the long life of the older fields and the newer discoveries that have been made. A thick Permian to Tertiary section, including an important development (about 5,500ft) of Lower Cretaceous reservoir beds, occurs in the Hanover region of the Lower

Saxony basin, between the Aller and Weser Rivers. Most of the oilfields lie along the northern margin of the basin and are associated with the flanks of salt structures. Some of these are mushroom-shaped diapirs below the overhanging edges of which oil accumulations have been found, as at *Eicklingen* and *Hanigsen*; while other fields are "super-cap" anticlines. The salt domes occur in belts running northwest from the Harz Mountains, in many instances along fault lines or at the intersections of faults. Most of the domes have cores of Zechstein salt and have undergone principal uplift in the Upper Cretaceous, after less intense movements in the Upper Jurassic and Lower Cretaceous. The salt is thought to have moved upwards as a result of lateral pressure, taking advantage of local folds, faults and unconformities, so that it tended to become more plug-like in shape the further upwards it travelled.

The older oilfields in this area lie to the northeast of the city of Hanover, and produce oil from reservoirs ranging in age from Rhaetian to Lower Cretaceous.

Nienhagen, with its extensions, was once the largest oilfield in Germany, and produced 2·2 MM brl in 1940. The main structural feature here is an elongated salt-stock around which is a complex system of peripheral faults. Oil was first found in shattered Triassic beds on the southern edge of the salt mass, then in a disturbed flank zone in Rhaetic and Upper Lias beds, and finally the best production of all was obtained from the northern flank, where an anticlinal trend intersects the salt. Valanginian sand lenses deposited near the salt mass have served as traps.

At *Wietze,* oil-sand mining and draining was carried out from 1874 to 1962, with about half the total production from this field being obtained in this way, and the balance by normal drilling. The salt here is in the core of an anticline overthrust to the north, with Triassic, Jurassic and Cretaceous beds upturned on its flanks. Oil has been produced from sandstones of Upper Dogger (Jurassic), Wealden (Lower Cretaceous), Senonian (Upper Cretaceous) and Rhaetian ages.

Oberg is a broad dome in the centre of which Upper Dogger beds outcrop, which is separated from an upthrust salt mass by an Upper Cretaceous graben. Oil has been produced for many years from Lower Dogger sandstones, complicated by unconformities.

At *Olheim-Eddesse* the beds surrounding the salt mass have been forced upwards into steep dips and broken by a series of peripheral and radial faults. There in an unconformity between the Wealden beds and the older strata. In the older *Olheim* field, production came from shallow Tertiary and Wealden sands, and in the *Eddesse* field from deeper Rhaetian reservoirs. A new and important discovery was made in this area at *Olheim-Süd* in 1968.

The elongated *Thonse* anticline produces gas from Dogger and Malm (Upper Jurassic) sandstones.

Postwar exploration has resulted in the discovery of a number of newer oilfields. Of these, *Suderbruch* (1949) is a deep, flat anticline situated between two salt uplifts, with Dogger and Valanginian sandstone reservoirs, and deeper Upper Jurassic oil in oolitic limestone lenses.

The Gifhorn trough runs north-northeastwards for about 60 miles between the towns of Celle and Braunschweig. It contains Carboniferous to Pleistocene sediments, and a number of anticlinal and block-faulted oilfields. *Hohne,* discovered in 1951, produces from part of a structural dome sealed between two sets of faults, running parallel with the Rhenish and Hercynian trends. Reservoirs are Middle and Upper Rhaetian (Upper Keuper) and Liassic (Lower Jurassic) sandstones.

Hankensbüttel (1954) is a similar structure, with Lower Cretaceous and Dogger sandstone reservoirs in down-faulted blocks. This is the largest oilfield in the Elbe-Weser area, with an annual output of about 4 million brl. Other important fields of the approximately 40 known accumulations in this area are *Ahrensheide, Lieferde,* and *Vorhop.* *Knesebeck* is an interesting example of oil in tilted fault-blocks sealed between two salt masses.

(iii) Between the Weser and the Ems

Diapiric salt structures are absent in this area, due presumably to the relative thinness of the Permian evaporites. The first oilfield to be discovered was *Quakenbruck* in 1950, followed by *Hemmelte-West.* They are heavily faulted anticlines, with oil in Wealden or Upper Jurassic beds. *Düste-Altdorf* has Dogger, Valanginian and Portlandian oil production, and deeper Triassic and Permian gas. *Bramberge, Bockstedt, Barenburg, Voigtei-Siedenburg* and *Wehrbleck* are other oilfields, and several of the *Süd-Oldenburg* group* of anticlines are large gas producers.

Rehden is a large gasfield which was discovered in 1952. It is a major anticline with reservoirs in Bunter and Upper Carboniferous sandstones and Zechstein dolomites.

The *Hengstlage* and *Buchorst-Siedenberg* anticlines have thick Bunter sandstone reservoirs.

(iv) West of the Ems and Mouth of the Ems

Asphalt impregnations were known in the Emsland area long before the first gas discovery was made at *Bentheim* in 1938. This is a broad surface fold with an east-west axis, divided by cross-faults into three blocks. The reservoir is a Permian Zechstein dolomite. Several similar gasfields were subsequently found at *Frenswegen, Itterbeck-Halle, Adorf-Dalum,* etc. All these have Zechstein dolomite reservoirs, and the individual well productivities are high. At *Hohenkorben* there is Valanginian gas.

---

* *Hengstlage, Dötlingen, Vistek, Barrien, etc.*

In the Ems estuary, near Groothuis and Bierum, several large gas accumulations were found in 1963, which are probably parts of one huge accumulation, an extension of the Dutch *Groningen* field. These fields are being jointly developed for the benefit of both Germany and the Netherlands.

The first oilfield to be discovered in this region was *Lingen*, in 1942, followed by *Georgsdorf* in 1944, and *Rühle* in 1949. The latter is an east-west anticline which is now the largest German oil producer, with an annual output of nearly 7 MM brl/a from its *Rühlermoor* and *Rühlertwist* pools, whose principal reservoirs are Lower Cretaceous (Valanginian) sandstones.

Other oilfields in this area are *Scheerhorn* (1949) with Malm, Wealden, Valanginian and Hauterivian oil, *Emlichheim* (1944) (Valanginian and Hauterivian) and *Hebelermeer* (1955) (Valanginian). All are productive from tilted fault-blocks. The output from the Emsland fields now amounts to nearly 27% of the national total.

The Upper Rhine Valley

In the German sector of the Rhine "graben" area, a number of small gas- and oilfields have been found on both sides of the river, lying between the Vosges mountains on the west and the Black Forest and Odenwald Mountains on the east. Most of the gas is found in the Darmstadt area in Miocene and Pliocene reservoirs; the oil occurs mainly in Lower Oligocene Pechelbronn sandstones, but there is also some deeper Eocene and Jurassic oil. On the west bank, *Landau* is the most important oilfield, while on the east the *Bruchsal* group of fields includes *Forst-Weiher* and *Weingarten* in tilted fault-blocks, with Tertiary and Jurassic reservoirs. At *Stockstadt* in the north, oil production comes from Oligocene sands in a dipping fault-block, with an output also of Plio-Miocene gas. The Upper Rhine fields account for less than $2\frac{1}{2}$% of the total German oil output.

The Alpine "Foreland" or Bavarian "Molasse Trough"

Several small oil- and gasfields have been found in the western section of the Central European "Molasse" Basin, which extends in Germany from Lake Constance to the Austrian frontier. For many years, a little oil had been produced in the Tegernsee area, and as many as 16 small oilfields and 13 gasfields have been discovered since 1954. The oil output provides only about 5% of the German total, but the gas production is locally important, being used to supply the city of Munich.

The area is one of complex tectonics, lying as it does in front of the zone of northern Alpine overthrusts. The hydrocarbon reservoirs are nearly all Tertiary in age, being Middle Oligocene (Rupelian) sands, as at *Isen* and *Mühldorf*, or Miocene (Burdigalian) sands, as at *Gendorf*. There is also some deeper Middle Jurassic (Dogger) production at

Mönchsrot. Assling and *Arlesried* are the largest producers of oil, with reservoirs of Lower Eocene and Middle Oligocene age respectively.

HOLLAND

Oil had been found in small quantities in wells drilled from time to time in different parts of Holland—mainly in the Carboniferous rocks in the south of the country—but it was not until 1943 that the first commercial oilfield was discovered at *Schoonebeek*, in the Permian basin of the Eastern Netherlands, which covers an area of about 2,000 sq miles and is an extension of the Northwest German Zechstein basin.

The Coevorden-Schoonebeek anticline is one of the most northerly of a group of east-west folds, some of which have formed oil accumulations across the frontier in Germany (p. 84). The structure plunges to the eastwards; and oil is found at relatively shallow depths, (2–3,000ft) in Valanginian (Lower Cretaceous) sands.[8] *Schoonebeek* is the largest oilfield yet discovered in the Netherlands, producing most of the country's crude oil output (11·72 MM brl in 1971). It also produces some non-associated gas from a deeper Permian reservoir. However, oil production is declining, and secondary and tertiary production methods are being employed to extend the life of the field for as long as possible.

The Mesozoic beds of the Western Netherlands Basin, which covers an area of about 4,000 sq miles, are folded into a series of roughly east-west anticlines which are of Variscan or younger origin. Oil is found here in Lower Cretaceous sandstones of Hauterivian and Valanginian age in several small, anticlinal oilfields at *Rijswijk, Pijnacker, Wassenaar, Ridderkerk* and *Zoetermeer*; and in Barremian sands at *De Lier, Ijsselmonde, Delft, Berkel* and *Moerkapelle*. There is a small accumulation in a Jurassic (Dogger) reservoir at *Werkendam*. Of these fields, *Ijsselmonde* and *Wassenaar* have produced the most oil.

However, the discovery of natural gas in Holland during the last few years has been of much greater significance than the production of oil, since this gas has now altered the whole balance of the Western European energy supply position and its sale to several neighbouring countries has brought considerable benefits to the Dutch economy.

Several minor gasfields were discovered in 1948 and the following years in Drenthe and Overijssel Provinces in the East Netherlands Permian Basin, with reservoirs in Zechstein (Upper Permian) dolomites. In the *Schoonebeek* oilfield there was also a separate Lower Cretaceous oil reservoir, but in the other small anticlines only non-associated gas

---

[8] W. Visser & G. Sung, *Habitat of Oil*, AAPG, 1958, 1067.

was found—e.g. the *Coevorden* culmination of the Coevorden-Schoone-beek structure, *Tubbergren, Wanneperveen*, etc.

In 1959, one of the largest gas accumulations in the world was found at *Slochteren* in Groningen Province. Extensions to this field have subsequently also been developed at *Delfzijl, Ten Boer, Nordbroek* and *Schildmeer*. The gas reservoir is a thick, highly permeable continental sandstone, which forms part of the Lower Permian Rotliegendes series. This sandstone rests unconformably on coal-bearing Westphalian "B" beds and is in turn covered by the basal red shales of the Zechstein evaporite section, which form an excellent seal. The underlying structure is believed to be a north-south trending Palaeozoic "high".

Gas has also been found in other parts of Holland, in the last few years, but on a smaller scale. Discoveries have been made in Friesland and North Holland, and also at Bierum on the Dutch side of the Ems estuary. The latter is a large accumulation, which together with the German discoveries in the estuary itself, is now being exploited jointly for the benefit of the two countries. This group of fields is probably a structural extension of the *Slochteren* accumulation.

The volume of gas in the *Slochteren* field and its extensions has been officially estimated at about 65 Tcf. These are original gas reserves of the same order of magnitude as those of the *Panhandle-Hugoton* field in the United States, which is, however, now half-depleted. With its other gasfields, Holland probably has onshore gas reserves of at least 83 Tcf, very much more than those of any other European nation.

ITALY

While oil has been produced in peninsular Italy since the 1860's, most of the accumulations discovered have been very small.

Thus, several minor oilfields occur in Emilia, in the foothill regions of the Apennine Mountains, which form part of the Alpine Tertiary orogeny and have been locally complicated by complex thrusting movements and the presence of the allochthonous "argille scagliose"—a mixture of clays and shales of varied ages which has slid outwards from the mountain arc.

The *Velleia* field in the Valle de Chero (Piacenza) was discovered in 1891, and the *Montechino* field soon after. These contained small volumes of oil in faulted and folded blocks of Eocene-Oligocene marls and sandstones surrounded by complex breccias. *Vallezza*, discovered in 1909, is a similar structure, and *Vizzola* (1939) has an Upper Oligocene sandstone reservoir.

Minor quantities of oil have also been found in Upper Miocene and Pliocene beds in several small anticlines in the Emilian Plain—notably at *Podenzano, Carpaneto* and *Fontevivo*.

Since 1944, exploration has been concentrated in the Po Valley basin where there is a thick sequence (up to 30,000ft) of Mesozoic, Tertiary and Quaternary strata above a Neogene-Mesozoic "basement", in an area which has been in a state of continuous tectonic unrest since the end of Cretaceous time, due to its position between the mobile belts of the Apennines and Alps. The Po plain is geologically divided[9] into a "Periapennic" trough in the south in which there is a great thickness of Oligocene to Quaternary neritic sediments folded into regular, long anticlines; and a "Pedealpine" region in the north where the Pre-Pliocene beds are strongly deformed, but the Pliocene strata occur in gentle structures over local buried uplifts. A number of gasfields have been found since 1944 in the Emilia-Romagna section of the basin.

The gas usually occurs in shallow anticlinal reservoirs in Quaternary, Pliocene and Miocene sands. *Cortemaggiore* is the largest field, with an output of about 15% of the total Italian gas production; it produces natural gas with a very high methane content from Pliocene beds, with some deeper Miocene oil.

There are a number of other gasfields in this area, all with sandy or silty Pliocene reservoirs. *Ravenna Mare* was the first offshore gasfield to be discovered; *Porto Corsini Mare, Porto Garibaldi, Agostino, Amelia* and *Riccione Mare* are offshore gasfields discovered in 1968/9 with Pliocene reservoir sandstones in structural and combination traps.

Exploration drilling has been carried out in the Bradano "trough" and in the Adriatic Basin, and gasfields have been found at *Cellino, Candela, Ferrandina-Grottole* (in Upper Cretaceous limestones and Pliocene sands) and near the Adriatic Coast at *Larino* (*Portocannone*), near Termoli. No commercial oilfields have been discovered in peninsular Italy, except for a small accumulation in Upper Cretaceous at *Pisticci-Dimora* in the extreme south.

However, there have been more successes in Sicily. For many years, asphalt-impregnated limestones of Lower Miocene age had been mined from the *Ragusa* anticline in the southeast of the island. This lies in a "foredeep" zone containing about 12,000ft of sediments, where exploratory drilling in 1953 discovered an accumulation of oil in fractured Lower Triassic (Keuper) dolomites. The structure there, and in the nearby *Gela* field which was discovered in 1956, is gentle and faulted. Oil is found in an Upper Triassic dolomite and in the overlying shales and carbonates.

Smaller volumes of oil occur in similar reservoirs in fields a little to the east of *Gela* (*Ponte Dirillo*) and to the north (*Cammarata-Pozzillo*).

---

[9] T. Rocco & D. Jaboli, *Habitat of Oil*, AAPG, 1958, 1153.

A Shell/Esso drilling rig complex shown operating in the British North Sea sector, 36 miles off the coast of Norfolk. The total weight of the complex is about 2,000 tons.

The million-barrel capacity reinforced concrete tank which will provide crude oil storage for the Norwegian Ekofisk fields seen in the course of construction near Stavanger. After completion in 1973, it was towed to the offshore location where it will rest in 230 ft of water. The perforated outer wall of the tank serves as a breakwater to protect the inner tank from wave action.

The *Gela* oil is very viscous and requires thermal production tech-niques and the injection of gas, which has been found in the north and northeast of Sicily at *Gagliano*, in early Miocene Collesano sands, inter-calated in a thick, shaly, "flysch" series, and in a similar horizon at *Bronte-San Nicola* nearby. Gas has also been found in smaller quantities in the west of the island at *Mazara-Lippone* (Upper Miocene) and in the east, near *Catania* (Pleistocene).

The Sicilian fields account for about 90% of the overall Italian output of 9·41 MM brl (1971). However, they are now declining, and the national reserves of crude oil are not more than about 271 MM brl. On the other hand, the output of natural gas is increasing, due to new offshore dis-coveries in the Adriatic.

SPAIN

The Hercynian "basement" outcrops extensively in Western Spain and has been subjected to intense tectonic disturbance. The search for hydrocarbons has therefore been concentrated in the Mesozoic and Cenozoic basins in the east and south of the country respectively.

Although exploration commenced as long ago as 1923, it was only in the 1960's that any success was achieved. Commercial gas was found at *Castillo*, west of Pamplona. The gas is now piped to Vitoria.

At *Ayoluengo*, near Burgos, and at nearby *Tozo* and *Hontomin*, small oilfields have been found with oil in a fractured limestone and several sandstone reservoirs of Upper Jurassic (Malm) to Lower Cretaceous (Neocomian) age. These fields were put on production for the first time late in 1966, and the total Spanish production for 1971 was 0·91 MM brl. Reserves are estimated at not more than 12 MM brl of recoverable crude.

Geologically, the *Ayoluengo* field is situated between the Polientes and Sedano sub-basins, which were connected with the main intracratonic Cantabrican basin to the north. Subsidence in these sub-basins during Upper Jurassic-Lower Cretaceous time, as the result of salt movement, led to the accumulation of a thick sedimentary series in which oil was generated and trapped in structures related to growing salt uplifts[10]. The structure at *Ayoluengo* is a domal culmination on a large faulted anti-clinal uplift.

The search for other hydrocarbon accumulations is continuing both onshore and offshore in Spain[11]. The most important offshore discovery is the *Amposta Marino* accumulation, south of the Ebro River delta in the Mediterranean. The reservoir here is a Cretaceous carbonate under-

---

[10] R. Sanz Sanz, 7 WPC Mexico, 1967, PD-2, 17.

[11] R. Querol, "Exploration for Petroleum in Europe & N. Africa", IP 1969, 49.

G

lying the unconformity at the base of the Upper Tertiary, which is thought to hold up to 200 MM brl of oil reserves. This field will begin production during 1973 at the rate of about 30,000 b/d. *Castellón* is another offshore discovery in the same area,

UNITED KINGDOM

The absence of other than small sedimentary basins and the complexity of the folding movements experienced make the land areas of the British Isles unpromising for major oil accumulations.

However, oil seepages are common in Carboniferous rocks, and oil-impregnated Mesozoic sandstones outcrop in Dorset and Sussex. A minor accumulation of natural gas was found as long ago as 1896 in Purbeck (Upper Jurassic) beds at Heathfield, Sussex; the gas was used to light the local railway station for a number of years. The first successful oil-well was drilled at *Hardstoft* in Derbyshire in 1919 and produced in all nearly 30,000 brl of oil from a small lens of Upper Carboniferous Limestone. Exploration was undertaken more systematically in the late 1930's, and several commercial oil accumulations were discovered in the Midlands Carboniferous Basin. The output from these fields was of special value during the 1939–45 war, but has since declined to only 657,000 brl in 1971.

The eastern flank of the broad anticline of Carboniferous Limestone, which makes up the central Pennine range, dips gently into the Nottinghamshire-Derbyshire coal basin. There it is unconformably overstepped by Permian and younger sediments. *Eakring* is a north-south fold in this area, about 10 miles northwest of Newark. Oil occurs here at a depth of about 2,000ft in the Rough Rock sandstone of the Carboniferous (Namurian) Millstone Grit.

Oil was also found (in 1941) in the same area, in the *Kelham Hills* anticline, in the topmost sandstone of the Millstone Grit, and in another culmination of the Eakring structure at *Duke's Wood*. A few years later, two more small anticlinal accumulations were discovered at *Caunton* and *Nocton*, with reservoirs in the Millstone Grit, and Carboniferous Limestone respectively.

Several other minor accumulations were subsequently found in the same area, notably *Plungar* (1953), *Egmanton* (1955), *Bothamsall* (1958), *Gainsborough* (1959) and *Beckingham* (1964). These are also small anticlines with Carboniferous Limestone reservoirs, but at *Beckingham* and *Gainsborough* the oil is found somewhat higher stratigraphically, i.e. in the Middle rather than the Lower Coal Measures.

The local relationship between coal and oil deposits is complex, and it is not clear whether the oil has been flushed away by meteoric waters from between the shallower coal seams, or whether it was in fact generated at progressively higher stratigraphic levels towards the north-eastern part of the basin of deposition.

A number of small Carboniferous gas accumulations have also been found in the East Midlands—notably at *Calow, Ironville* and *Trumfleet*, which are anticlines with gas in Coal Measures or Millstone Grit beds. It has been suggested[12] that these gas accumulations, which lie west and north of the oilfields, have been produced by a regional effect related to the Carboniferous basin of deposition, although the cause of this is not certain.

Another basin in which commercial oil has been found in Britain is in the Dalkeith area of Midlothian, east of Edinburgh, where the *D'Arcy-Cousland* anticline has produced a modest volume of oil and gas from Calciferous sandstones of Lower Carboniferous age. Gas has also been found in the *Pumpherston* anticline in East Lothian, where the local oil-shales were mined for many years. Scottish shale oil production commenced in 1873 and reached a maximum output of about 1·8 million brl in 1913; thereafter it declined until its final abandonment as un-economic in 1962.

A third basin, probably also of Carboniferous age, has produced a little oil in Lancashire, at *Formby*, in an area of local seepages. The structure here is thought to be a faulted homocline, the reservoir being in Keuper (Upper Trias) beds, sealed by glacial boulder-clay. Since the Keuper is a continental-type formation, and therefore devoid of source material, the oil has probably come from a deeper, presumably Carbon-iferous, reservoir, which may yet be found by further exploration, pos-sibly offshore.

In the south of England, outcrops of oil-impregnated Mesozoic sand-stones occur in the Wealden and Hampshire Basin, which is a continu-ation of the Continental Paris Basin. Several exploratory wells have been drilled to test possible anticlinal accumulations, so far without notable success. Some heavy oil was found in basal Purbeck rocks in a well drilled on the Portsdown anticline, and small accumulations in Upper Dogger (Jurassic) beds at *Kimmeridge* and *Wareham*.

Another area of interest is in the eastern part of Yorkshire, where Permian limestone outcrops. The Permian beds in this area, which probably forms a western extension of the Zechstein Basin, are developed in the form of interbedded anhydrites and dolomitic limestones of facies similar to the equivalent Zechstein beds of Holland and Germany. Gas found in an anticline at *Eskdale* was used to supply Whitby and other towns. In 1966, other and larger accumulations of gas were found in the neighbouring *Lockton* and *Ralph Cross* anticlines near Scarborough, in Magnesian Limestone reservoirs.

---

12 R. Brunstrom, *Proc. 6th World Pet. Congr.*, 1, 11, 1963.

Finally, some gas "shows" have been recorded in exploration wells drilled in the Palaeozoic basin in the South Midlands—notably in Berkshire, Oxfordshire and Bucks.

Indigenous production in Britain has never been able to supply more than a tiny fraction of the requirement for crude oil and oil products. This has grown at the rate of about 10% p.a. over the last decade, and reached nearly 750 MM brl in 1971. The remaining proved onshore reserves are only about 18 MM brl. However, this picture has been radically changed, by the remarkable offshore discoveries, first of gas and subsequently of oil, that have been made in recent years in the British *North Sea* sector. These are described on p. 97.

OTHER COUNTRIES OF WESTERN EUROPE

BELGIUM

Although there is a considerable thickness of sediments available and some drilling has been carried out, no commercial oil or gas accumulations have yet been found in Belgium. There are reasonable petroleum prospects in the folded Devonian strata of the Namur Basin and in the faulted Carboniferous and Devonian beds of the Campine Basin.

GREECE

Exploration for oil has been carried out sporadically in Greece for many years, but without commercial success. Test wells have been drilled in northeastern Thrace, and in the Aetolia-Akarnia province of western Greece. A small amount of crude oil has been produced at *Keri* on the southwest of the Ionian island of Zante (Zakinthos) where heavy oil and asphalt are found at shallow depths in Plio-Miocene beds. Offshore exploration has been carried out without success in the seas surrounding Thassos Island.

IRELAND

At least four sedimentary basins are known in the Republic of Ireland, and a number of test wells have been drilled in the last few years west of Dublin, in County Cavan, in County Kilkenny, etc; but only non-commercial "shows" of gas have been reported. Offshore exploration in the Celtic Sea discovered a gas accumulation in 1971 about 27 miles off County Cork.

MALTA

The first offshore wells to be drilled off the coast of Malta in 1972 were abandoned without success.

PORTUGAL

A large part of Portugal consists of the highly metamorphosed western part of the Hesperian Massif, so that oil prospects must be confined to the coastal strips.

Several wells have been drilled in the Torres Vedras basin northeast of Lisbon, where non-commercial "shows" of oil and gas were found in Miocene formations. There is also some asphalt production. Offshore exploration is expected to begin in 1973.

SPITZBERGEN (SVALBARD)

A well was reported to have been drilled on an anticline on Edge Island early in 1972 without success.

SWEDEN

There are fairly thick Mesozoic beds in Skaane province and on the island of Gottland. Several exploration wells have been drilled in this area, but with only small oil "shows" reported in Mesozoic and Palaeozoic beds. Here again, offshore prospects in the Baltic appear to be more promising.

Sweden possesses extensive deposits of oil-shale from which locally important quantities of oil are obtained by distillation, using the Ljunstrom method of *in situ* distillation.

SWITZERLAND

A number of small oil and gas seepages are known in various parts of Switzerland, and several exploratory wells that have been drilled (mainly in Canton Vaud) have found non-commercial "shows" in the "Molasse" Basin formations which are productive in Austria. However, no commercial oil accumulation has yet been found. The Cretaceous asphaltic limestone of Val de Travers near Neuchatel has been worked for many years; the oil from which the asphalt has been derived is thought to have come from underlying Chattian (Oligocene) oilsands, which outcrop at La Plaine and Dardagny, near Geneva.

THE NORTH SEA

THE discovery of hydrocarbons in the North Sea Basin has been one of the most remarkable developments in international exploration operations during the last few years. It has meant that a huge volume of oil and gas has been found almost in the centre of one of the world's largest petroleum-consuming areas.

The potentialities of the North Sea were emphasised by the discovery of the large Groningen gasfields in 1959, and the search was stimulated by the knowledge that oil had been discovered in several of the neighbouring countries in reservoir formations which might well be expected to thicken out to sea. Initial submarine gravity and aeromagnetic surveys showed that the 220,000 sq miles area of the main basin was divisible into several sub-basins in which up to 20,000ft of varied sediments occurred. There was also evidence of uplifted structures and piercement salt domes.

The ratification by Britain in 1964 of the Geneva Convention on the Territorial Sea was the beginning of the first phase of North Sea exploration. By this time also, the technology of offshore drilling first developed in the Gulf of Mexico, California, Venezuela and the Caspian Sea, had become sufficiently advanced to make possible drilling operations by a variety of special rigs in the stormy North Sea waters.

The 1958 Convention, which came into force in 1964 when it had been ratified by 22 of the original 46 states represented at Geneva, defined for the first time the Continental Shelf* and laid down rules by which the Shelf areas could be equitably divided between the coastal states concerned. The principle employed was that where two or more states have coasts opposite to each other, the boundary of their Continental Shelf sovereignty is determined by application of equidistance measurements, i.e. the boundary is the median line equidistant from the nearest points on the base lines (low-water lines) from which the breadth of the territorial seas is measured.

These principles enabled a number of agreements to be made which delineated the different North Sea sectors, but in some cases disputes still arose about these boundary delimitations. In particular, West Germany maintained that the definition quoted was unacceptably disadvantageous in view of the inward-curving coastline of that country. The dispute with her neighbours was taken to the International Court in the Hague, which ruled in 1969 that the German-Dutch and German-Danish offshore sector boundaries should be revised.

In the North Sea area, a Mesozoic-Tertiary basin some 500 miles long

---

*"All the sea bed and the subsoil of the submarine areas adjacent to the coast but outside the territorial waters to a depth of 200 metres, or beyond that limit to where the depth of the superjacent waters still permits the exploitation of natural resources in such areas".

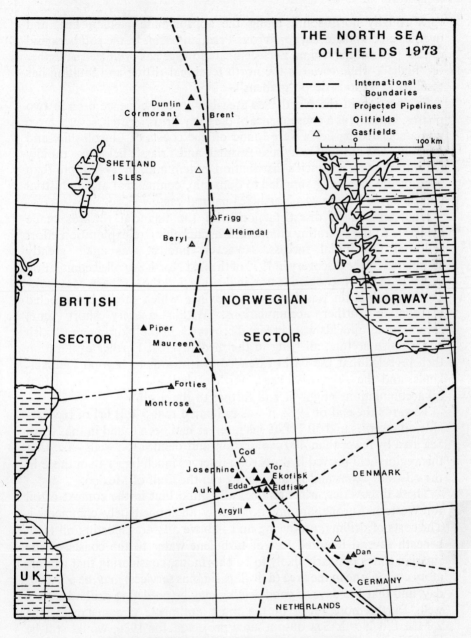

Fig. 20. OILFIELDS OF THE NORTH SEA BASIN

runs roughy north-south along the approximate median line, and overlies a composite Carboniferous-Permian basin. In the southern and central parts of the region, Zechstein salt flowage has produced a number of "highs", while towards the north tensional rifting and faulting has resulted in various deep "grabens".

Exploration in the North Sea area has in consequence been in two phases. Initially, as a consequence of the early British search, several large non-associated gasfields were found off the coasts of Lincolnshire and Norfolk. The output from these Permian and Triassic fields was rapidly fed into the Gas Council's distribution system and, by 1972 more than 90% of the gas being supplied to domestic, commercial and industrial consumers in Britain was North Sea natural gas.

Due perhaps to dissatisfaction with the financial rewards made available to the operating companies, a slackening of exploration effort followed this initial success. However, interest was again greatly stimulated by the discovery in 1970 of the first North Sea oil accumulation –*Ekofisk* in the Norwegian sector, indicating that the northern Mesozoic-Tertiary sub-basin was a new oil province which might be expected to contain further accumulations. Within a very short time, several other oilfields were in fact discovered, in the Norwegian, British and Danish sectors, all lying in the median area. Further exploration then revealed that the basin extended northwards to beyond the Shetlands, and the *Brent* discovery, north of latitude 61 °N, showed that oil accumulations might lie still further to the north.

Towards the end of 1972, it was estimated that 5–6 B brl of recoverable oil reserves and 50 Tcf of gas reserves had been found in the North Sea area by fewer than 400 exploration and delineation wells—i.e. that this region held proved hydrocarbon reserves much larger than those in the offshore Texas and Louisiana area in the Gulf of Mexico.

These discoveries, while strategically important in the context of oil for Western Europe, are not necessarily all immediately commercial. The costs of drilling, producing and—above all—transporting oil from beneath several hundred feet of turbulent water to the coasts of the riparian states is clearly likely to be enormous, considering that exploratory costs alone in the area (a drilling rig plus services) may be $50,000/day, and that 125 days of good drilling may be needed to drill a 12,000ft well. Furthermore, even in the most optimistic circumstances, the 2 MM b/d North Sea output now projected for 1980, would still be insufficient to have met even the 1970 UK requirement for imported oils. Also, since North Sea oil is generally light, it will have to be blended with imported heavier oils to form suitable industrial fuels. Again, from the European point of view, if a further 1 MM b/d becomes available for Continental use over and above the British consumption, North Sea oil will still provide only a fraction of the 23 MM b/d oil consumption predicted for 1980. Nevertheless, these discoveries will greatly assist the balance of payments of the states which control them, and materially reduce their requirements for oil imports.

THE BRITISH SECTOR

The first offshore hydrocarbon accumulation to be found in the North Sea basin (at the end of 1965) was the elongated, faulted dome of the *West Sole* gasfield, some 42 miles east of the Humber. The reservoir here is basal Permian (Rotliegendes) sandstone, lying at a depth of about 9,000ft, beneath the southwest flank of a large salt "pillow",[13] and containing reserves of about 1·0 Tcf of recoverable gas. This was also the first North Sea gasfield to pipe gas to the mainland, at Easington, and thence into the National Grid.

The *Leman Bank* field was discovered in 1966, about 30 miles from the Norfolk coast. It is a large, flattish anticline, 18 miles long by five miles wide containing the largest gas accumulation so far found in the British sector—probably one of the largest offshore gasfields in the world, with recoverable reserves as high as 12 Tcf. The reservoir is again a thick section of Rotliegendes sandstone, lying here at a depth of about 6,000ft, and sealed by Zechstein evaporites.

Other smaller, but important offshore gas accumulations were subsequently found in the period 1966–70—*Hewett, Ann, Indefatigable, Deborah, Viking* and *Rough*, the largest of which were connected to Bacton in Norfolk by a major pipeline system, and thence to the National Grid.

The reservoir formations in these fields are generally Rotliegendes sandstones, but this formation is not as consistent in the British North Sea sector as it is, for example, in the Groningen area of Holland. There is a tendency for permeability to be reduced by the occurrence of shale or clay streaks. Zechstein-age evaporite deposits form efficient seals over most of the area, and there are also post-Zechstein gas reservoirs (Middle and Lower Bunter sandstones) sometimes present—in *Hewett*, for example. The potential output from these fields is 4·5 Bcf/d, and their recoverable reserves up to 35 Tcf.

The *Ekofisk* oil discovery in the Norwegian sector in 1970 (p. 99) then stimulated the exploration of the British sector of the northern Tertiary Basin. Success was surprisingly rapid; the *Forties* field, with a basal Paleocene sandstone reservoir, was the first major oilfield to be discovered in the British sector. The trap is a broad, low-relief anticline holding recoverable reserves of about 2 B brl, and production is planned to start in 1974 via a 133-mile-long, 32in pipeline to Peterhead and thence to

[13] P. Kent & P. Walmsley, *Bull.AAPG*, 54(1) January, 1970, 168.

Grangemouth and the Firth of Forth. Production will begin at the rate of 90 MM brl/a, building up to 150 MM brl/a later. An investment of well over £360 MM is envisaged as necessary.

Auk was the second oilfield to be found in the British sector, drilled in 1971–2 on block 30/16. It lies about 160 miles off the Scottish coast, due east of Dundee, in an average water depth of 275ft. The reservoir here is a Permian dolomite overlain by basal Tertiary beds.

Production from this field will begin in 1974, initially into tankers loading from two single-buoy moorings. The capacity of the field is thought to be up to 50,000 b/d and its recoverable reserves perhaps 1·0 B brl.

Several other smaller oil accumulations have also been found in the northern basin; notably *Montrose*, which may hold 2–300 MM brl of recoverable oil, *Piper*, *Argyll* and *Josephine*, which is probably sub-commercial under present conditions.

In mid-1972 the most northerly discovery yet made was found in water 460ft deep about 120 miles northeast of the Shetland Islands in the Northern North Sea "graben" area. This *Brent* accumulation is thought to hold up to 1·0 B brl of recoverable oil in Jurassic sandstone and is a discovery of great importance, both for its own value and also as an indication of the northern potentialities of this sedimentary basin. Its ultimate production rate may be 100,000 b/d, depending upon the provision of adequate outlets. Initial production is expected to begin in 1975/6 using the "spar loading" system* into tankers; a pipeline to a Shetlands terminal will probably follow.

Finally, towards the end of the 1972 exploration season, the announcement was made of two other oil discoveries—the *Beryl* prospect on block 9/13 about 100 miles east of the Shetlands in 384ft of water, and *Cormorant*, 25 miles south of *Brent*, in 500ft of water. It therefore seems likely that future exploration will discover further large oil accumulations in this area in which other large structures are known to occur.

Estimates made in the latter part of 1972 agreed that about 4·0 B brl of commercially recoverable oil had been found by that time in the British sector in four structures, out of the 25 or so which had been tested. Four others among these structures probably contained a further 1·0 B brl or so of sub-commercial oil. On the basis that about 125 potential oil-bearing structures are known in British waters, the ultimate commercial potential of the British sector might therefore be 31 B brl of commercial and 8·5 B brl of sub-commercial oil[14]. However, in such waters, depths and weather conditions, the technical problems involved in exploration and production operations are very great, and enormous

---

* Designed in Holland, this consists of installing 300,000-brl capacity storage tanks floating on and below the surface, flexibly anchored to the seabed. Tankers would load directly from the storage—itself connected with the producing platforms—from loading facilities installed on top.

[14] Dr. J. Birks, speaking at "Financial Times" North Sea Conference, London, Sept. 1972.

capital investments would be necessary. Thus, it is estimated that the development costs of an oilfield in northern waters could well be of the order of £250 MM[15].

It certainly seems likely that by the early 1980's North Sea production of 2·0 MM b/d will account for at least half the UK consumption at that time. Again, some of the British Sector output might not be consumed directly in Britain but treated and exported, thus earning foreign currency to pay for the purchase of the heavier oils required for blending. This could result in balance of payments savings of the order of £500 MM/a. (In 1971, Britain's oil imports of about 100 MM tons cost about £1,000 MM). However, to achieve this result an investment of at least £2,000 MM on capital equipment will be necessary, i.e. about £1,000 for each b/d of producing capacity.

THE NORWEGIAN SECTOR

Most of the land area of Norway is made up of crystalline rocks of the Fenno-Scandian "Shield" and is therefore devoid of petroleum accumulations. Offshore, however, the Norwegian sector of the North Sea basin includes a thick sequence of Tertiary and Mesozoic beds which have been shown to hold important oilfields.

The first Continental Shelf discovery, made in 1968, was *Cod*, about 145 miles from land, a gas and condensate accumulation in a Tertiary sandstone reservoir. Then, during 1970, a much larger accumulation of good quality, low-sulphur oil was found in the *Ekofisk* structure, which lies about 185 miles southwest of the coast of Norway, near the boundary with the British sector. The reservoir here is reported to be a fractured chalk of Danian (Upper Cretaceous) age. This momentous discovery—the first major oilfield to be found in the North Sea—was rapidly followed by several others in the same area, which have since been named *West Ekofisk, Ergfisk, Eldfisk, Edda, Albuskjell* and *Tor*.

The *Ekofisk* field itself was put into production in June, 1972 at the rate of 40,000 b/d, the oil flowing directly into tankers from a loading buoy. To obviate delays due to bad weather, a million-barrel concrete storage tank which was constructed at Stavanger has been successfully towed to the field area and there settled on the sea floor. Built in the form of a cluster of concrete tanks surrounded by a perforated skirt to break the force of the sea, the structure, which floats when the tanks are empty was sunk on the seabed by filling the tanks with water. A deck across the top will take process and separation equipment.

The production from the Norwegian offshore oilfields is expected to be about 180 MM brl/a in 1976, rising perhaps to 350 MM brl/a by 1980. Gas production could be 200-300 Bcf and 1.0 Tcf in these years

---

[15] Sir D. Barron, speaking at Scottish Council's Aviemore Offshore Forum, 1972.

respectively. Reserves are probably at least 2 B brl of oil and condensate and 7 Tcf of gas (Table 31).

There are obviously considerable technical as well as financial problems involved in deciding the best way to produce and market oil and gas derived from a group of oilwells drilled in several hundred feet of water in the middle of the North Sea. The productive area is separated from the coast of Norway by the 60-mile wide 2500-ft deep Norwegian "Trench" which makes direct pipeline laying extremely difficult.

It was eventually announced (June, 1973) that a 220-mile oil pipeline would be constructed from the *Ekofisk* fields to Teesside in Britain, while a rather longer gas pipeline would be built to a terminal at Emden in West Germany.

In addition to these oilfields, a very large non-associated gas accumulation (*Frigg*) has been found, further to the north, and probably straddling the boundary between the Norwegian and British sectors. Reserves of gas here are said to be of the order of 12 Tcf, or sufficient to support a production rate of up to 1·8 Bcf/d. The field will probably be developed jointly by British and Norwegian interests if an economic outlet can be decided—perhaps to the Shetlands, 130 miles to the northwest—for liquefaction and eventual dispatch by LNG tankers, or by piping the gas all the way to the British mainland.Here again, crossing the Norwegian "Trench" by a subsea pipeline 120 miles long appears to be an unacceptable alternative, particularly in view of the relatively small eventual market for the gas. Another oil and gas accumulation, *Heimdal,* to the south of *Frigg,* awaits evaluation.

TABLE 31

THE NORWEGIAN OFFSHORE FIELDS

	Recoverable Reserves		
	Oil MM brl	Condensate MM brl	Gas Bcf
Ekofisk	1,000	140	3,500
West Ekofisk	340	177	2,500
Tor	109	19	300
Cod	18	32	700
Total	1,467	368	7,000

Source: Ekofisk Committee Report. A number of other oilfields have since been found—*Edda, Ergfisk, Eldfisk, Albuskjell*—and an extension of the *Frigg* gasfield.

THE DANISH SECTOR

Many of the sediments which lie between the western edge of the Fenno-Scandian "Shield" and the Variscan folds of central Germany are similar to those of northwest Germany, and there are also many salt structures. However, no commercial accumulation of hydrocarbons was found in the various onshore exploration wells drilled in Denmark since 1935.

On the other hand, the offshore exploration which has been carried out since 1966 has discovered at least six structures which contain gas or oil in a chalky Upper Cretaceous (Danian) reservoir horizon. The estimated reserves of "M", since renamed the *Dan* field, which lies about 120 miles off the Danish coast, are thought to be more than 275 MM brl, and production from it is expected to begin in 1973, oil flowing directly into tankers from a production platform at the initial rate of 15,000 b/d. Eventually, the output from this group of structures may reach 10 MM brl/a of oil, with an additional substantial volume of gas.

OTHER OFFSHORE AREAS

Offshore exploration was carried out off the coasts of **Holland** between 1961 and 1963. Subsequently, due chiefly to difficulties arising from the German claim in the International Court for a revision of the boundary of their North Sea sector, there was considerable delay before exploration was resumed in 1968.

A number of so far non-commercial gas discoveries have been made in Rotliegendes sandstone reservoirs lying in an arc about 40–50 miles west of The Hague and northwest of Den Helder. An oil accumulation has also been reported about 50 miles north of the island of Vlieland, but this is not known to be commercial.

Several offshore wells have been drilled since 1964 off the coast of **Germany**, in the general region of Borkum Island. The gas which was discovered was mainly non-commercial nitrogen, due perhaps to a stratigraphic change in the nature of the Rotliegendes formation to a saline facies in this area.

The large *Bierum* gasfield discovered in 1963 in the Ems estuary is being jointly developed for the benefit of both Germany and Holland.

CHAPTER 4

Eastern Europe and the USSR

EASTERN EUROPE

THE petroleum accumulations of Eastern Europe outside the Soviet Union fall into two groups—those, the largest, which lie in a Tertiary "foredeep" belt along the outer side of the great Alpine-Carpathian mountain arc, and the smaller fields lying on the inner side of the arc.

The "Neogene"* Pannonian Basin, extending across parts of six countries (Romania, Hungary, Austria, Czechoslovakia, USSR, and Yugoslavia), is the most important sedimentary area. Within it, some 6,000ft of freshwater Pliocene strata were deposited and subsequently affected by faulting and folding processes. Whereas on the outer side of the mountain arc the Pliocene strata were strongly folded, on the inner side the folds are gentle, due to the proximity of the "basement", which lies here at a relatively shallow depth.

The main basin can be differentiated into two sub-basins—the Danube basin in the west, and the Transylvanian basin in the east, which experienced differential epeirogenic movements with consequent variation in the lithology of their constituent beds[1]—most notably shown by the presence of a great thickness of salt in the Transylvanian basin which is absent in the west. In the Danube basin, hydrocarbon accumulations occur in Pliocene beds in gentle anticlines which are probably the result of compaction rather than lateral folding. In Transylvania, most of the gas accumulations are in the underlying Middle and Lower Miocene strata.

---

*"Neogene" is a European term used to include undifferentiated younger Tertiary (i.e. Miocene and Pliocene) as opposed to older Tertiary ("Paleogene") beds.
[1] M. Pauca, Bull.AAPG 51(5), May 1967, 696.

Eastern Europe is one of the few areas of the world, where, until recently, petroleum consumption showed only very slow growth. However, the five-year economic plans announced for the period 1971–75 now show that the trend away from the traditional dependence on solid fuels is likely to accelerate.

TABLE 32

EASTERN EUROPE—ENERGY CONSUMPTION

	Total energy consumption MM tons c.e.	Market shares %			
		Solid fuels	Liquid fuels	Nat. gas	Hydro-nuclear
1950	154	93	5	—	—
1958	269	90	6	5	—
1969	430	76	14	10	—
1975*	590	58	21	21	—

*Projected (Totals do not exactly make up 100% due to rounding).
Source: UN Economic Commission for Europe Economic Survey, 1971.

Hydrocarbon fuel production in Eastern Europe is rising slowly but is still generally insufficient to meet expanding domestic and industrial demands, so that both oil and gas are being imported in increasing quantities by pipeline from the Soviet Union.

TABLE 33

EASTERN EUROPE—OIL PRODUCTION, 1971

	million brl
Albania*	9·7
Bulgaria	2·2
Czechoslovakia	1·4
East Germany*	0·4
Hungary	14·1
Poland	2·9
Romania	99·4
Yugoslavia	21·3
Total	151·4

*Estimated figures

TABLE 34

EASTERN EUROPE—ESTIMATED OIL RESERVES

	million brl
Albania	630
Bulgaria	245
Czechoslovakia	245
Hungary	267
Poland	37
Romania	867
Yugoslavia	460
Total	2,751

TABLE 35

EASTERN EUROPE AND USSR—OIL PRODUCTION 1870-1970
(*thousands of brl*)

Year	Albania	Bulgaria	Czecho-slovakia	E. Germany	Hungary	Poland	Romania	USSR	Yugo-slavia	Total
1870	—	—	—	—	—	—	84	261	—	345
1880	—	—	—	—	—	287	115	2,761	—	3,163
1890	—	—	—	—	—	631	383	24,609	—	25,623
1900	—	—	—	—	—	2,347	1,629	65,955	—	69,931
1910	—	—	—	—	—	12,673	9,724	65,970	—	88,367
1920	—	—	69	—	—	5,607	7,435	31,752	—	44,863
1930	—	—	157	—	—	4,905	41,624	125,555	—	172,241
1940	1,497	—	163	—	1,881	3,891	43,168	218,600	10	269,210
1950	2,800	—	292	—	3,700	1,205	32,000	266,200	780	306,977
1960	4,857	1,460	929	—	9,270	1,442	85,712	1,079,371	6,671	1,189,712
1970	10,950	2,190	1,460	440	14,600	3,070	98,920	2,566,000	20,080	2,717,800

ALBANIA

A belt of Tertiary sediments which lies parallel with the coast of Albania has been folded along a north-northwest—south-southeast axis by the Alpine orogeny. The folds are intense in the east, dying out towards the west, where block-faulting occurs.

Oil was first discovered near *Valona* (*Vlore*) in 1917, and several small accumulations were subsequently found near Scutari and between Durazzo and Tirana. Of these, the most important is the *Devoli* (*Kuchova*) field (now renamed *Stalin*).

The structure here is a buried, eroded anticline of Lower Eocene-Upper Cretaceous limestones covered by Miocene and Pliocene beds which reflect the buried structure. Oil comes from a number of sand lenses in clays of Upper Miocene and Upper Tortonian age. Two unconformities are distinguishable, one at the base of the Pliocene and one in the Tortonian.[2]

The *Pathos* field is a monocline dipping north-northwest, with a productive Tortonian sandstone reservoir. The most recently discovered accumulation is *Marinze,* south of Valona. The maximum production of oil during the Italian occupation was 1·3 million brl in 1943; since the war, the output has steadily increased to a peak of 9·7 MM brl in 1971, with about 7 Bcf of mainly associated gas.

Two small Tertiary gasfields, at *Bubullime* and *Divjake*, about 30 miles north of Valona, are awaiting pipeline outlets.

Crude oil reserves in Albania were estimated[3] as about 630 MM brl in 1960; little recent information is available.

---

[2] A. Mayer Gürr, *Sci.Petr.*, 6, 1, 52, 1953.
[3] Report by US National Petroleum Council (Impact of Oil Exports from Soviet Bloc), 1960.

Disaster at Campina in Romania! A "gusher" out of control has ignited and the derrick is collapsing. Photographed about 1925.

(Source: Petroleum Information Bureau.)

Liquid crude oil seepage at Maidan-i-Naftun in Iran, about 1926. A primitive form of refining was used in this area, the lighter fractions of the oil which seeped into the pit being periodically burned off; the resultant liquid was then drawn off through an underground channel.

A "mud volcano" erupting with the semi-explosive evolution of high-pressure gas and mud near the Seria oilfield in Brunei. *(Source: Shell Photograph.)*

BULGARIA

After some years of sporadic exploration dating from 1927[4], the first commercial oilfield was found on the Black Sea coast at *Tyulenovo* in 1954. Production comes from Oligocene strata at a depth of about 2,600ft and an offshore extension of this field is being developed.

In 1962/3, the *Dolni Dabnik* field was discovered near Pleven, and has since become the country's largest oil producer. Other small oilfields have been found at *Gigen*, south of the Danube in the same area, and at *Gorni Dubnik* (in 1967). The overall oil output reached a peak of 3·3 MM brl in 1968, but fell to 2·2 MM brl by 1971.

Minor accumulations of natural gas have been discovered at *Cheren*, between Vratsa and Pleven, and at *Staro Oryakhovo*, south of Beloslav, near the Black Sea coast.

CZECHOSLOVAKIA

A small volume of oil has been produced over a period of many years from the embayments of Neogene deposits lying between the Bohemian massif and the Carpathian Mountains. These accumulations are on the updip edge of the Vienna basin and hence are less productive than the neighbouring Austrian fields.

At *Gbely* in Slovakia, shallow oil has been obtained since 1913 from sand lenses of Sarmatian age in a locally faulted anticline. A number of other small oilfields have been found since 1953 in East Slovakia, in structures related to Carpathian basin folds (*Smilov, Olka, Mikova*).

In Moravia, the faulted *Hodonin* anticline is a similar structure, giving shallow oil and also production from deeper sands, which are thought to be of the same age as the reservoir of the *Zistersdorf* field in Austria. An oil discovery was also made in 1963 at *Hannakei*, near Kromeriz, in the northeast extension of the Vienna basin. Total Czechoslovakian oil production is about 1·4 MM brl/a. The balance of the country's requirements of some 70 MM brl/a is imported by the "Druzhba" pipeline system from the USSR.

A gasfield with an Oligocene reservoir was discovered before the 1939–45 war at *Vacenovica*, and a more important gas accumulation has been developed in recent years at *Vysoka*, on the Austrian border, which is structurally the continuation of the *Zwerndorf* field (p. 74). The recoverable reserves of gas in the Czech section of this field have been estimated to be 280 Bcf. Smaller gasfields have also been found in the Cerky-Tesin area, near the Polish frontier (*Strahava, Pribor, Zukow*). Czechoslovakia produced about 29.7 Bcf of gas in 1970, but had to import a further 31.4 Bcf from the West Ukraine gasfields of the USSR, a volume likely to increase to 122 Bcf/a by 1975.

---

[4] *Inst.Pet. Review*, June 1966, p. 199.

I

EAST GERMANY

Exploration for hydrocarbons in East Germany began in 1951, and some oil was subsequently found at *Stralsund* on the Baltic Coast, and gas at *Storkov*, southeast of Berlin. By 1964, natural gas production was being obtained from several minor gasfields in Thuringia (*Langensalza, Mülhausen,* etc.). A small oilfield was discovered in 1961 near *Reinkehagen* in Pomerania. This field has a partly dolomitic limestone reservoir and the associated gas has a high nitrogen content. The annual output of oil has been estimated at about 440,000 brl/a.

About 95% of the petroleum requirements of East Germany are imported from the Soviet Union via the northern branch of the "Druzhba" pipeline.

HUNGARY

Present-day Hungary lies wholly within the Pannonian Neogene Basin. The first oilfield to be discovered (in 1936) was *Bukkszak*, which gave production from the crestal part of a faulted dome. The reservoir was a porous andesitic tuff in mudstones of Mid-Oligocene age, but the field was soon exhausted.

In 1937, oil was found in the *Budafapuszta* anticline in the Lispe area, in lenticular sands of Lower Pliocene age. Between 1940 and 1942, several more fields—*Lovaszi, Ujfalu* and *Hahot*—were discovered in the same area; their reservoirs are Miocene-Pliocene sandstones. The Lovaszi and Ujfalu structures are parallel domes on the western end of the Peklenica-Budafapuszta uplift, whereas Hahot is on a separate Tertiary structure, which reflects a deeper Mesozoic uplift.

After the war, *Nagylengyel* was discovered (in 1951) in the Zala highlands. This field now produces about 70% of the total output of Hungarian oil, which amounted to about 14·1 MM brl in 1971. There are Mesozoic (Triassic-Cretaceous) dolomitic limestone reservoirs, in addition to shallow Pliocene sandstones.

Oil was subsequently also found in the southeast corner of the Great Hungarian Plain, at *Kunmadaras, Pusztafoldvar* and *Battonya.*

In the east of Hungary, there is a group of large gasfields. The most important of these are *Hadjuszoboszlo, Ebes,* and several recent discoveries in the *Bekes* area. Gas production from these fields is expected to increase to exceed 165 Bcf/a by 1975.

The most recent hydrocarbon discoveries have been made in the Szeged basin in southern Hungary, where several important gas- and oilfields have been found at *Algyo, Ullés, Szank, Tazlar* and elsewhere. These fields, in which the main accumulations are in Upper Pannonian strata, are being intensively developed in an effort to reduce Hungary's dependence on oil imported from the Soviet Union (about 34 MM brl/a in recent years). The gas reserves at *Algyo* alone have been estimated as 882 Bcf, nearly as large as those of *Hadjuszoboszlo*, and its oil reserves are said to be equivalent to 50% of those of the whole country[5].

Hungarian oil reserves are estimated to be about 267 MM brl. Non-associated gas reserves are of the order of 2·9 Tcf, and associated gas reserves are about 1·4 Tcf, so that total Hungarian gas reserves are at least 4·3 Tcf.

POLAND

Oil and gas have been obtained since the 1860's from a number of small fields in Western Galicia, where the reservoirs are Boryslaw (Oligocene) sandstones. Tectonically, this area is a continuation of the overthrust "nappe" zone of eastern Galicia (whose oilfields now lie within the boundaries of Western Ukraine—p. 118), with Cretaceous and Lower Tertiary folds thrust over a Miocene foreland. However, the structures here are usually less complex than in the east, often being relatively simple anticlines.

The post-war search for oil in Poland has been concentrated on discovering deeper reservoirs outside the Carpathian zone, notably on the "pre-Sudeten" homocline between Zielona Gona and Wielun. Oil-bearing Zechstein (Permian) dolomites have been found at *Rybaki*, but the most important accumulation is the *Grobla* field, near Bochnia, which now produces almost half the Polish oil output of about 3 MM brl/a. The reserves of this field alone are said to exceed those of the thirty or more old Carpathian fields which still lie within the Polish frontiers.

The proportion of oil now obtained from the non-Carpathian areas of Poland exceeds 70% of the total output, and overall reserves are about 37 MM brl. The production of indigenous crude oil does not account for more than 7% of consumption, and the balance has to be imported, mainly from the USSR.

A considerable volume of natural gas is being produced, mainly from several gasfields in the Sanok area. The *Debowiec-Simoradz* gasfield in the western Carpathians supplies Cracow, Tarnow and Warsaw with natural gas, and new fields have been found near Lubaczow and Przemysl in the Carpathian foreland area. The overall output of natural gas from the Polish gasfields was about 140 Bcf/a in 1969, and is likely to increase significantly within the next few years. Reserves are estimated at probably about 3·5 Tcf.

Gas is also being imported by pipeline from the USSR to make up the volumes currently needed.

---

[5] V. Dank & J. Kokai, "Expl. for Petroleum in Europe & N. Africa", IP, 1968, 131.

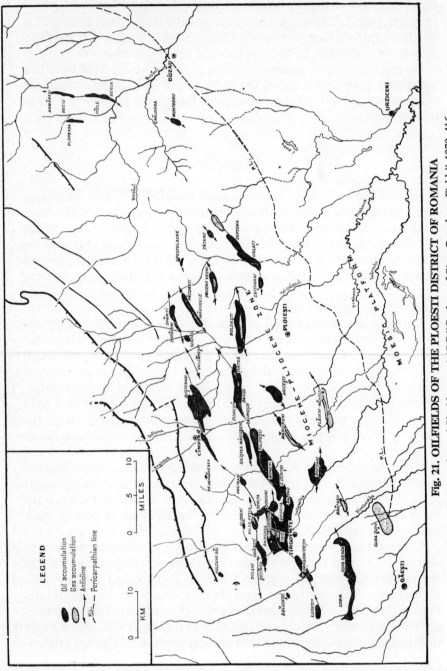

Fig. 21. OILFIELDS OF THE PLOESTI DISTRICT OF ROMANIA

Source: D. Paraschiv and Gh. Olteanu, AAPG "Geology of Giant Petroleum Fields", 1970, 416.

ROMANIA

Oil had been obtained from shallow wells for many years before exploration drilling began in Romania in 1860. A number of accumulations were found in the "pre-Carpathian depression" in Wallachia, in structures related to the foothill folds of the Carpathian Mountains.This area was the original centre of the Romanian oil industry. Oil was subsequently found in the Bacau area, and in Oltenia, and large volumes of gas have been discovered in the inner Transylvanian basin.

Romanian oil production declined from its pre-war peak of 53 MM brl due to exhaustion of the older fields, lack of exploration, and wartime destruction, to only about 24 MM brl in 1945. However, there has again been expansion in the post-war years, and the 1971 output reached a peak of 99·4 MM brl of oil and about 850 Bcf of gas, making Romania easily the largest European hydrocarbon producer outside the USSR. Oil reserves are about 867 MM brl and gas reserves at least 8·8 Tcf.

The core of the Carpathian mountain chain, which curves in a great arc through the heart of Romania, consists of metamorphosed Palaeozoic and Mesozoic rocks which have been subjected to Hercynian and early Tertiary folding and uplift. Between the end of the Lower Cretaceous and the end of the Oligocene, typical shallow-water sediments were deposited around this central core, followed by a great thickness of Miocene and Pliocene beds. Late Tertiary folding has produced many structures containing oil, mainly in Pliocene (Dacian and Meotian) sandstone reservoirs. The principal petroleum areas of Romania can be considered in the following groups:

Pre-Carpathian Depression
(a) Southern Sub-Carpathians
The Ploiesti oilfields provided 99% of the Romanian oil output before 1938, but now account for only about 30%.

The most important local tectonic features are the salt uplifts and the structures they have produced, which vary with the distances from the Carpathian range. From north to south, overthrusts are succeeded by folds with cores of diapiric salt, and then by anticlines associated with salt "pillows" at depth. The oilfields occur along lines of uplift related to salt movements; the age of the salt is probably Mid-Miocene. The petroleum source beds in this area may be the Aquitanian shales which lie close to the Pliocene sandstone reservoirs. There seems to be a general relationship between oil quality and reservoir age—the oil from Meotian reservoirs being paraffinic, while that from Dacian beds is uniformly non-paraffinic[6].

At *Moreni* there is an interesting example of an intruding and over-hanging salt mass, whose upward motion has caused the soft surround-

---

[6] R. Walters, *Bull.AAPG*, Oct. 1969, 2160.

ing Tertiary beds to be warped upwards and thus form reservoir structures sealed against the salt mass.

A single large salt uplift has produced a number of traps in the surrounding beds—the pools of *Pleasa, Stavropoleos, Bana, Piscuri, Cricov, Tuicani* and *Pascov,* with oil in Dacian and Meotian sands.

Ochiuri, west of Moreni, has a trap which is the result of a similar, though simpler, salt intrusion. Production again comes from Dacian and Meotian sands.

At *Boldesti,* no salt mass is known, but one is suspected to be present at depth. The structure is a broad, elongated, somewhat asymmetrical dome, locally productive at *Boldesti, Scaeni* and *Harsa.* Only gas is produced from the Dacian, but below four gas sands in the Meotian there are two oil reservoir sandstones.

The *Ceptura* field at Urlati on the River Cricov is a long, narrow, anticlinal fold which runs northeast-southwest. Principal production (of highly paraffinic oil) comes from the Meotian.

Bucsani was the first field to be discovered by a seismic survey on the gravel-covered southern plain of Romania. It is structurally a broad fairly long buried anticline, over a deeper salt mass, which gives oil from the Meotian.

At *Bustenari,* near the Carpathians, there is oil in the Meotian and also in the Oligocene. Here, the structure is complex and the Oligocene has been thrust southwards over the Lower Miocene Salifère beds, with the southerly dipping Meotian again overlapping the Oligocene. The continuation of the Bustenari field is *Runcu,* an elliptical east-west dome, overthrust to the south.

Filipesti is a long, broad anticline with Meotian oil. At *Campina,* one of the oldest Romanian fields, an upthrust salt mass has arched the overlying Meotian, which is productive on the southern flank. There is also some Dacian oil here.

The *Baicoi-Tintea* field is structurally the result of a salt plug which reaches the surface, and the salt at *Floresti* is probably connected at depth with the Baicoi-Tintea mass.

A number of small oilfields occur near Buzau in the southern sub-Carpathian zone, where the main structural feature is the asymmetrical *Arbanasi* anticline—about 18 miles long with a north-northeast—south-southwest axial trend. Oil accumulations are found at several culminations on this fold, the reservoirs being Sarmatian sandstones. "Mud volcanoes" also occur in this area, where attempts have been made to mine the shallow oil sands, with little success.

Several oilfields have been discovered in recent years in other parts of the southern plain, in the Arges area and west of the River Olt. The most important of these are *Ticleni* and the *Balteni* stratigraphic trap; this area now accounts for 22% of Romanian oil output, mainly from Miocene reservoir beds.

(b) Eastern Sub-Carpathians

In a region of complex tectonics and many seepages near Bacau and Moinesti in Moldavia, a number of oil accumulations (*Zemes-Tazlau, Lucacesti, Stanesti,* etc.) have been developed in an area where the "flysch" of the eastern Carpathian range has been thrust over the Oligo-Miocene Salifère foreland and subsequently folded and faulted. Production comes mainly from Oligocene sands in complex local fold and fault structures, and now provides about 20% of the total national production.

Transylvanian Basin

This is an intermontane depression formed at the end of the Cretaceous. Salt uplifts on the inner (western) side of the mountain arc have produced anticlinal traps in which large gas accumulations occur in Miocene and Pliocene reservoirs. The gas has a remarkably high proportion of methane—more than 99%, and production from these fields accounts for nearly 80% of the total Romanian gas output, the balance being associated oilfield gas.

The gasfields lie near the Mures River, in a line paralleling the eastern Carpathian range; they include *Teg-Mures, Saros, Sin Miclaus, Sarmasel, Medias,* and *Gurghiu.*

Other Areas

Considerable exploration effort is being directed to finding new oil provinces, e.g. along the edge of the Pannonian Basin in the west and on the Moesian or "Pre-Balkan" Platform south of Bucharest, in both of which areas some accumulations of pre-Tertiary oil and gas have recently been found.

The *Clejani* field, near Videle, is the most important oil discovery so far made on the Moesian Platform, where a considerable thickness of Mesozoic and Palaeozoic sediments occurs. Structures in this area are controlled by a system of longitudinal faults which approximately parallel the Carpathian range. Reservoirs have been found in Triassic, Jurassic, Cretaceous and Neogene beds.

The search for oil is also being carried out on the Moldavian Platform, in the Birlad Depression and on the North Dobrujan promontory[7], while offshore exploration in Black Sea waters was initiated in 1969, so far without announced success.

With increasing consumption, Romania became an importer of crude oil during the late 1960's; by 1970 oil imports were about 15 MM brl/a, mainly from the Middle East and Algeria.

---

[7] N. Grigoras & N. Constantinescu, "Exploration for Petroleum in Europe & N. Africa", IP, 1969, 147.

YUGOSLAVIA

Oil in small quantities was obtained from shallow mineshafts in Croatia as long ago as 1856. The first small oilfields were discovered in 1885 near the present Hungarian frontier at *Peklenica* and *Selnica*. These are crestal accumulations on a gentle east-west fold in an extension of the Pannonian Basin; the reservoirs are Lower Pliocene sands[8]. During the 1939-45 war, a larger field was found at *Peteshava* in the same area; this is a stratigraphic trap on the flank of the same uplift which has produced the *Lovaszi* and *Ujfalu* accumulations in Hungary (p. 106).

In general, the oil and gas accumulations of modern Yugoslavia are mainly the result of stratigraphic traps formed in several "graben" areas. Most of the reservoirs are Miocene or Lower Pliocene sandstones. Nearly 80% of the annual oil output of about 21 MM brl comes from Croatia, where the fields lie along the northern bank of the Sava river[9]. The most important of these are *Struzec* and *Zutica*, which each produce about 10,000 b/d. Other locally important fields are *Klostar* in the Sava depression, *Sandrovac* in the Drava depression, and *Kikinda* and *Elemir* in the Banat area. *Kikinda*, discovered in 1959, is of particular interest in that it produces from Pliocene sandstone overlying fractured crystalline schists, which also act as a reservoir[10]. *Mokrin* and *Okoli* are the largest gasfields; gas output is about 26 Bcf/a and reserves have been estimated at 3·5 Tcf.

Little exploration has as yet been carried out in the Adriatic Coastal Basin, where prospects of hydrocarbon accumulations have been described as good[11].

Proved oil reserves have been estimated at about 460 MM brl.

THE USSR

PRODUCTION of oil began in the Baku area of Azerbaijan in 1873, in a region where petroleum had been recovered from hand-dug pits and wells for many centuries. In the years prior to 1918, Russian oil output was second in the world only to that of the United States—and in fact sometimes even surpassed it. After a period of dislocation in the early 1920's, output once more increased until it reached a peak of 238 MM brl in 1941. The destruction brought about in the course of the ensuing war, during which many of the Caucasus oilfields were destroyed, resulted in a fall in production to only 167 MM brl in 1946.

8 A. Mayer-Gurr, *Sci.Petr.*, 6, 1, 60, 1953.

9 R. Filjak *et al.*, "Expl. for Petroleum in Europe & N. Africa", IP, 1969, 113.

10 M. Vujkov, *Nafta*, Nov. 1969, 20, 11, 529.

11 R. Markovic, 7 WPC, Moscow, 1968, AI, 92, 9.

However, in the postwar years, the development of the petroleum industry was very rapid as new productive areas were opened up, so that by 1960 the Soviet Union stood once again second only to the USA as a world oil producer.

In 1971 output was 2,701 MM brl—more than eleven times the pre-war peak. By 1975, it is planned to reach 3,500 MM brl/a.

A striking trend has been the continuing eastward movement of the focus of production, the combined result of the depletion of the early discoveries, the desire for wartime security and the development of previously unexplored territories. Thus, Azerbaijan produced 70% of Soviet oil in 1939, but less than 4% (of which more than half came from offshore fields) in 1970; whereas the fields lying between the Volga and Urals, which yielded only 7 MM brl in 1938, provided 71·4% of the national output by 1965, i.e. about 1,500 MM brl, or more than 200 times the output of only 30 years before. By 1970, this proportion had decreased to 58·8%, since during the last few years the new oilfield areas in West Siberia have been rapidly developed; in 1970 they accounted for 8·8% of the national output with the production from less than a dozen huge fields, and West Siberia is expected to produce 850 MM brl of oil in 1975.

A further new and promising producing area has now been discovered in East Siberia, which awaits development; the eastward trend of the centre of gravity of Soviet oil production is therefore likely to continue in the years ahead.

The USSR is a major exporter of petroleum, in the forms of crude oil, finished products and natural gas, mainly to the Comecon countries of Eastern Europe. Over the 10-year period up to 1971, more than 170 MM tons of oil were delivered via the 5,300 km-long "Druzhba" pipeline system to Czechoslovakia (65 MM tons), East Germany (45 MM tons), Hungary (25 MM tons) and to other Eastern European states. As the capacity of the pipeline system grows, it is planned to increase these deliveries by 39% over the 1971–75 period, to attain a 50 MM ton/a delivery rate by 1975.

However, it is interesting to note that since 1969, the USSR has also been *importing* oil, mainly from the Middle East (about 157,000 b/d in 1971). Also, in spite of the record domestic output of natural gas (7780 Bcf in 1972) gas is also being imported by pipeline from Afghanistan (72 Bcf/a) and Iran (220 Bcf/a rising to 350 Bcf/a in 1973). Presumably, therefore, there are certain deficiencies in the Soviet petroleum producing and distributing industry, resulting from strategic requirements or the rapid increase in domestic consumption. The growth in demand for hydrocarbon fuels is shown by the fact that, in 1970, about 37·9% of the primary energy requirements of the USSR were provided by oil, and 18·8% by natural gas, compared with the equivalent proportions of 23% and 6% in 1958.

TABLE 36

USSR—TOTAL ENERGY CONSUMPTION

	Total	Percentages			
	MM tons c.e.	Solid fuels	Liquid fuels	Natural gas	Hydro/ nuclear electricity
1950	304	73	23	4	—
1958	598	70	23	6	—
1969	1,011	43	32	24	1
1975*	1,434	33	35	30	2

Source: ECE, 1971
*planned.

TABLE 37

USSR—CRUDE OIL PRODUCTION BY AREAS

	1950	1955	1960	1965	1970
Total (million brl)	273	510	1065	1750	2540
Percentage shares:—					
Ural-Volga	29·0	58·2	70·5	71·4	58·8
Transcaucasus/North Caucasus	55·0	30·8	20·2	17·5	15·7
Central Asia/Kazakhstan	11·8	8·0	6·1	5·6	8·8
West Siberia	—	—	—	—	8·8
Ukraine/Byelorussia	4·2	3·0	1·5	3·3	5·2
Other	—	—	1·7	2·2	2·7
	100·0	100·0	100·0	100·0	100·0

Source: "Petroleum Industry of the USSR", 8 WPC Moscow, 1971.

TABLE 38

USSR—PRIMARY FUEL OUTPUT (percentages of total)

Year	Coal	Oil	Natural gas*	Other fuels†
1950	66·1	17·4	2·3	14·2
1956	63·2	23·3	3·0	10·5
1958	58·8	26·3	5·5	9·4
1960	53·9	30·5	7·9	7·7
1962	48·8	34·2	10·9	6·1
1966	40·7	36·7	16·5	6·1
1968	37·9	39·3	17·9	4·9
1970‡	38·5	37·9	18·8	4·8

*including associated gas
†peat, shale, wood
‡planned
Source: V. Vasilyev, 8 WPC Moscow, 1971.

Up to the end of 1969, the USSR had produced about 27 B brl of oil in the course of its petroleum history. One estimate for the remaining crude oil reserves* at the end of 1965 was 29·8 B brl, of which 23·8 B brl was in the Ural-Volga area, 1·8 B brl in Azerbaijan and 4·2 B brl was in other areas.

Since 1965, there have, of course, been many new oil discoveries, particularly in the Mangyshlak Peninsula of Kazakhstan and West

*For an explanation of the methods used to classify reserves in the USSR, see p. 335.

Siberia. "Proved" oil reserves were therefore probably about 60 B brl at the end of 1971, or double the 1965 total. In addition, there are important oil-shale reserves in Estonia, near Lake Peipus, and in the Ukhta-Pechora and Kuibishev-Saratov areas. Proved natural gas reserves are estimated to be some 630 Tcf, at least 80% of which total lies east of the Urals.

"Ultimate" oil reserves have been estimated to amount to 210 B brl on land, with perhaps a further 39 B brl of offshore oil (mainly from the inland seas).

Offshore oil and gas prospects are of great importance, since it is estimated that the Continental Shelf area of the USSR covers an area of more than $2\frac{1}{2}$ million sq miles. There are already about 20 oil- and gas-fields in production around the Caspian Sea (some of the Baku fields in the west, the Emba area in the northeast, the Mangyshlak Peninsula in the east and the Nebit Dagh fields in the southeast).

Immediate offshore prospects are thought to lie in other parts of the shallow waters of the Caspian Sea; off the Black Sea and Azov Sea coasts where there are likely continuations of the Stavropol-Krasnodar fields; on the southern Baltic Sea shelf; in the Sea of Okhotsk to the north of Sakhalin Island; and in the huge northern Arctic Ocean shelf areas.

PETROLEUM PROVINCES OF THE USSR

The land and inland sea area of the Soviet Union covers 8·6 MM sq miles, or one-seventh of the Earth's surface. About 65% of the total area is made up of sedimentary rocks.

The two principal tectonic elements are the Russian Platform in the west and the Siberian Platform in the east. The former comprises a large area of semi-horizontal Palaeozoic, Mesozoic and Tertiary strata, while the latter is composed of semi-horizontal Palaeozoic and Mesozoic rocks. In both cases, the underlying "basement" is an extension of exposed, extensively metamorphosed, Pre-Cambrian "Shields"—on the west, the Baltic (Fenno-Scandian) massif and the smaller Ukrainian "Shield", and on the east the Angara "Shield" of Siberia and its subsidiary massifs. Mobile geosynclinal belts with their associated mountain chains garland these platform areas. They can be divided into the following groups:

(1) Palaeozoic geosynclines, forming the Ural, Tian Shan, Altai, West Arctic, and East Kazakhstan mountain chains.

(2) Mesozoic-Tertiary geosynclines, which include:

(a) the Mediterranean geosyncline, with the Carpathian, Caucasus and Pamir ranges; and

(b) the Far East geosynclines, with the Transbaikal, northwest Siberia, Kamchatka, and Sakhalin Island mountain chains.

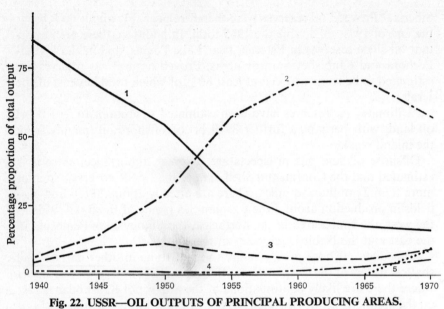

Fig. 22. USSR—OIL OUTPUTS OF PRINCIPAL PRODUCING AREAS.

1—Transcaucasus and North Caucasus; 2—Volga-Urals; 3—Central Asia and Kazakhstan; 4—Ukraine and
Byelorussia; 5—West Siberia

TABLE 39

USSR—PETROLEUM PROVINCES AND PRODUCTION, 1971

MM brl

I West Russian Platform
    (a) *Ukraine and Byelorussia*
        (i) West Ukraine (Carpathians)
        (ii) East Ukraine (Dnieper-Don Basin) } 103
        (iii) Prepiatsky Basin   38
    (b) *Pechora-Ukhta Basin*   43
    (c) *Ural-Volga Region*   1,540

(i) Perm Province	132	
(ii) Tartar ASSR	727	
(iii) Bashkir ASSR	292	
(iv) Kuibishev area	253	
(v) Orenburg area	63	
(vi) Lower Volga	73	

II The Caucasus
    (a) *Azov-Kuban Basin*   265
        (NW, NE Caucasus and Georgia)
    (b) *South Caspian Basin*
        (Azerbaijan)   145

III East Caspian Area
    (a) *Pre-Cambrian Depression (Emba)*
    (b) *Mangyshlak Peninsula* } 115
    (c) *Nebit Dagh area*   110

IV Central Asia   15

V West Siberia   310

VI East Siberia (awaiting outlets)   —

VII Soviet Far East   17

                            TOTAL   2,701

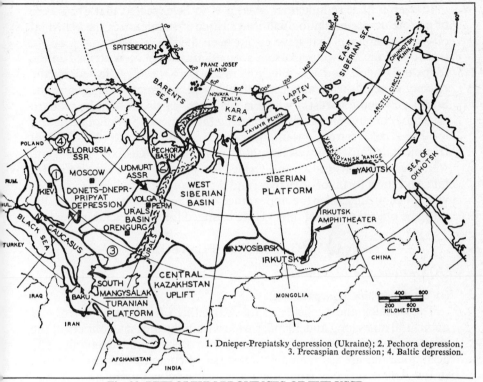

1. Dnieper-Prepiatsky depression (Ukraine); 2. Pechora depression;
3. Precaspian depression; 4. Baltic depression.

Fig. 23. PETROLEUM PROVINCES OF THE USSR

Source: N. Eremenko, G. Ovanesov, V. Semenovitch, *Bull. AAPG* 56(9), Sept 1972, 1719.

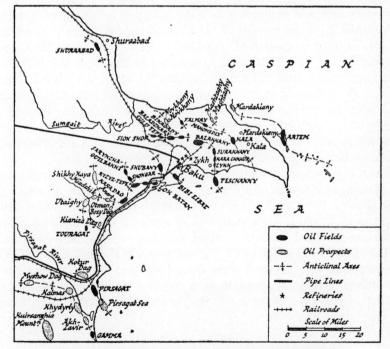

Fig. 24. OILFIELDS OF THE APSHERON PENINSULA

In such a huge sedimentary area, it is to be expected that every type of petroleum accumulation will exist. Table 39 summarises the principal oil-producing areas that have so far been developed. Obviously, only a fraction of the known petroleum accumulations can be described in the following pages, and the selection made is inevitably affected by the paucity of published statistical and technical information regarding the Soviet petroleum industry which is available, particularly in its modern aspects. Problems also arise in rendering the names of Soviet oilfields and local geological formations into English. Some are caused by the different treatments afforded in western publications to the Russian "soft" sign, which has been variously transliterated; others arise from the alternative long and short forms of adjectives, and the use of the genitive case, factors which can greatly alter the appearance of a proper name. Occasionally, also, the official names of oilfields are changed for unpredictable reasons.

I. WEST RUSSIAN PLATFORM

(A) *Ukraine and Byelorussia*

(i) West Ukraine (Carpathians)

The oilfields and gasfields of the Carpathian foothills lie in an area about 150 miles long and 30 miles wide within the present frontiers of the Western Ukraine, with prolongations into Poland and Romania. The accumulations are associated with a series of complex "nappes" formed in the Alpine orogeny from the Tethys geosynclinal sediments. Several overthrust folds of "flysch" sediments have been driven outwards over each other and over a Miocene "foreland". In general, oilfields are found in an inner belt, and gasfields in an outer belt, paralleling the mountains.

Boryslaw is on the margin of the Central nappe, where it overlies a further large recumbent fold. Production has come since 1860 from Oligocene and Upper Cretaceous sandstones on the upper side and crest of the recumbent fold, which is considerably affected by local faults. There is also a little oil in the beds of the overlying Central nappe, which have provided plastic cover for the lower accumulation. The oil is paraffinic, and ozokerite wax has been mined for many years from steep veins which occur over the crest of the fold. These veins are often filled with fault breccia, for which the ozokerite acts as a cement.

Other nappe-type oilfields in this area which are structurally and stratigraphically similar are *Dolina, Bitkow, Maidan* and *Ripensk*. Of these, *Bitkow* is of special interest in that it produces from three fault blocks of shale and is associated with an almost horizontal thrust plane.

The outer belt of gasfields around *Dashava* provides non-associated gas for export by pipeline to Czechoslovakia and Poland. Kiev has also been supplied with gas from here since 1948, and Moscow since 1951. The largest of these accumulations is *Bilkhe-Volykhia* with reserves of 780 Bcf of gas. *Uhesko* (with 600 Bcf) is an anticline near Striy, which was discovered in 1947. Production here comes from a number of Lower Sarmatian and Tortonian (Middle Miocene) sandstone reservoirs in a structure bounded by several faults. *Opary* is a gentle dome with 12 productive horizons; other gasfields in the same area are *Kalusz*, *Kadobninskoe* and *Kossow*.

(ii) East Ukraine (Dnieper-Don Basin)

The Dnieper-Don sedimentary trough is about 300 miles long with a northwest-southeast axis, and is separated by the Pre-Cambrian Ukraine massif from the Carpathian Tertiary nappes on the west, and the Azov portion of the Mediterranean geosyncline on the south. The basin is mainly filled with Upper Palaeozoic and Mesozoic strata. A number of small oilfields have been found (*Romny, Katschanovo,* etc.) with Carboniferous and Triassic reservoirs.

Towards the centre of the basin, several large gasfields have been discovered, with Carboniferous and Jurassic sandstone and Permian carbonate reservoirs. Of these, *Shebelinka,* found in 1956 a few miles southeast of Kharkov, is the largest, with original reserves of some $16\frac{1}{2}$ Tcf. It is an anticlinal fold over an underlying Devonian salt-pillow, with maximum dimensions of 24 miles by 8 miles. The structure is markedly asymmetric with dips of 25–30° on the southwest flank and 8–10° on the northeast flank. There are at least a dozen reservoirs in Carboniferous and Permian dolomites, anhydrites and sandstones.

Other important gasfields in this area are *Zapadno-Krestischche, Poltava, Chergimov* and *Radshenko.* In all, there are at least 23 gasfields, of which nine produce condensate. Gas is piped from here to the industrial centres of the Ukraine and also to Moscow via Briansk, at an annual rate of at least a trillion cu. ft.

(iii) Byelorussia (Prepiatsky Basin)

The Prepiatsky Basin in the south of the Byelorussian SSR is one of the newer oil provinces of the USSR. The basin covers an area of about 14,000 sq miles, lying between the crystalline Voronezhsky and Byelorussian massifs on the north and the Ukraine "Shield" on the south. Oil and gas have been found since 1941[12] in Jurassic to Upper Devonian fractured limestone reservoirs.

Faults in the "basement" seem to have controlled the development of lines of uplift in the overlying beds; accumulations of oil are found in faulted anticlines and fault-blocks, many associated with salt domes.

The most important of these oilfields are *Rechitsa, Ostashkowizi, Semilusko-Petinsky* and *Zadonsko-Eletsky.*

(B) *Pechora-Ukhta (Timan) Basin*

The Pechora-Ukhta basin in the northeastern corner of European Russia extends over some 120,000 sq miles west of the Urals. It contains

---

[12] V. Avvakumov *et al.,* 7 WPC Mexico, 1967, PN 2–4.

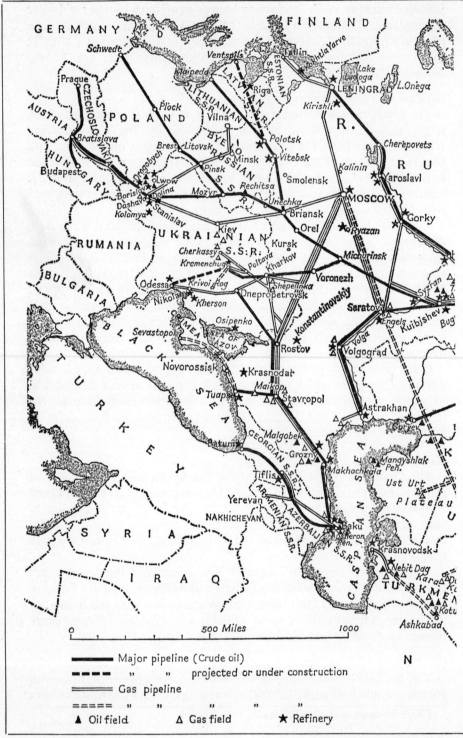

Fig. 25. OILFIELDS AND

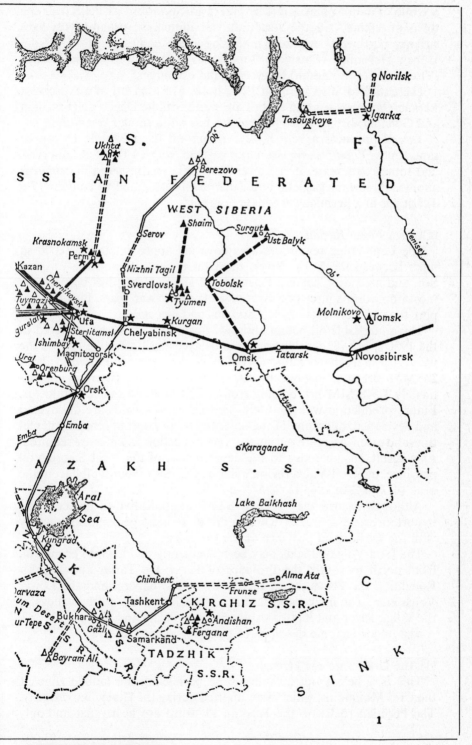

GASFIELDS OF THE USSR

Source: Petroleum Information Bureau.

a group of minor oilfields and several more important gasfields near the town of Ukhta. All the structures are anticlines, usually with axes striking northwest-southeast, and production comes principally from Upper Devonian (Frasnian) sandstones and Lower Carboniferous (Visean and Tournaisian) limestones and dolomites.

The *Usa* field, discovered in 1968, holds 614 MM brl of oil reserves; *Verkne-Omra* has gas and oil in Upper and Middle Devonian (Frasnian and Givetian) beds; *Nishne-Omra* also has oil in similar reservoirs.

Yarega produces a very heavy viscous oil from Middle Devonian sandstones; *Vukhtylskoye* is a major gasfield, with 18 Tcf of gas reserves, and some condensate, in a Lower Carboniferous reservoir. *Dzehbol* is another important gasfield, and *Layavozh* is estimated to contain 17·5 Tcf of gas in a Permian carbonate reservoir.

(C) *Ural-Volga Region*

The Ural-Volga region, an area covering approximately 350,000 sq miles between the Volga River, the Ural Mountains and the Caspian Sea, and extending beyond Perm to the Kama River, has become the most important source of Soviet oil in the post-war years, far outstripping the older oilfields of the Caucasus and Georgia.

The remarkable oil potentialities of this area were first discovered in the 1930's, but its development was delayed until after the end of the war. Little more than 7 million brl of oil was produced in 1938, and about 28 MM brl in 1946—but by 1971 the output had risen to approximately 1,540 MM brl, coming from a belt of more than 500 oilfields. Future production will no doubt increase as new fields are discovered and developed, but many of the older fields are now approaching depletion, and extensive repressuring and water injection schemes are required. As a result, and due also to the development of the West Siberian oil province, the Ural-Volga region's share of USSR oil output fell from the peak of 71·4% in 1965 to 58·8% in 1970.

Although termed the "Second Baku" as indicative of its economic importance as a source of Russian oil, there is no geological similarity with the Tertiary oil province of the Caucasus.

The Ural-Volga oilfields may be conveniently divided into groups in, from north to south, the Province of Perm, the Tartar and Bashkir Republics, the Provinces of Kuibishev and Orenberg, and the Lower Volga area. The most important accumulations occur in the valleys of the Volga River and its two tributaries, the Belaia and Kama.

The major tectonic divisions of the Ural-Volga area are:

(1) **The Ural "foreland" trough**

This is a belt about 50 miles wide paralleling the mobile zone of the Ural Mountains, which were uplifted during the Hercynian orogeny. The Permian rocks on the Russian Platform are nearly flat and only moderately thick; but as they pass eastwards towards the west flank of the present Ural Mountains, they thicken rapidly and are often found in a fossil reef facies which can act as a highly permeable oil reservoir. The Permian section varies in thickness from 12,000ft in the south to only 4,200ft in the centre, thickening again to 9,000ft in the north. Most of the oil is found in a north-south line of reefs of Artinskian (Lower Permian) age, which were probably formed along older fault lines. Carboniferous, Devonian and Silurian oil accumulations are also found, often associated with local folds.

(2) The Ural-Volga arch

The broad regional arch, whose southern flank dips towards the Caspian depression, is made up of two independent Pre-Cambrian uplifts—those of Samarski Luka (Bashkiria) and Tartaria. Several thousand feet of marine Devonian and Lower Carboniferous clastics and carbonates were laid down upon these uplifts, and they include the principal hydrocarbon source rocks and reservoir beds. Accumulations are also found in Middle Carboniferous and Permian strata. The Devonian structures are thought to be the result of the uplift of "basement" fault blocks, while the Carboniferous anticlines were probably the result of late-Palaeozoic folding movements. This area has provided the major oilfields of the "Second Baku", usually in the form of relatively small, flattish anticlines occurring as local culminations on lines of deep-seated uplifts, which can often be traced by geophysical methods over distances of several hundred miles.

THE URAL-VOLGA OILFIELDS

(i) Perm Province

The town of Perm has given its name to the rocks of the Permian system which outcrop extensively in this area. (After the 1939–45 war, the name of the town and the Province was changed to Molotov, but it has subsequently reverted to Perm). Oil was first discovered in the east of the region in the early 1930's in the *Chusovaskiya-Gorodki* field, where the reservoir is a buried limestone reef of Permian age. Since the war, a number of oilfields have been developed, mainly in the west and south of the Province. They are generally small culminations along lines of gentle regional uplift, which are probably the result of sedimentary deposition on pre-existing reefs. Oil is produced from Permian dolomites, Lower Carboniferous sandstones and limestones, and Devonian limestones; the Lower Carboniferous (Visean) sandstones are the most prolific reservoirs.

There are at least 40 oilfields in Perm Province, whose numbers are being added to as the many anticlines that have been discovered are drilled and developed.

Yarino in the north can be considered as typical of the fields in this area. It is a gentle anticline with a slightly steeper west flank, a north-northeast trending axis, and maximum dimensions of six miles long by 2½ miles wide. This field was discovered in 1956 and produces oil from the Lower Carboniferous Visean "Bobriki-Stalinogorsk" sandstone formation, which is petroliferous over much of the Ural-Volga area.

Tanipo in the south of the Province is a rather smaller faulted anticline, which contains oil in Middle Carboniferous (Westphalian) limestones as well as in Visean sandstones.

Polazna is another, slightly steeper, rather more asymmetrical anticline (with flank dips of 4/5° and 2°). Other similar fields are *Kuedino, Pavlovo, Severokamsk, Krasnokamsk, Overiata* and *Shilovo*. The output from the Province was 132 MM brl of oil in 1971.

(ii) Tartar ASSR

The Tartar Republic, which lies roughly in the centre of the Ural-Volga area, accounted for more than 727 million brl of oil in 1971, or more than 27% of the oil production of the whole of the Soviet Union in that year. The oilfields, which mainly lie east of the Volga and have been discovered only in the last 20–25 years, are gentle anticlines formed as local culminations on the southern flank of the main Tartar uplift, which itself reflects an underlying Pre-Cambrian "basement" flexure. The principal axis of this roughly oval uplift runs northwest-southeast for about 180 miles. There are also a number of subsidiary fold axes, of which one of the most important is the 70-mile long "Sok-Scheshma trend".

The largest oilfields are gentle structures which have flank dips of only a few degrees, becoming even flatter at depth. Oil is generally found in Carboniferous sandstones, and there is also Permian and Devonian carbonate production.

The *Romaschkino* field, west of Bugulma, will probably turn out to be the largest oil accumulation in the whole Ural-Volga area when it is fully developed; it is certainly one of the largest in the USSR. The field was discovered in 1948 and about 2,000 producing wells have subsequently been drilled into a large number of reservoir sandstones of Upper Devonian (Frasnian and Givetian) age, folded in a huge gently-dipping domal structure. There is also some less important Lower Carboniferous oil. A vigorous peripheral water injection pressure maintenance programme was initiated soon after production was started in 1953, and the field has been divided into at least 23 separate productive areas, which are being methodically developed. Some of these pools have individual names—*Minnibaevo, Pavlovo, Selenogorsk,* etc.

The *Bavly* field is a similar large, gentle structure (discovered in 1946) which produces principally from Frasnian sandstones, but also from

Carboniferous (Visean) beds. It is situated in the extreme southeast of the Tartar Republic.

Sulino, near *Bavly*, is an example of Permian production in this area, with oil in Permian Sakmara limestones and dolomites; structurally, it is another large, flattish anticline.

Yamashin is another large oilfield in this area, which was developed in 1969 and is expected to produce 35 MM brl/a.

(iii) Bashkir ASSR

The Tartar uplift is separated by the Kama-Bielaya depression from the Bashkir uplift, which again reflects a deep Pre-Cambrian "basement" flexure. Oilfields occur in minor anticlinal structures along a number of roughly parallel northwest-southeast trends within the depression; on the northeast flank of the Tartar uplift which extends into Bashkiria; on the southwest flank of the Bashkir uplift itself; and in the Ural Mountain geosynclinal zone, southeast of Ufa.

The *Arlansk* field, discovered in 1959, lies on one of several lines of uplift within the Kama-Bielaya depression. The reservoirs here are Visean (Lower Carboniferous) sandstones.

The *Oriebasch* (*Kaltasy*) field lies a few miles to the northeast on a parallel trend, with similar reservoirs.

The most important accumulation in Bashkiria is the *Tuymazy* field (near Bugulma), on the northeast flank of the Tartar uplift. This is an unfaulted, asymmetric, gently folded anticline, with its southeastern flank steeper ($2\frac{1}{2}°$) than its northwest flank ($1°$). There are three culminations, and the maximum dimensions of the field are about 18 miles long by nine miles wide; there is a closure of some 250ft in the Devonian D^1 and D^2 reservoir horizons, and an oil column of 120ft. The field was originally developed as a Carboniferous producer in 1934, but subsequent drilling to the Devonian in 1944 proved that the deeper sandstones at depths of about 5,200ft were much more prolific. The 1970 production was about 100 MM brl/a of oil, with about 0·5 Bcf/a of gas. Extensive water injection pressure maintenance schemes have been adopted in this field, whose output is roughly one-third of that of the whole Bashkirian ASSR—itself the second-largest oil-producing region of the Soviet Union.

A few miles east of the Tuymazy uplift is the east-west Serafimov-Baltaev trend, with which are associated the four oilfields of *Serafimovka Leonidovka, Konstantinovsk* and *Nikolaeka*. Productive horizons here are similar to those at Tuymazy. *Serafimovka* is the largest field of this group, with a northwest-southeast axis, and maximum dimensions of six miles by three miles. The north flank is slightly the steeper, although dips are generally gentle ($2–3°$).

Schkapovo is another important oil accumulation, a few miles to the southeast of these fields. It is an anticline with a northwest-southeast axis, and prolific production from Devonian (Givetian and Frasnian) sandstones, with some production also from Lower Carboniferous sands.

The *Kusch-Kul* (*Sedyasch*) field is one of the relatively few oilfields so far discovered on the southeastern flank of the Bashkirian uplift, the other flank of which in Perm Province is associated with many more accumulations.

The Ural foreland trough, which extends for some 240 miles roughly north-south in Bashkiria was the area in which the oilfields of the Ural-Volga area were first developed. *Ishimbayevo* was discovered in 1932, with production from Lower Permian (Artinskian) reef limestones which have since been developed as a group of individual domes with dips of 25–40°. A few miles north of Ishimbayevo, the *Kinsebulatovo* dome has flanks dipping as steeply as 60°, with oil in Artinskian limestone whose permeability is controlled by local faulting. *Malyschevo* is another asymmetrical dome in this area, with an east flank dip of 10°–12° and a west flank dip of 50°.

Kantschurink and *Musinsk*, to the south of Ishimbayevo, are condensate fields with a common water drive and gas reserves of some 210 Bcf.

(iv) Kuibishev Province

This region is third only to the Bashkir and Tartar Republics as an oil-producing area in the Soviet Union. The loop of the Volga at Kuibishev coincides with the east-west axis of the Shiguli-Puhatschev uplift, which is about 180 miles long and is associated over much of its length with a number of important oilfields in local anticlinal accumulations.

At least 108 oil- or gasfields are known in this Province, of which 77 are oilfields, 15 condensate fields and 16 gasfields. Reservoirs are mainly Permian limestones and Carboniferous and Devonian sandstones[13]. There is also some Mesozoic production. The greater part of the oil output comes from the limestone reservoirs, and fault tectonics have played an important part in controlling local accumulations.

As in the other oilfield areas of the Ural-Volga region, most of the structures are relatively small, gentle anticlines, usually 6–12 miles long and a few miles wide.

The *Syzran-Saborovsk* field was discovered in 1935 and is one of the most important oilfields of this Province. It lies on a gentle faulted anticline, about six miles long with an axis striking northwest-southeast. Some oil is produced from Lower Carboniferous limestones, but principal production is from the underlying Devonian sands. There is also some Mesozoic-Tertiary output. Outcropping asphaltic limestones of Upper Carboniferous age and Permian asphaltic dolomites have been extensively mined in this area.

---

[13] W. Lobow, G. Alkeseew & M. Sajdelson, *Geol. Nefti.*, 1958, 5, 8.

Only a few of the most important fields can be mentioned here. The *Muchanovo* field is probably the largest oilfield so far discovered in the area, an anticline about 11 miles long and nearly three miles wide with gentle flank dips, and a thick reservoir development of Lower Carboniferous (Visean) sandstones and Devonian (Givetian-Frasnian) limestones.

The *Soliny Ovrag* field, in the west, was one of the earliest discoveries. It is a faulted anticline, with major production in Lower Carboniferous (Visean and Tournaisian) sandstones, and in Devonian (Fammenian and Frasnian) limestones and sandstones.

The neighbouring *Tablonovo* field is also a flat, asymmetrical faulted anticline with similar reservoir horizons. Further to the east, *Shigulevskoye* is a faulted anticline with an east-west axis, an axial length of about 2 miles, and production from Devonian, Carboniferous and Permian strata.

(v) Orenburg Province

In the part of Orenburg Province lying west of Kuibishev and south of the Tartar ASSR, there are several important oilfields which are geologically related to structural trends in the neighbouring areas. Thus, in the north, the anticlinal *Baytugon* field lies on the southern flank of the Tartar uplift, and has a Visean reservoir. Further south, there are two lines of oilfields running roughly northwest-southeast along the Bolschoy-Kinel uplift and the Samarkino fault zone. Oil occurs in Devonian, Carboniferous and Permian reservoirs. Other accumulations worthy of mention are *Stepanovo, Buguruslan* and *Sultangulovo,* which are all gentle, asymmetrical anticlines with similar reservoirs, ranging from Upper Permian limestones to D^2 and D^3 Devonian sandstones.

A major gasfield was found in 1967 at *Krasny Kholm,* which is thought to contain as much as 35 Tcf of gas in a thick Lower Permian limestone reservoir with several structural culminations.

(vi) Lower Volga Region

The Region comprises three petroleum-producing areas:

(a) *Saratov area*

A number of large gasfields and smaller oilfields lie near Saratov, on both sides of the Volga. Discovered during the 1941–45 war, these fields were linked by pipeline to Moscow in 1947, and supply gas for the capital at a rate of 17/18 Bcf/a.

The area is structurally on the gently sloping southeastern flank of the Russian Platform. Tectonic movements were more intensive here than further to the north, and there is a general thickening of Palaeozoic sediments towards the southeast.

Oil and gas are produced from a number of anticlines with Carboniferous and Devonian reservoirs, the gas being mainly produced from the clastic beds of Middle Carboniferous (Namurian) age, and the oil from Carboniferous and Devonian reservoirs.

A typical example is the *Stepanovo* field, about 50 miles from Saratov, which is one of several culminations along an east-southeast trend. The structure here is a large faulted anticline, which contains reserves of some 900 Bcf of mainly non-associated gas in Devonian (Givetian) clastics. Similar anticlinal fields are *Sovetskaya, Sokolova Gova, Suslovskaya,* etc.

The *Elshan-Kurdjum* gasfield lies to the northwest of Saratov and is structurally a faulted anticline with a steep southeast flank. The gas (with some oil) is produced from Middle Carboniferous (Visean and Tournaisian) formations.

(b) *Volgograd area*

The Volgograd group of gas- and oilfields lies south of Saratov, on the right bank of the Volga, in an area which is tectonically one of transition between the Russian Platform and the Mediterranean geosyncline.The fields are mainly asymmetrical anticlines lying in the zone between the river Don and its left bank tributary, the Medveditza. Upper Devonian, Carboniferous and Jurassic reservoirs have been found—with most of the gas in Carboniferous limestones.

Examples of the 25 or more fields so far discovered in this area are the most northerly pair—the *Bachmetievo* and *Shirnovo* fields, which are culminations on the same north-south striking faulted anticline, some 15 miles long, which has a slightly steeper (5°) west flank than east flank (2°). The *Linevo* field lies to the south, but with a northwest-southeast axis. A number of Middle Carboniferous (Visean) sandstone horizons are productive in each of these fields, which may together provide an example of differential entrapment, with the fluids moving from the southeast to the northwest so that the gas has accumulated in the first structure encountered (*Linevo*), leaving the heavier fluids to flow on to fill the further structures with gas and some oil (*Shirnovo*), and finally mainly oil (*Bachmetievo*).

Korobkovo is an important gasfield which was discovered in 1952 on an anticlinal structure about nine miles long. It is asymmetrical, but in contrast to most of the other neighbouring structures, its east flank is steeper than its west flank. The gas reserves of this field are quoted as 3·6 Tcf. The reservoir formations here are Carboniferous (Visean, Tournaisian and Namurian) sandstones, in which gas and condensate are found; and also a higher Jurassic (Bajocian) sandstone in which gas only occurs.

Abramovo is another gasfield, with a Middle Carboniferous limestone reservoir, in which important gas reserves have been developed.

(c) *Astrakhan Province*

Some relatively minor gas- and oilfields are found in Astrakhan Province, southeast of Volgograd. Most of these occur in flattish, northwest-southeast striking anticlines, with Mesozoic reservoirs. Thus, at *Olejnikovo* and *Promyslovoe*, near the coast of the Caspian Sea, gas and some oil has been found in Cretaceous (Lower Albian) sandstones.

TABLE 40

VOLGA-URAL AREA—PALAEOZOIC STRATIGRAPHY

Permian—Upper	P_2	Tartarian
		Kazanian
		Ufimian
—Lower	P_1	Kungurian
		Artinskian
		Sakmarian

Carboniferous—Upper	C_3	Stephanian
		(Uralian)
—Middle	C_2	Westphalian
		(Moscovian)
		Bashkirian
—Lower	C_1	Namurian
		Visean
		Tournaisian

Devonian—Upper	D_3	Fammenian
		Frasnian
—Middle	D_2	Givetian
		Eifelian
—Lower	D_1	Bavly

II THE CAUCASUS

THE Caucasus is the land junction between Europe and Asia. The area was originally part of the Tethys geosynclinal depression in which a great thickness of Jurassic and Cretaceous sediments was deposited from two land masses that lay on its flanks. During the Tertiary orogeny, the sediments were compressed and uplifted into a fan-like complex of folds, with volcanoes along the principal lines of stress, and some overthrusts to the south.

The great oilfields which made Russia pre-eminent as an oil producer in the last century were associated with the foothill regions of the Caucasus Mountains, generally on the northern flank.

Many oil seepages occur along the northern foothills, and traces of oil and bitumen are also found along the southern mountain flank. "Mud volcanoes" are common in the southeast, rising to a remarkable height and often ejecting large quantities of gas and rock detritus. Near the

Caspian Sea, the mud cones of Touragai, Kianzi Dag, Kalmas and others, rise to heights of as much as 1,200ft.

There are two principal sedimentary basins, both of which contain major petroleum accumulations:

(a) *The Azov-Kuban (Middle Caspian) Basin*

This stretches from the Sea of Azov to the east side of the Caspian Sea, and includes a number of important oil and gas accumulations, all in Tertiary reservoirs. These fall into three main groups—those around Maikop in the northwest, Grosni in the northeast, and the fields in Georgia, near the mountain axis.

(i) Northwest Caucasus (Krasnodar-Stavropol area)

This was for many years one of the most important petroleum provinces of Russia. Oil has been produced since 1907 at *Maikop* from coarse gravels and grits of Lower Miocene-Oligocene age laid down in erosional cavities in the underlying marls, and also from sand lenses.

Klutschevskoe, south of Krasnodar, is a small fold with a roughly east-west axis. The structure is gentle and asymmetric, and the reservoirs here are Upper Oligocene sands.

Novo-Dimitrievskoe, discovered in 1953, is another asymmetrical east-west fold in the same area with Upper Oligocene production.

Anastasievsk-Troizkoe is the largest of this group of fields; it has an east-west axis some 15 miles long, with two culminations. The reservoirs are of Upper Miocene age. *Achtyrsk-Bugundyrskoe*, some 40 miles west of Krasnodar, has a more complex structure, the fold being overthrust and faulted. Production comes from the Paleocene beds on the north flank.

Other fields in this area are *Schirokaya-Balka, Zentralnoe Pole* and *Sybsa*, all of which are important oil or gas producers with Upper Oligocene or Paleocene reservoirs.

At *Kaluzkaia,* the reservoir is a Lower Miocene-Oligocene brecciated dolomite; sandstone lenses are also productive.

Near Stavropol, gas has been found in the *Severo-Stavropol* anticline, where the reserves have been estimated at $6\frac{1}{2}$ Tcf. This fold is about 21 miles long and a maximum of 10 miles wide, with gentle (3°) dips. There are in all 16 gasfields in this area, all of which have gas in Oligocene and (to a lesser degree) in Eocene reservoirs.

(ii) Northeast Caucasus

The Grosni district of Circassia, in the northeastern foothill zone of the Caucasus Mountains, was an important oil-producing area in the USSR before the last war, but suffered much wartime damage. Oil had been obtained here from hand-dug pits for many years before the first well was drilled in 1893.

The *Grosni* anticline is an asymmetric fold about nine miles long paralleling the mountain range. The southern flank has dips of 30°–50°, while the northern flank is frequently overthrust. Production came originally from the Miocene Chokrak sandstones, but now comes mainly from deeper oil and gas reservoirs which are mainly of Upper and Lower Cretaceous age.

The *Novo-Grosnenskoe* (*Oktrabeskoe*) field was discovered in 1913 a few miles from the main field, but on a more symmetrical structure.To the northwest of Grosni, the *Vosnessenski* anticline was found in 1924, and the *Malgobek* anticline in 1934. These fields still produce important volumes of oil, mainly from Miocene sandstones, but with some additional Pliocene, Oligocene and Cretaceous production.

The newest field of this group is *Karabulak-Atschaluki*, which has an Upper Cretaceous limestone reservoir.

The Kuma-Terek sub-basin contains a number of buried folds paralleling the mountains. Several of these are associated with anticlinal oil accumulations, as for example, *Osek-Suat*, discovered in 1951, which has oil in Oligocene, Lower Cretaceous and Jurassic beds. *Velischaevka* is a slightly smaller fold which has productive Jurassic and Lower Cretaceous sandstone reservoirs, as has also *Simnaia-Stavka*.

Southeast of Grosni, a number of minor accumulations lie along the Caspian Sea coast of Daghestan. These are small, sharply folded structures with northwest-southeast axes and Miocene or Oligocene sandstone reservoirs. Among them, *Makhatch Kala, Izerbash* and *Kaiakent* are oilfields; and *Daghestan Ogni* and *Duzlak* (near Derbent) are gasfields.

(iii) Georgia

The oilfields of Georgia occur in the west of the Kura Valley, in a number of complex folds related to the main mountain chain. At *Mirsaani* and *Patri-Shiraki*, oil is obtained from Pliocene and Miocene reservoir sands. The fields are all relatively small, their total output being only about 560,000 brl/a.

(b) *The Southern Caspian Basin (Azerbaijan)*

This contains a great thickness of sediments ranging from Permian to Tertiary in age. The southeastern end of the Caucasus range plunges down to the Apsheron Peninsula, projecting into the Caspian Sea. It is in this area that the most productive and famous of the older oilfields have been found, centred round Baku in Azerbaijan. Although no longer of such importance as in past years, Azerbaijan produced 145 MM brl of oil and 194 Bcf of gas in 1971.

The stratigraphy here is complex; major oil production has come from the sands of the Upper Pliocene "Productive" series, which can be divided into upper, middle and lower groups, all of which form reservoirs in different fields. The source of the oil has been variously ascribed to the Eocene shales, to the Lower Miocene Oligocene bituminous clays, or to the Middle and Upper Miocene diatomaceous clays.

The oilfields of the Apsheron Peninsula lie on a series of long fold lines, which roughly parallel the mountain axis, but with many branches and loops. Thus, Baku is almost encircled by a line of folds, and there is another almost complete ring to the west.

The curious crescentic and hook-shaped trends of these fold lines are probably due to the "slumping" of sediments towards the tectonic depression of the Caspian Sea. Many of the structures show an increased dip near their axis, and a more complete succession in the depressions than the uplifts, indicating that they have been rising through long periods of geologic time. The cores of Tertiary clays which are found piercing younger formations seem to indicate the diapiric nature of the structures.

The most important of the many fold lines is the structural trend running from Kirmaku and Binagady via Balakhany and Surakhany to Zykh and Peschanny Island. This fold line probably swings round from Peschanny Island westward to Bibi-Eibat. At culminations along this line are a number of important oilfields producing from as many as 34 individual oil sands in the Upper Productive series, and also from deeper horizons in the lower part of the same series. Oil is also produced in some of the fields from sands in the overlying Apsheronian and Akchaghylian beds.

Binagady is a broad, faulted structure that has produced oil since 1901. The *Balakhany-Sabunchy-Romany* east-west anticline is also broad and faulted, plunging eastwards. *Surakhany,* which has been one of the richest fields in the world for its size, is a broad, flat arch, running north-south, much faulted at its crest. *Kara-Chukur* is another north-south anticline, cut by two longitudinal branching faults. The *Zykh* dome, which was discovered in 1935, has a major axis striking north-northeast-south-southwest.

Bibi Eibat is a broad dome, with several cross-faults; it has perhaps had a larger yield per acre than any other field in the world, and still gives oil from 18 sands in a thickness of about 5,000ft of the Productive series, in an area much of which has been reclaimed from the Caspian Sea.

Lok-Batan was discovered in 1932 and is thought to have a diapiric core of Maikop clays piercing the lower strata of the Productive series and thus producing an anticlinal structure. Oil is found in nine sands at a depth of 2,700ft. *Puta* and *Ker-Gez* are faulted anticlines. *Sulu Tepe* is an asymmetrical anticline with an eastern flank dipping almost vertically and a gentler western flank. *Kala* is a north-south fold which is much faulted.

The *Karadag* field is another north-south striking fold, with large gas production from Apsheronian-Akchaghylian and Productive sands.

In recent years, the development of many of the Apsheron Peninsula oilfields has been continued into the shallow waters of Ilytch Bay, and several new fields have also been discovered on extensions of the fold trends which sometimes outcrop on local islands (e.g. Peschanny Island) or may be reached by offshore drilling in the Caspian Sea. The offshore oil production of Azerbaijan currently accounts for 55% of its overall output, and this proportion is increasing as exploration is carried eastward.

The most important offshore oilfield so far discovered (in 1949) is *Neftyane Kami*, which lies some 25 miles from the coast. It is an anticlinal fold with a northwest-southeast axis about seven miles long; part of the field lies on a group of local islands (joined by trestle bridges) and part is below the water level of the surrounding sea. There are about 20 sandstone reservoir horizons of the Productive series.

Three important oil accumulations have been found on the local culminations of a single asymmetric, partly overturned fold; the central field is on *Artyom* (*Holy*) *Island*, and the other two fields are offshore at *Darvin* (to the northwest) and *Gyurgyany* (to the southeast), respectively. The reservoirs in these fields are the usually Productive series sands.

South of Baku there is a belt of Tertiary sediments in which oil is found in a number of small accumulations lying between the Apsheron Peninsula and the mouth of the Kura River. Impregnated oilsands have been mined here successfully on the southern flank of the *Tscheldag* fold.

III EAST CASPIAN AREA

(a) Nebit Dagh Area

The South Caspian basin extends eastwards across the Caspian Sea to the Nebit Dagh area of the Turkmenian SSR. Oil was first found here in 1913 on *Cheleken Island*, just off the coast. This is a faulted dome which is still productive. *Nebit Dagh* is a large fold with an east-west axis some 18 miles long. Surface dips in Quaternary strata are 10°–12°, but the fold flanks become steeper at depth, being 20°–25° in the underlying Pliocene sands, which provide numerous reservoir horizons. *Kotur Tepe* (also called *Leninsk*), which was discovered in 1956, is a somewhat larger and steeper fold, with dips in the Pliocene beds as high as 40°.

Chikishlyar, Okarem and *Kamischlyar* are smaller fields on the coast, near the Iranian frontier; *Kizyl-Kum* is a large gasfield. All these fields produce from multiple Pliocene sandstone reservoirs.

(b) Pre-Caspian Depression (Emba area)

The southeast corner of the Russian Platform was encircled by old Caledonian and Hercynian mobile belts which connected the Ural and Caucasus uplifts, and now lie below the Caspian Sea. This pre-Caspian depression is filled with Palaeozoic, Mesozoic and Tertiary sediments, thickening towards the southwest, which are pierced to varying degrees by many diapiric salt domes—only about 15 of which are visible at the surface, although more than 1,500 are known to exist in the subsurface. Oil accumulations occur in Mesozoic beds flanking salt domes, or less frequently in Palaeozoic and Mesozoic strata arched over buried salt masses.

The oilfields lying between the Ural and Emba rivers, in Kazakhstan and also south of the Emba, were first developed commercially in 1911. *Dossor* was the first field to be discovered, a faulted, elliptical structure above a salt mass, which contains oil in shallow Jurassic "super-cap" sandstones. Similar salt dome fields have been found at *Makat, Baitchunas Iskin,* and *Koschagil* in the coastal area near Gurjev; and at *Shubarkduk, Jacksymai* and *Barankul.* Production comes principally from Cretaceous and Jurassic sands lying on the flanks or over the caps of buried salt masses. The age of the salt is probably Permian.

Makat also produces from the Neocomian and from a deep Permo-Triassic reservoir.

(c) Mangyshlak Peninsula

An important group of oilfields was discovered about 1960 on the Mangyshlak Peninsula, which is part of the Ustyurt basin of Kazhakstan on the east coast of the Caspian Sea. These are gently folded anticlinal accumulations—about a dozen fields in all, with Middle Jurassic sandstone oil reservoirs and Lower Cretaceous gas reservoirs. *Uzen* is an anticline about 21 miles long, with flank dips of $2°$ on the north and $3°$ on the south, and a gross Lower Cretaceous reservoir thickness of some 3,000ft. The oil produced is very paraffinic, with a high wax content which makes processing difficult.

Zhetybai is one of the largest oilfields in the Soviet Union; it is a fold with axes 15 miles long by three miles wide, and some 15 Jurassic reservoir sands totalling 3,300ft in thickness. Other important oil accumulations in this area are *Tengin, Tasbulat-Aktas, Yushni,* etc.

These Mangyshlak oilfields have been termed the "Fourth Baku". Their output in 1966 was only 14 MM brl, but probably reached a rate of 95 MM brl/a in 1971.

IV CENTRAL ASIA

Soviet Central Asia is a political and economic concept rather than a separate geologic province. The term is used to describe the area of the republics of Uzbekistan, Turkmenistan, Tadjikistan, Kirghiztan, and the southern part of Kazakhstan—in all about 600,000 sq miles, much of it desert, which lies to the southeast of the Aral Sea, and is bounded on the south and east by the mountain ranges of the Afghan and Chinese frontiers. Geologically, this is part of the Angaran Variscan Platform, with (in the south) the eastern extension of the Mediterranean Tertiary geosyncline.

Hydrocarbons have been found in two sedimentary basins—those of Ferghana and Karakum. In the Ferghana intermontane basin there are more than fifty anticlinal structures, many of which contain oil in Oligocene, Eocene or Upper Cretaceous limestones and sandstones. The principal fields are *Palvantash, Majlisu, Isbaskent* and *Andizhan*, close to the frontier with China.

In the eastward extension of the Karakum Basin, which includes the Amu Darya river valley, along the border with Afghanistan, there are four small Tertiary oil accumulations near Termez, just north of this frontier. These are all anticlines discovered between 1934 and 1938—*Khaudag, Uch-Kyzyl, Dzhar-Kurgan* and *Kirovabad*. In the main basin, between Bokhara and Samarkand in Uzbekistan, a group of about 25 Mesozoic gasfields has been discovered, including several very large accumulations at *Gazli, Uchkyrsk, Farab, Bokhara* and *Mubarek*.

Of these, *Gazli*, which was discovered in 1953, is the most important, with reserves of about 17·6 Tcf of gas. It is an anticline about 24 miles long by six miles wide, with Cretaceous (Cenomanian, Albian and Neocomian-Aptian) reservoir sandstones, which have an overall thickness of nearly 300ft. Some oil, which is probably condensate, is also produced from the deepest reservoir.

The output from the Central Asian gasfields was about 1·3 Tcf in 1970, most of the gas being sent via the transcontinental gas pipeline running via Saratov to Moscow and the industrial areas of the western USSR. The gas reserves of these fields are probably not less than 25 Tcf.

V. WEST SIBERIA

The West Siberian Platform forms an extensive plain between the Urals and the Yenisei River, which is mainly covered by featureless Quaternary deposits. Structurally, this is an enormous flat cratonic basin*, some 1,200 miles across, containing sediments which reach their maximum thickness of 20–30,000ft in the northern and central parts of the basin and may be 9,000–12,000ft thick elsewhere.

Thick, marine and shallow-water deposits are dominant in the western and northwestern parts of the area, which have been subjected to prolonged and steady downwarping, while the thinner continental facies in the south and southeast show that the vertical movements here were variable in direction and intensity.

The "basement" consists of folded and metamorphosed Palaeozoic

---

* *More correctly a northern (Ob-Taz) and a southern (Irtysh) basin, separated by a line of uplifts.*

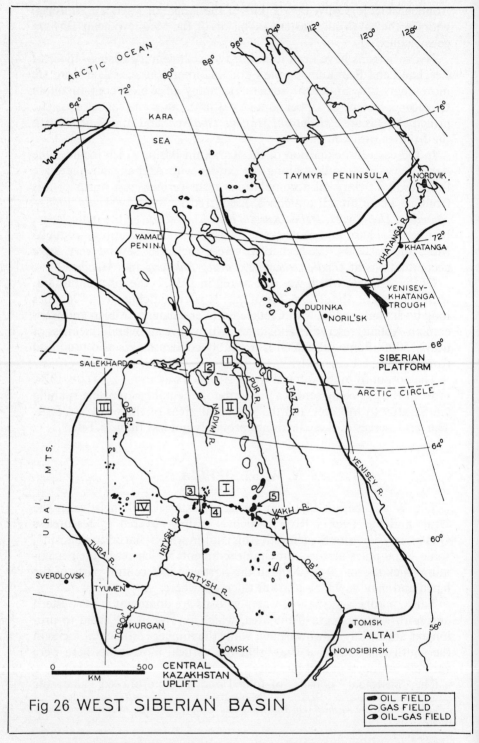

Fig 26 WEST SIBERIAN BASIN

Key: (1) Urengoy field; (2) Medvezh'ye field; (3) Ust'-Balyk field; (4) Mamontovo field; (5) Samotlor field. I = Middle Ob' oil province; II = Pur-Taz gas province; III = Berezovo gas provinces; IV = Shaim oil province.

Source: N. Eremenko, G. Ovanesov, V. Semenovitch, Bull. AAPG, 56(9) Sept. 1972, 1719.

rocks, probably of the same age as those exposed on the eastern flanks of the Urals, underlying a thick Mesozoic section. This is succeeded by Paleocene, Eocene and Quaternary beds.

In the west, along the eastern foothills of the Urals, oil and gas accumulations usually occur in Upper Jurassic sandstones, often in stratigraphic traps resulting from the pinching-out of reservoir beds against local uplifts. In the central area, the principal oil production comes from Neocomian sandstones. Upper Cretaceous sands form particularly important gas reservoirs in the north.

Most of the hydrocarbons of West Siberia are thought to have been formed in the middle or at the end of the Lower Cretaceous and consequently to have been trapped in structures which had already been formed in Valanginian-Neocomian time.

Until relatively recently, little was known of the oil prospects of this area; only a few seepages had been reported from East Siberia (Lake Baikal area and in the delta of the Chatanga River). The first successful exploratory well was drilled near Tomsk in 1954.

Since then, very rapid progress has been made, and at least 114 oilfields (containing reserves of 28 B brl) have been found[14] in Tyumen and Tomsk Provinces, mainly large anticlines and domes with thick Mesozoic sandstone or siltstone reservoirs. Nearly 124 MM brl of oil was produced from only a few of these fields in 1968, and 310 MM brl in 1971. The area has been termed the "Third Baku", and its output will certainly expand rapidly as new fields are put into production and further pipeline outlets completed. Output planned for 1975 is 850 MM brl and a potential 1985 output of as much as 2,800–3,500 MM brl/a has been suggested. Some of the largest gasfields in the world have also been discovered in West Siberia, and the production of natural gas from this area was some 220 Bcf in 1966 and probably reached about three times this volume in 1970.

The petroleum-producing areas of West Siberia may be considered in three main groups:

(a) The Ural Foothills

In this outer belt, gas and oil have been found mainly in Jurassic reservoirs, as in the group of about 20 gasfields discovered around Berezovo on the eastern end of the North Sosva uplift. Of these, the *Berezovo* field itself is the largest, being an anticline with a northwest-southeast axis, about $4\frac{1}{2}$ miles long and three miles wide, with gently dipping flanks and several Jurassic sandstone reservoirs. The *Punga* gasfield, discovered in 1961, may hold 3·2 Tcf of recoverable gas. *Igrim* and *Chuelskoye* are other important gasfields of this group.

Oil has also been found in stratigraphic traps with Jurassic reservoirs

---

14 P. Dickey, *Bull.AAPG*, March 1972, 454.

L

in the Khanty-Manskii and Nadym basins. At *Shaim*, on the Kondin uplift, there are a number of oil accumulations (22 had been discovered by 1970) which all have Upper Jurassic reservoirs, just below a pre-Cretaceous unconformity.

(b) Middle Ob Region

In this central area, east of the junction between the Ob and Irtysh Rivers, a large number of hydrocarbon accumulations have been discovered since 1961. Most of the fields are structural domes with Lower Cretaceous (Neocomian) or Jurassic sandstone reservoirs.

One of the largest oilfields is *Meghion (Meghionsky)*, on the Surgut arch, and another is *Ust-Balik (Ust-Balikskoye)* (an anticline 15 miles long by eight miles wide). This field has a number of very prolific Jurassic reservoirs—mainly slightly arkosic sandstones. Its reserves may exceed 3 B brl. Probably the largest accumulation of all is *Samotlor (Samotlorskoye)*, covering nearly 300 sq km, which is believed to hold as much as 15·1 B brl of oil reserves in very thick Lower Cretaceous reservoir sandstones. This field is due to produce 270 MM brl in 1973, via a new 48-in pipeline to the west. Another very large oilfield is *Mamontovo (Mamontovskoye)*, south of Surgut, with reserves estimated to be not less than 3 B brl. *Zapadno-Surgut* is an irregular anticline, with sandstone and siltstone reservoirs which may hold 2 B brl of oil.

East of Surgut, an important group of oilfields (*Sosninskoye, Medvedevskoye,* etc.) have Neocomian and Jurassic reservoirs, while *Okhteuresk* is a large Cretaceous gasfield.

To the south, in the Vasyngan area, several Jurassic oil and gas accumulations are associated with the Parabelsk and Mezhorsk uplifts; of these, *Pudinsk* and *Meyovsk* are the most important.

(c) Northern Region (Tyumen Province)

In this previously little-explored area, a number of very large gasfields have recently been discovered. In the northwest, the South Yamal uplift on the Yamal Peninsula has produced the *Novoportovskoye* and *Archticheskoye* gasfields, the latter of which is said to contain 63 Tcf of gas in Lower Cretaceous beds.

In the northeast, two large gasfields have been found on the Taz uplift. The *Taz (Tazofskoye)* field itself has reserves of 40 Tcf of gas. *Zapolvarnoye* is also one of the world's largest gasfields, with 45 Tcf of gas in reservoirs consisting of thick, silty sandstones of Cenomanian age.

By far the largest field in this extraordinary area is *Urengoy*, just south of the Arctic Circle. This is probably the largest gas accumulation in the world. It was discovered in 1966, and is an enormous fold with maximum dimensions of 87 miles by 55 miles, which is believed to hold some 210 Tcf of gas in a very thick and permeable Cenomanian (Middle Cretaceous) reservoir.

The *Medvezhye* field in the Nadim River valley is another enormous gas accumulation in the same area with reserves of about 35 Tcf. Other large gasfields are *Yamburg* (30 Tcf) and *Gubinskoye* (*Gubkin*), (12·3 Tcf).

In all, there are said to be at least 43 important gas accumulations in the Tyumen area, of which four are condensate fields, as well as 41 oil accumulations. Individual well productivity is very high, due to the thick and highly permeable reservoir formations, and ultimately the output from this area may reach 15 Tcf/a. The gas reserves of Tyumen Province account for 56·4% of the total USSR "industrial reserves". The oil has generally been found in Lower Cretaceous and Jurassic beds. Discussions have taken place as to the possibility of exporting up to 280 MM brl/a of Tyumen oil to Japan via a 4,000-km pipeline via Irkutsk to Nakhodka.

In the Far North, near the mouth of the Yenisey, several large gas accumulations have been found on the Taymyr Peninsula. The first to be discovered was *Messoyakha* (in 1967) from which gas is now piped to Norilsk. The proved and potential reserves of these lower Yenisey gasfields is thought to be at least 15 Tcf, and the reservoirs are Lower Jurassic and Cretaceous (Valanginian to Cenomanian) sandstones.

VI. EAST SIBERIA

The Eastern Siberian Platform is a huge area of Pre-Cambrian "basement" rocks onto which have been laid great thicknesses of nearly horizontal Palaeozoic and Mesozoic beds. The sediments cover an area of nearly 1½ million sq miles, and the section differs notably from that of the Russian Platform in the west in the relatively minor development of Devonian and Lower Carboniferous strata. The "basement" is exposed to form several "Shields"—e.g. the Aldan Massif in the centre, and the Anabar Massif in the north.

The western boundary of the Platform is approximately the line of the Yenisey, and the Lena and Aldan rivers can be taken as forming an approximate eastern boundary over much of their lengths.

Caledonian and Hercynian folding movements have affected the sediments in the pericratonic basins lying around the edges of the Platform, resulting in the development of anticlinal structures favourable for oil accumulation. The major basins include the Yenisey (Tunguska) Basin in the west, the Irkutsk (Angora) Basin in the south, the Khatanga Basin in the north, and the Vilyuy Basin in the east.

In the Yenisey Basin, gas has been found at *Bystryansk* and *Novomikhailovskoye*, in gently dipping Jurassic and Cretaceous beds.

In the Irkutsk Basin, northwest of Lake Baikal, Lower Cambrian (or possibly even Pre-Cambrian) gas and oil accumulations have been found in sandstone and dolomite reservoir rocks in anticlines at *Markovo* and *Balychtinsk*.

In Yakutia, important hydrocarbon discoveries have been made in the Vilyui Basin, in culminations on the large *Sredne-Vilyuy* uplift, which has maximum dimensions of 130 miles by 40 miles. The oil and gas reservoirs here are Permian, Triassic and Lower Jurassic sandstones. Gas and condensate have also been obtained from Lower and Middle Triassic and Lower and Middle Liassic beds, while there is also Middle-Upper Triassic, Upper Jurassic and Cretaceous oil[15].

In the area to the south and west of the junction of the Vilyuy and Lena rivers, important gas accumulations have been found at *Ust-Vilyuisk* and *Sobo-Khainsk*. These are faulted, asymmetrical folds with Lower and Middle Liassic sandstone reservoirs. Further to the south-east, minor oilfields occur in similar structures at *Oloisk* and *Kangalassy*.

In the Far North, oil has also been discovered in the Nordvik Bay area, in the Khatanga Basin on the edge of the Laptev Sea.

VII. SOVIET FAR EAST

Narrow troughs of folded Cenozoic strata lie on the west side of the Kamchatka Peninsula in the Soviet Far East, and on the east side of the island of Sakhalin (Karafuto), between Siberia and Hokkaido. Small oil and gas production has been obtained from *Voiampolka*, *Totschili* and *Chromovo* on Kamchatka, which are minor anticlinal accumulations with Miocene sandstone reservoirs.

Before the 1939–45 war, the Japanese developed a few small oilfields in the southern part of the island, below latitude 40°N. The yield from these was poor, however, and the principal production came from concessions in the northern (Russian-owned) part of the island. This area is part of the Sea of Okhotsk geosynclinal trough. Most of the oil has come from elongated north-south folds in which Upper Miocene sands lenses overlie an organic Mid-Miocene shale, which is the probable source rock. Some Pliocene oil has also been found.

The principal oilfield is *Okha* (*Okhinskoye*) on the northeast coast of the island (originally discovered in 1925). This is a north-south trending anticline with dips of 50°/60° on the faulted east flank and 20°/25° on the west. Production comes from numerous Upper Miocene sandstones.

Ekhabi, a few miles to the south, is a sharper fold with Miocene and Pliocene reservoirs. *Piltunskoye*, about 50 miles south of *Okha*, produces oil from Pliocene and Upper Miocene beds; it is a large fold which is cut by faults.

The total output of oil from Sakhalin is about 17 MM brl/a. In addition, a considerable volume of non-associated gas has been found at

---

15 *E. Bazanov et al., 7 WPC, Mexico 1967, PD 2-3.*

Kolendo, *Tungor* and *Maloe Salo*, so that the proved gas reserves were estimated to be 2·8 Tcf at the beginning of 1969. Various proposals are being examined for exporting some of this gas, either by pipeline to Komsomolsk on the Siberian mainland, or as LNG by tanker to Japan.

TABLE 41

USSR—AGE OF OIL AND GAS RESERVES
(Categories $A + B + C_1$)
in percentages of totals

	On Jan. 1, 1966		On Jan. 1, 1970	
	Oil	*Gas*	*Oil*	*Gas*
Palaeozoic	61·0	21·8	37·0	10·0
Mesozoic	24·2	58·5	50·0	80·0
Cenozoic	14·8	19·7	13·0	10·0

Source: N. Eremenko et al., Bull.AAPG, 56(9), Sept. 1972, 1717.

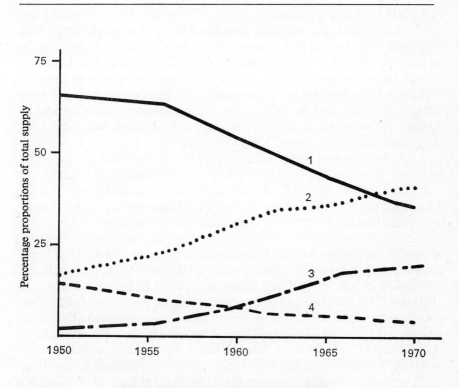

Fig. 27. PRIMARY FUELS IN THE USSR.

1—Coal; 2—oil; 3—gas; 4—oil-shale, peat, wood.

CHAPTER 5

The Middle East

WHILE oil and bitumen in small quantities have been obtained from many parts of the Middle East throughout human history, it was not until 1908 that the first commercial accumulation was discovered, in Iran. After the 1914–18 war, and the subsequent political and concessionary rearrangements, commercial oil was also found in Iraq in 1927, but the relative remoteness of the producing fields from centres of consumption meant that by 1938, when further discoveries were made in Qatar, Kuwait and Saudi Arabia, the output from Iraq and Iran together still did not exceed 120 MM brl/a.

Little development was possible during the 1939–45 war, so that at its end only seven fields were actually in production, although a further 14 oil accumulations had been discovered by that time, and it was already clear that the Middle East would become a major petroleum producing and exporting area when adequate outlets became available.

The construction of additional pipeline links between the oilfields and a number of outlets on the Mediterranean coast and in the Gulf area, coupled with the rapid increase in world oil demand, led to a remarkable expansion in Middle East petroleum production in the post-war years. This is still continuing, and even accelerating. In 1971, the Middle East countries produced a third of all the oil produced anywhere in the world (about 6 B brl), and the overall rate of increase in their output is about 12% p.a., i.e. nearly 70% greater than the average world rate of increase. Furthermore, the area is now known to contain as much as 61% of world proved oil reserves (347·2 MM brl at the end of 1971), so that in the future it is likely to account for an ever-increasing proportion of global output.

TABLE 43

MIDDLE EAST—ESTIMATED PROVED OIL RESERVES—end-1971

		B brl
Abu Dhabi		15·1
Bahrein		0·5
Dubai		1·5
Iran		60·5
Iraq		33·1
Israel		0·0*
Kuwait		75·0
Neutral Zone		13·0
Oman		4·8
Qatar		4·8
Saudi Arabia		137·4
Syria		1·3
Turkey		0·2
	Total	347·2

*2 MM brl
Source: World Oil, Aug. 15, 1972, p 60.

Iran and Saudi Arabia are the leading oil-producing countries, followed by Kuwait and Iraq; together, these four states account for nearly 85% of the oil output of the whole region. However, production from Abu Dhabi is also increasing rapidly, while Qatar, Oman and the Neutral Zone are each important oil exporters. Offshore accumulations have also been discovered under the waters of the Gulf; the full development of these fields will be accelerated by the resolution of various boundary disputes now pending.

The Middle East oilfields comprise more than 150 accumulations, lying in a roughly rectangular area between southern Turkey and Dhofar. These are shown on pp. 144-5; but all are not necessarily commercial.

There are three principal geotectonic components of the region:
(a) *Pre-Cambrian Massifs*—blocks of intensely deformed igneous and metamorphic rocks, principally of Pre-Cambrian age, which occupy most of the western and central parts of Arabia. On the south, they form part of a crystalline mass extending into Somaliland, and on the west are separated from the adjacent Nubian "Shield" by the Red Sea rifts.
(b) *The Shelf area* to the north and northeast of the Arabian massif, which is the result of its intermittent slow subsidence under shallow seas between Cambrian and Cretaceous times. A thick sequence of continental and shallow-water sediments forms a wide "foreland" area which dips gently towards the Mesopotamian basin and that of the Empty Quarter.

The Shelf can be divided into an almost flat "interior homocline", bordering the Arabian massif, and an "interior platform" further to the east, which forms a belt of more sharply dipping strata. The irregularities of dip in this area are related to the existence of a number of major north-south anticlinal "trends" or "swells" which rise above the general

KEY TO MAP OF MIDDLE EAST OIL ACCUMULATIONS

(Compiled from published data: all accumulations listed are not necessarily commercial.)

Turkey
1 *Beykan*
2 *Kurkan*
3 *Sahaban*
4 *Kayokoy*
5 *Selmo*
6 *Silivanka*
7 *Celikli*
8 *Magrip*
9 *Kurtalan*
10 *Garzan/Germic*
11 *West Raman*
12 *Raman*

Syria
13 *Sokhne*
14 *Djibissa*
15 *Rumelan*
16 *Karatchok*
17 *Derik*
18 *Souetde*

Iraq
19 *Ain Zalah*
20 *Butmah*
21 *Khurmala*
22 *Avanah*
23 *Baba*
24 *Bai Hassan*
25 *Qasab*
26 *Jawan*
27 *Najmah*
28 *Qayarah*
29 *Jambur*
30 *Naft Khaneh/Naft-i-Shah*
31 *Abu Ghurab*
32 *Buzurgan*

Iran
33 *Maleh Kuh*
34 *Sarkan*
35 *Chesmeh Kush*
36 *Lab-i-Safid*
37 *Lali*
38 *Karun*
39 *Par-e-Siah*
40 *M-i-S*
41 *Naft-i-Safid*
42 *Haft Kel*
43 *Karanj*
44 *Faris*
45 *Gach Saran*
46 *Agha Jari*
47 *Kupal*
48 *Mulla Sani*
49 *Marun*
50 *Shadegan*
51 *Ahwaz*
52 *Susangerd*
53 *Ab Teymur*
54 *Mansuri*
55 *Ramshir*
56 *Rag-i-Safid*
57 *Pazanan*
58 *Bibi Hakimeh*
59 *Kilur Karim*

60 *Bushgan*
61 *Gulkhari*
62 *Binak*
63 *Darius-Kharg*
64 *Norouz*
65 *Bahregansar*
66 *Hendijan*

Iraq
67 *Seeba*
68 *Nahr Umr*
69 *Zubair*
70 *Rumaila*

Kuwait
71 *Raudhatain*
72 *Sabiriyah*
73 *Bahra*
74 *Khashman*
75 *Magwa*
76 *Ahmadi*
77 *Burgan*
78 *Minagish*
79 *Umm Gudair*

Neutral Zone
80 *South Umm Gudair*
81 *Wafra*
82 *Fuwaris*
83 *Arq*
84 *Khafji*
85 *Hout*
86 *Dorra*

Iran
87 *Cyrus*

Neutral Zone
88 *Lulu*

Iran
89 *Esfandiar*
90 *Fereidoon*

Saudi Arabia
91 *Marjan*
92 *Zuluf*
93 *Safaniya*
94 *Manifa*
95 *Karan*
96 *Jana*
97 *Abu Hadriya*
98 *Khursaniyah*
99 *Fadhili*
100 *Berri*
101 *Abu Safah*
102 *Qatif*
103 *Dammam*
104 *Abqaiq*
105 *Fazran*
106 *Juraybi'at*
107 *Khurais*
108 *Qirdi*
109 *Mazalij*
110 *Ghawar fields*
111 *Shutfah*
112 *Kidan*
113 *Hamaliyah*

Bahrein
114 *Awali*

Qatar
115 *Dukhan*
116 *Idd-el-Shargi*
117 *Maydan Mahzan*
118 *Bul Hanine*

Iran
119 *"W"*.
120 *Rostam*
121 *Raksh*
122 *Sirri*

Sharjah
123 *Mubarek*

Dubai
124 *Fateh*
125 *S.W. Fateh*

Abu Dhabi
126 *Mandoos*
127 *Nasr*

Iran
128 *Sassan*

Abu Dhabi
129 *Abu Khoosh*
130 *Bunduq*
131 *Umm Shaif*
132 *Zakum*
133 *Raz Boot*
134 *Dalma*
135 *Mubarraz*
136 *Addalkh*
137 *Jarn Yaphour*
138 *Zubbaya*
139 *Bab*
140 *Bu Hasa*
141 *Rumaitha*
142 *Sahil*
143 *Asab*
144 *Shah*
145 *Zarrara*

Saudi Arabia
146 *Shayba*

Oman
147 *Lekhwair*
148 *Yibal*
149 *Al Huwaisah*
150 *Fahud*
151 *Natih*
152 *Maradi*
153 *Al Ghubar*

Dhofar
154 *Marmul*

level of the platform and are probably related to basement uplifts. Many major oilfields occur in structural culminations on these north-south "swells"—the most important of which is the En Nala (Hasa) arch of Saudi Arabia, with which are associated many major oilfields—notably those of *Ghawar, Abqaiq, Qatif, Bahrein, Dukhan, Burgan,* and some of the Iranian offshore accumulations.

(c) *The Geosynclinal Zone* lies to the north and east of the Shelf area and includes the mountain chains of the Taurus, Zagros and Oman mountains. Sediments deposited in a series of broad basins were folded by tangential pressures during the Alpine orogenies, the direction of movement being northeast to southwest, with the folds gradually broadening and becoming less intense towards the southwest. Thus, in the northeast of Iran there is a series of overthrusts, followed to the southwest by tightly folded and faulted mountain ranges, and then by folds which become more and more gentle across Iraq, before dying out towards Arabia.

THE OILFIELDS

The oilfields of the Middle East vary in their structural and stratigraphic characteristics according to their position in the main geosyncline. Thus, accumulations in Jurassic reservoirs occur in the typically broad, gentle structures which are probably related to salt uplifts at depth trending roughly north-south in the "interior platform" area, nearest to the Arabian massif. These include the fields of Saudi Arabia and Qatar. Towards the centre of the basin, the folds are stronger and the reservoirs are dominantly Cretaceous in age, while further to the east, the mainly Tertiary oilfields of Iran occur in the strongly-folded and sometimes overthrust structures of the Zagros Mountains foothill belt.

The analysis[1] of the age distribution of the approximately 250 commercial reservoirs discovered up to 1967 has shown that more than 200 of these are Middle Jurassic-Upper Cretaceous in age. A further substantial group of some 32 reservoirs are in Miocene-Oligocene rocks, but these are mainly the Asmari Limestone fields. All other stratigraphic levels accounted together for the balance of only about 14 accumulations then known.

The similarity in the physical and chemical properties of many Middle East crudes is remarkable. Thus, they have high API gravities (i.e. they are mainly light oils), appreciable sulphur contents and nearly parallel correlation index curves. This supports the theory of a common time of origin and initial accumulation, and the most likely source rocks are considered to be Middle-Lower Cretaceous dolomites and limestones.

These "basinal" source rocks are thought to have provided the oil which passed by lateral migration into initial contemporaneous reservoirs during Upper Cretaceous—Early Tertiary migration. Vertical

---

[1] H. Dunnington, *J.Inst.Pet.*, April 1967, 129.

TABLE 44

MIDDLE EAST—OIL PRODUCTION 1920–1970
(thousands of brl)

Year	Abu Dhabi	Bahrein	Dubai	Iraq	Iran	Is-rael	Kuwait	Neutral Zone	Oman	Qatar	Saudi Arabia	Turkey	Total
1920	—	—	—	—	12,230	—	—	—	—	—	—	—	12,230
1930	—	—	—	913	45,828	—	—	—	—	—	—	—	146,741
1940	—	7,074	—	24,225	66,317	—	—	—	—	—	5,075	—	102,691
1950	—	11,016	—	49,726	49,2457	—	125,722	—	—	12,268	199,547	108	640,862
1960	—	16,500	—	353,748	385,748	932	594,278	49,829	—	63,088	456,453	2,624	1,923,285
1970	253,231	27,973	30,710	569,100	1,396,125	547	998,110	183,435	121,310	132,357	1,386,400	25,231	5,154,079

TABLE 45

MIDDLE EAST—OIL PRODUCTION, 1971

	MM brl
Iran	1,657·0
Saudi Arabia	1,641·6
Kuwait	1,067·8
Iraq	621·8
Abu Dhabi	341·3
Neutral Zone	198·0
Qatar	156·9
Oman	107.4
Dubai	45·7
Syria	40·1
Bahrein	27·3
Turkey	24·0
Israel	0·4
Total	5,929·3

TABLE 46

MIDDLE EAST—LARGEST OILFIELDS

	Original Reserves (MM brl)	Production in 1969 (MM brl)
Greater Burgan, Kuwait	72,000	803·0
Ghawar fields, S. Arabia	45,000	543·4
Safaniya fields, S. Arabia	27,000	155·0
Kirkuk, Iraq	15,000	394·1
Rumaila, Iraq	13,600	97·8
Gach Saran, Iraq	10,000	264·0
Agha Jari, Iran	9,500	314·2
Abqaiq, S. Arabia	9,000	231·1
Wafra, N. Zone	8,150	30·3
Marun, Iran	6,000	214·2
Ahwaz, Iran	6,000	87·3

Source of data: Bull.AAPG, 54/8, 1524, 1970; and Int. Petroleum Enc., 1970.

migration occurring at a later stage thereafter filled the Asmari anticlines of Iran and Iraq. The large Mesozoic fields lying on the stable side of the present basin are also believed to have received their oil by lateral migration from similar source rocks.

Due to the plasticity of the shallow Fars clays and their salt content, the surface folds often differ substantially from those at depth. Seismic surveys have therefore been of special value in delineating the sub-surface structures.

Oil and gas "shows" and seepages of every type and stage of evolution are common in the platform and geosynclinal areas, including major deposits of residual asphalt and large areas of oil-stained gypsum. Since most of the crude oils are sulphurous, the evolution of hydrogen sulphide gas and deposition of sulphur are also common. However, in spite of these evidences of periodic leakage from the reservoirs, there are relatively few major faults which have affected both cap rocks and reservoir beds, and few examples also of igneous activity detrimental to petroleum accumulations. The sedimentary sequence contains many excellent plastic sealing beds in the form of shales, clays and—best of all —bedded evaporites, which have provided very effective protection for underlying reservoirs. The Lower Fars salt series and the Jurassic Hith anhydrites are notable examples.

Nearly all known Middle East accumulations are the result of structural trapping in major anticlines. Some of these are very large, having amplitudes of 10,000ft or more and axial lengths of 10–100 miles, often with several pools on crestal culminations. These huge structures have probably resulted from the interaction of several tectonic systems— lateral compression paralleling the mountain belts, vertical movements over underlying faults, and deep-seated salt uplifts.

Many intrusive salt masses are exposed in southwest Iran, or form islands in the Gulf. Some of them are several miles in diameter and make up mountain masses 4–5,000ft high. The age of the salt is probably Cambrian (Hormuz) and the most active period of intrusion was the Upper Tertiary. From the point of view of oil accumulation, these salt plugs and stocks appear (as yet) to be of less consequence than their equivalents in Germany or the US Gulf Coast, presumably because of the nature of the salt and the sediments in the area (p. 50). Although very few oilfields are actually associated with piercement-type salt plugs (*Nahr Umr* in Southern Iraq is one such example), many accumulations have resulted from the bulging of underlying salt "pillows", which has caused arching of the overlying sediments.

The size of the anticlines, coupled with the other factors discussed below, has resulted in many individual oil accumulations of truly "giant" dimensions; in fact, the "typical" Middle East oilfield is usually considered to be an accumulation with not less than 1·0 B brl of recoverable oil reserves.

The fields in this area are also characterised by the excellent qualities of their reservoir rocks, which are usually thick, homogeneous, very porous and permeable beds. The majority of these are Tertiary or Cretaceous limestones, usually possessing both primary and secondary porosity and permeability. Their primary properties are sometimes due to reefal facies (e.g. at *Kirkuk* and *Bu Hasa*), or more often result from the fact that the intergrain pore-spaces of many of the bedded limestones have remained largely uncemented (e.g. the Kuwait Minagish oolites and the Saudi Arabian pellet-bioclastic grainstones)[2]. This preservation of pore-space is probably the consequence of the mode of formation of the limestone beds in relatively undisturbed depositional conditions. As a corollary, dolomite reservoirs are relatively rare, being known only in a few of the Middle East fields (e.g. the Arab/Darb reservoir of *Umm Shaif*). Secondary porosity and permeability due to local fracturing have made even the relatively dense limestone beds (e.g. the Asmari Limestone at *Gach Saran, Agha Jari*, etc.) very prolific reservoir rocks. There are also some notable Cretaceous sandstone reservoirs—e.g. the thick, loosely cemented beds of *Rumaila* and *Zubair* in Iraq, or of the largest Middle East field of all—*Burgan* in Kuwait.

Because of their thick, uniform, permeable reservoir beds, Middle East oilfields usually require relatively few wells to drain them effectively, so that they have a remarkably high oil output per well. Thus, only 240 wells were needed to produce 1,040 MM brl of oil from Iran in 1968, and 107 wells produced 550·1 MM brl from Iraq in that year. Throughout the whole of the Middle East, there are probably not more than 3,000 active oil-wells, compared with 11,600 in Venezuela and 512,500 in the United States. For the whole area, the average production rate is about 5,500 b/d, but many individual wells are capable of sustained rates of production of 20,000 b/d or more*. This high per-well productivity is one important factor in making Middle East crude oil relatively cheap to produce; another is the relative shallowness of many of the fields.

On the other hand, the generally high sulphur content of the crudes and the distances of the producing fields from their markets are adverse economic factors when compared with some other oil areas, e.g. Libya or Nigeria.

The extraordinary concentration of petroleum in this relatively small area of the world must be due to the co-existence here of all the factors favourable for oil accumulation[3]—abundant source beds laid down in suitable environments, thick, permeable reservoir beds, very large structures formed at the right time to provide traps, and preservation under impermeable and plastic cap rocks.

---

[2] L. Illing *et al.*, 7 WPC Mexico, 1967, PD(4) 4.

[3] J. Law, *Bull.AAPG*, Jan. 1957, 41, 1, 51.

Indeed, one well in the Gach Saran field in Iran is reported to be capable of a steady flow of nearly 100,000 b/d.

ABU DHABI

Abu Dhabi is the largest Trucial Coast state and one of the newer of the world's major oil producers. The first onshore discovery of oil was made in a well drilled in 1954 near Tarif, but the presence of noxious H_2S gas delayed the development of the Bab (*Murban*) field until 1959. In 1962, a second oilfield was discovered about 40 miles to the southwest, at *Bu Hasa*, and in 1966 other fields, as yet undeveloped, were found to the southeast. Offshore exploration has also been successful and several oilfields have been discovered in shallow contiguous waters. Oil was first exported in 1962, and by 1971 output had reached 341·3 MM brl, making Abu Dhabi the fifth largest Middle East oil producer.

An important difference between the Abu Dhabi onshore oil accumulations and those of the older oil-producing areas to the west is the age of the principal reservoir formation. This comprises a number of permeable zones in the thick Lower Cretaceous Thamama limestone, in contrast to the dominant Jurassic Arab Zone reservoirs of Qatar and Saudi Arabia. Production comes mainly from the Thamama B reservoir |in *Bab,* and the Thamama A (Shuaiba)—a highly permeable reef limestone—in *Bu Hasa*[4].

The *Bab* dome is ovoid in shape with a northeast-southwest longer axis. Maximum dimensions are 33 by 13 miles, and flank dips are very gentle—as little as 2° on the southeast or steeper flank. The fold has been subject to a long period of structural growth, from Upper Cretaceous to Middle Miocene time, followed by a regional tilt to the northeast during the late Miocene. There is minimum structural closure of some 800ft. It is of interest to note that the field provides an example of horizontal layering as a result of solution-pressure diffusion processes and the growth of stylolites[5].

(Stylolites are clay-filled partings between blocks of rock—usually limestones—produced as pressure-solution phenomena, whose development can adversely affect reservoir rock capabilities.)

The *Bu Hasa* structure grew along a different north-south trend, and began its growth later than *Bab*; however, it is probably also the consequence of deep-seated faulting and the movement of the Cambrian (Hormuz) salt, which is exposed as diapiric domes in some parts of the Gulf.

The *Abu Jidu* anticline (with its two culminations at *Asab* and *Shah*) was discovered to be oil-bearing by a test well drilled in 1966, and awaits development. The reservoir here again comprises the porous zones of the Thamama limestone, of which the B is the most important. Production in 1971 from the Abu Dhabi onshore fields totalled 209·9 MM brl.

---

[4] G. Hajash, 7 WPC Mexico, 1967, PD2–5.

[5] H. Dunnington, 7 WPC Mexico, 1967, PD3, 4. p. 7.

Umm Shaif is the largest of the offshore fields, with oil in the Jurassic (Arab Zone) Darb and Araej limestones. (The Darb formation has been correlated with the Arab No. 4 limestone of Qatar, while the Araej reservoir is thought to be equivalent in part to the Uwainat of Qatar[6].) The output from *Umm Shaif* was 36·4 MM brl in 1971.

A second offshore accumulation has been found at *Zakum*, 50 miles to the southeast, where the oil reservoirs are a series of porous Thamama limestones and there is gas in the Araej. Production from *Zakum* was 96·7 MM brl in 1971. Both these offshore fields are linked to Das Island, which serves as an operational and storage base.

A third offshore oilfield, *El Bunduq*, awaiting production facilities, has an Arab Zone (Darb) reservoir. This field will be developed for the joint economic benefit of Abu Dhabi and Qatar.

The newest offshore accumulation—*Mubarraz*—was discovered in 1969 about 40 miles west of Abu Dhabi City on a "trend" which may extend from Zakum to Murban. The reservoir here is again a Lower Cretaceous Thamama limestone, thought capable of producing as much as 200,000 b/d when fully developed.

The oil and gas reserves of Abu Dhabi have not been published; but they are probably at least of the order of 15–16 B brl of recoverable oil, and perhaps 10 Tcf of gas[7]. Most of the gas produced with the oil still has to be flared for lack of commercial outlets, but some of the gas from the offshore fields will be liquefied and exported to Japan as LNG, beginning in 1976.

BAHREIN

Bahrein Island has been producing commercial oil for longer than any other state in the Arabian Gulf, an accumulation having been discovered at *Awali* in 1932. This is an anticline about $7\frac{1}{2}$ miles long and $2\frac{1}{2}$ miles wide, with a northwest-southeast axis and gentle dips. The reservoir beds are Wasia (Middle Cretaceous) limestones, and there is also gas and some oil in Jurassic Arab Zone limestones, and still deeper gas in Permian Khuff beds—one of the few examples of a Palaeozoic hydrocarbon accumulation so far known in the Middle East. Some of the Arab Zone gas is currently being used for pressure maintenance of the overlying oil reservoir. Production in 1971 was 27·3 MM brl of oil and 33·4 Bcf of gas, and remaining recoverable reserves are estimated to be about 541 MM brl.

Exploration in other parts of Bahrein and on the adjoining islands has not so far discovered any other commercial oil accumulation. The *Abu Safah* offshore field is being developed jointly with Saudi Arabia.

---

6 H. Dunnington, *J.Inst.Pet.*, April 1967, 143.

7 *International Petroleum Encyclopaedia*, 1970, 106, and *World Oil*, Aug. 15, 1972, 60.

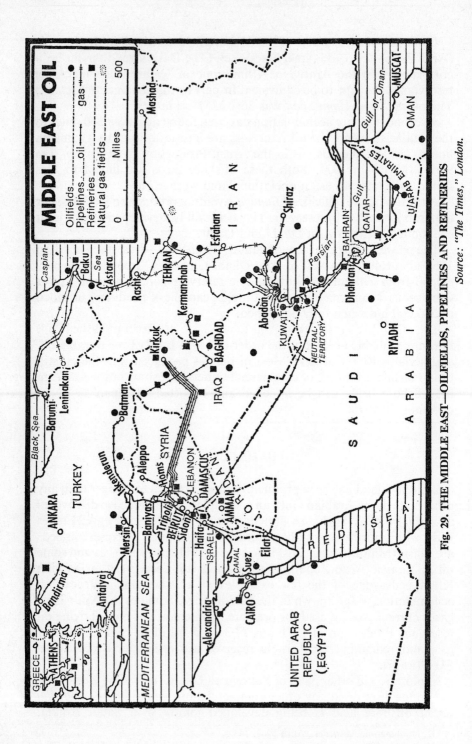

Fig. 29. THE MIDDLE EAST—OILFIELDS, PIPELINES AND REFINERIES

Source: "The Times," London.

DUBAI

After unsuccessful onshore exploration, an offshore discovery of commercial oil was made on the Dubai Continental Shelf in 1966; this was put into production late in 1969. The output from this *Fateh* field was 45·7 MM brl in 1971, the reservoir beds being of Wasia and Thamama (Middle and Lower Cretaceous) age. The reserves here and in the nearby *Southwest Fateh* accumulation, which was discovered in 1970, have been estimated at about 1·5 B brl.

Production storage is in three huge submerged tanks which have been lowered to the sea-bed in 158ft of water. Each has a 0·5 MM brl storage capacity.

IRAQ

The asphalts and seepage oils obtained from various places in Mesopotamia were highly valued in the ancient world and widely distributed.

Modern Iraq covers a major part of the Mesopotamian sedimentary basin in which conditions for petroleum generation, accumulation and preservation have been outstandingly favourable. Oil and gas shows are common, while the many bitumen deposits and shallow impregnations, particularly in the Hit-Awasil area and in Kurdistan* are evidence of the past dissipation of enormous quantities of hydrocarbons.

Northern Oilfields

(a) *Kirkuk*. The discovery well on the Kirkuk structure was drilled in 1925 near the gas seepages of the famous "Eternal Fires". Surface thrust faulting, and incompetence in the overlying Tertiary beds, made the interpretation of the subsurface structure initially difficult. However, upon development, it was found to be a simple, very large anticline, with an axis running northwest-southeast, a wide, flat top, and three major domal culminations—*Khurmala, Avanah* and *Baba*.

The principal reservoir is the so-called "Main" Limestone, a reefal, highly permeable bed of Oligocene–Upper Eocene age. Oil is also found in higher "transition" limestones, and in a deeper Upper Cretaceous reservoir. The overlying Lower Fars salt and anhydrite beds have formed an adequate but imperfect seal, since there are seepages of oil and gas near the crest of the southeastern culmination. There has been considerable debate as to the source of the oil, but the discovery of the deeper Upper Cretaceous reservoir has lent support to the belief that the oil originated in the thick succession of underlying Middle-Lower Cretaceous dolomites and limestones, before migrating upwards in at least two stages[8].

Production from Kirkuk was 33 MM brl in 1946; with the availability of greatly increased pipeline outlets, it had risen to more than 364 MM

---

*e.g. at Berat Dagh, Aqra, Pir-i-Mugrun, Bekhme, etc.

[8] H. Dunnington, *op. cit.* April 1967.

M

brl in 1971. This is one of the world's largest oilfields, with original recoverable reserves which have been estimated at not less than 15 billion barrels[9], and its development has been a remarkable example of the scientific control that is possible when a single operator develops a major reservoir. An extensive hydraulic repressuring system is in operation. By the end of 1971, the field had produced a total of some 5·6 B brl of oil.

(b) Two smaller folds near Kirkuk produce oil from similar structures and reservoirs. At *Bai Hassan,* a few miles to the southwest, there is both a "Main" Limestone development and deeper Cretaceous oil. Production from this field amounted to 20·1 MM brl in 1971. *Jambur,* about 45 miles to the southeast and on the same trend as *Bai Hassan,* has an equivalent Tertiary reservoir which is permeated by bituminous residues, although the oil produced is free from asphaltenes. This has been taken as an indication that light oil arrived in the structure by lateral migration, causing the deposition of bitumen from a heavier oil already present.

(c) *Fields West of the Tigris.* A large accumulation of heavy, high-sulphur oil was found before the war on the west bank of the Tigris, in Lower Miocene and Upper Cretaceous limestone reservoirs. However, the quality of this oil is such that it is not at present commercially refinable. The *Qayarah, Najmah* and *Jawan* anticlines are culminations on the same fold axis, while *Qasab* is a parallel structure.

In the same general area, but further to the northeast, an anticlinal field with a lighter oil content was discovered in 1940 at *Ain Zalah.* The reservoirs here are two Upper and Middle Cretaceous limestone beds, and production, which is declining, is about 7·2 MM brl/a. *Butmah,* another minor anticline in the same area, has Cretaceous oil and also some deeper Triassic production.

Southern Oilfields

The *Nahr Umr* and *Zubair* accumulations were found by seismic surveys in the Basrah area of southern Iraq in 1948/49, and an even more important discovery was made at *Rumaila* in 1953. *Nahr Umr* appears to be associated with a diapiric salt uplift, and the other two fields are large subsurface anticlines. There is some shallow Tertiary heavy oil, and a Middle Cretaceous "Mishrif" reservoir, but the most important reservoir beds are the permeable Lower Cretaceous "Third Pay" and "Fourth Pay" sandstones. The Zubair structure extends for at least 22 miles and has flank dips of about 3°; Rumaila is even larger.

Production from *Zubair* was about 35 MM brl in 1971 and from *Rumaila* more than 190 MM brl. The northern part of the *Rumaila* field was expropriated in 1960; total reserves in this field have been estimated at 13·6 B brl[10]. These accumulations owe their richness to the rapid, but intermittent, subsidence of the local basin, which was filled partly with

---

9, 10 *International Petroleum Encyclopaedia,* 1970, 106.

organic reef and shoal limestones and partly with reservoir sandstones and cap rock shales.

In 1971, the Iraq National Oil Co. reported that the *Buzurgan* field, located in the Tayeb region, near the frontier with Iran, was capable of producing 100,000 b/d from a Mishrif limestone reservoir. *Abu Ghurab* is an associated structure.

The overall recoverable oil reserves of Iraq have been estimated at 33·1 B brl, with about 19 Tcf of mainly associated gas[11].

TABLE 47

IRAQ—PRODUCING OILFIELDS, 1971

	(MM brl)
Northern Fields	
Kirkuk	364·6
Bai Hassan	20·1
Ain Zalah	7·2
Jambur	2·3
Butmah	2·0
Southern Fields	
Rumaila	190·6
Zubair	35·0
	621·8

IRAN OIL FIELD

The site of the first human workings for petroleum was probably in Iran. According to Herodotus, at least one hand-dug pit for oil was excavated at Shush in the time of Darius the Great (about 500 B.C.), but many such pits and shallow wells were no doubt dug in the southern oil seepage area at various times throughout human history.

The first commercial oilfield was discovered in 1908, and by 1951, when the Iranian Oil Exploration Co (the "Consortium") commenced operations, six large anticlinal oilfields had been found in the southwest of the country, the reservoirs in each case being thick Asmari (Oligo-Lower Miocene) limestones.

A number of other large accumulations have since been discovered in the Consortium concession area (totalling about 30), and the output from only 17 of these oilfields accounted for nearly 92% of the overall Iranian output of more than 1,650 MM brl in 1971—a production which was greater than that of any other Middle East country including Saudi Arabia; this huge output came from less than 300 producing wells. The remaining 8% of production was derived either solely or jointly by the State-owned National Iranian Oil Co (NIOC), from one accumulation which lies across the Iraq frontier and has produced oil for

[11] *International Petroleum Encyclopaedia*, 1970, 106.

many years (p. 159), and various newly discovered fields in the relatively shallow offshore waters of the Gulf. Iranian production is increasing rapidly year by year (1971 output was 18·6% above that of 1970) and may reach 2,800 MM brl/a by 1980.

The total proved oil reserves of Iran are more than 60 B brl[12]. About 95% of the annual production, which multiplied by a factor of six between 1950 and 1970, is exported; the balance of 5% provides in turn at least 95% of the total energy requirements of the whole country.

About 60% of the 1·65 MM sq miles area of present-day Iran is made up of sedimentary basins. The major oilfields lie in the province of Khuzestan in the Dezful geosynclinal embayment.

The principal tectonic feature of Iran is the folded Zagros mountain belt, the result of the Alpine orogenesis of a mobile geosynclinal trough, containing late Triassic to Neogene sediments, superimposed on the Arabian platform[13]. During the later phases of geosynclinal development, there is evidence that the trough axis shifted southwestwards.

Paralleling the mountain belt, a series of long, sinuous asymmetrical folds with a dominant northwest-southeast trend, contrasts sharply with the north-south trends of the principal Arabian shelf structures.

The major reservoir formation in nearly all the oilfields is the Asmari Limestone, a development of Oligo-Lower Miocene shallow-water and reef-type limestones some 1,000–1,600ft thick, which have a characteristically high secondary permeability due to fissuring and fracturing. The limestone is often locally dolomitic, and is conformably overlain by Lower Fars evaporites, which have provided generally efficient seals, although oil and gas "shows" of nearly every type are nonetheless common throughout the region.

The Asmari Limestone interfingers to the west with continental sandstones (Ghar formation) and these beds also contain oil in some fields. The Ilam (Upper Cretaceous) and Sarvak (Middle Cretaceous) limestones of southwest Iran also act as oil reservoirs, but generally to a lesser degree than the Asmari Limestone.

Consortium Fields

Table 48 lists the currently productive oilfields of the Consortium area. All these fields are anticlines of the same general type—asymmetric folds paralleling the Zagros mountain front. Several of the accumulations have been found in culminations occurring along the same structural axis. Within an area less than 200 miles long by 60 miles, there are 18 large oilfields—most of them "giants".

(a) *Masjid-i-Sulaiman*

This field was discovered in 1908 and has been one of the world's

---

12 *World Oil*, Aug. 15, 1972, 60.

13 However, recent reconsideration of the tectonics of Iran (J. Stöcklei, *Bull.AAPG*, 52(7) July 1968, 1229) has modified this oversimplified concept.

largest oil producers. The first well was located in an area of seepages on a local Lower Fars fold, but this was later shown to be structurally unrelated to the subsurface Asmari Limestone anticline, because of the plasticity of the intervening evaporites and marls. The main structure is a broad, flat anticline containing an oil accumulation about 18 miles long by 4 miles wide. Dips on the southwest flank are nearly vertical and on the northeast flank are 20°–30°. The reservoir limestone outcrops to form the Asmari Mountain only 12 miles to the southeast of the oilfield.

The reservoir bed in the Upper Asmari Limestone is of fairly low porosity but has zones which are highly permeable due to fissuring. The original oil column was 2,200ft, but as production progressed an expanding gas cap developed, while the oil-water level remained almost stationary.

Smaller volumes of deeper oil have been found in Eocene and Middle Cretaceous limestones, and also a large volume of gas in Lower Cretaceous and Jurassic beds. It is interesting to note that due to a failure in the cementing of the deep test well which made the latter discovery, high-pressure gas repressured the overlying Asmari Limestone at the rate of about 350 MM cf/day.

It is likely that the Asmari oil of *Masjid-i-Sulaiman* (and in fact of all the Asmari oilfields of Iran) originated in the Cretaceous (probably the Middle Cretaceous) and subsequently migrated upwards into its present reservoirs.

(b) *Haft Kel*

This field was discovered in 1928 by drilling on another surface anticline on a line of folding parallel to *Masjid-i-Sulaiman*, in an area of active oil and gas seepages. The structure here is an anticline 17 miles long by nearly 3 miles wide, which is slightly asymmetrical and has two culminations. Asmari Limestone forms the reservoir.

(c) *Naft-i-Safid*

The axis of the Haft Kel fold continues to pitch downwards gently to the northwest for many miles. A lower culmination separated from it by a deep saddle has acted as an Asmari Limestone trap for hydrocarbons migrating southwestwards along the axis of the fold. A light condensate seepage at the surface has given this field its name which means "white oil springs". There is also a deeper Sarvak limestone reservoir in this field.

(d) *Agha Jari*

Currently one of the most important oil producers in Iran (and also one of the few oilfields in the world to have produced more than 3 B brl of oil), *Agha Jari* was discovered before the war, but production did not begin until 1945. Its output in 1970 was nearly 300 MM brl of oil, so it is still one of the world's major producers.

The field is a strong surface structure in Lower Fars rocks on the edge of the foothill belt, with many gas seepages. The main fold is some 15 miles long and 3 miles wide, and most oil has come from the upper part of the highly porous and fissured Asmari and deeper Sarvak limestones.

(e) *Pazanan*

This is an accumulation in a structural culmination lying on the same axis as the Agha Jari fold, from which it is separated by a deep saddle. A high-pressure gas column of some 2,000ft overlies the oil in the Asmari Limestone reservoir.

(f) *Gach Saran*

Discovered in 1928, this is the second largest oil producer in Iran; the oil comes from Asmari and Sarvak limestones and is heavier than most Iranian crudes. The structure is a relatively simple asymmetrical fold with a steeper southwest flank, entirely in disharmony with the Fars surface structure. The productive area is some 14 miles long by 4 miles wide, and there are extensive gas shows, with local deposits of "sour" gypsiferous earth and heavy oil seepages. Output in 1971 was 322 MM brl. There is a 600 ft column of hydrocarbons present.

(g) *Lali*

Although discovered in 1938, first production from this field did not commence until 1948. It is a limestone fold about 25 miles from *Masjid-i-Sulaiman,* along the same anticlinal trend but slightly offset to the northeast. The anticline is about 15 miles long by 4 miles wide in the central area, with a steep and partly overturned southwest flank. The porosity and permeability of the Asmari and Sarvak limestone reservoirs are poorer than at *Masjid-i-Sulaiman.*

(h) *Other fields*

In recent years, the most striking increases of production have come from the *Marun* and *Bibi Hakimeh* fields (Asmari and Sarvak limestone reservoirs) which produce light and heavy crudes respectively. Output from *Marun* increased between 1966 and 1971 from 45,000 b/d to 893,200 b/d; and from *Bibi Hakimeh* in the same period from 70,000 b/d to 445,600 b/d. Production from the *Ahwaz* field, which now supplies the Teheran and Abadan refineries, has also notably increased over a short period of time.

Exploration in the Consortium area has been very successful in finding further accumulations. Thus, three new fields were put on production for the first time in 1966—*Faris, Ramshir* and *Rag-i-Safid*; while *Binak* was a new producer in 1967, from a Bangestan reservoir. Other important recent discoveries are *Cheshmeh Khush,* some 100 miles northwest of Ahwaz, where the Asmari limestone was found to be highly productive at a depth of about 11,300ft; *Par-e-Siah* where Asmari gas

and a relatively small oil column were found; the Asmari limestone
accumulation of *Mulla Sani*, northwest of Marun; and *Maleh Kuh* and
Sarkan in Lurestan, with Ilam and Sarvak reservoirs.

The *Kharg Island* field is interesting in that it has a roughly north-
south axis, like the other Gulf ,offshore 'fields, rather than one paralleling the Zagros mountains, and is a continuation of the same offshore
trend which is productive in the *Darius* field (p. 160), with oil in Asmari
and Lower Cretaceous limestones.

TABLE 48

IRAN—OIL PRODUCTION, 1971

Consortium (IOPEC) fields

	Net crude oil M b/d
Marun	893·2
Gach Saran	882·2
Agha Jari	867·5
Bibi Hakimeh	445·6
Faris	347·5
Ahwaz	265·6
Karanj	195·7
Binak	52·3
Haft Kel	44·1
Rag-i-Safid	36·1
Naft-i-Safid	32·4
Pazanan	26·8
Kharg	22·2
Par-e-Siah	13·1
Masjid-i-Sulaiman	7·8
Kupal	6·2
Ramshir	4·4
Lali	1·6
Total IOPEC	4,144·3
Other fields	395·3
Total output	4,539·6

The NIOC* onshore fields

(a) *Naft Khaneh—Naft-i-Shah*. This field lies on both sides of the Iraq-
Iran frontier, about 80 miles from Baghdad, and is an asymmetrical
anticline, the southwest flank of which is the steeper and is affected by a
reverse fault. The subsurface axis is considerably displaced relative to the
surface axis, along which there are many oil and gas shows. In this
respect, the field therefore resembles the northern Iraq accumulations
rather than those of Iran. The producing formation is some 250ft of
Kalhur limestone, which is the local name for the Asmari. Output was
4·4 MM brl of oil and 3·3 Bcf of gas in 1971.

---

* It was announced in 1973 that all future petroleum production operations in Iran are to
be brought under the control of NIOC.

(b) *Central Plateau Fields*

Several large surface structures in the Central Iranian Basin—
Alborz, Sarajeh and (far to the northeast) *Gorgan*—contain hydrocarbons in the Oligo-Miocene (Qum) limestone-shale series, which is
roughly equivalent in age to the Asmari Limestone.

Complex faulting and salt tectonics complicate the structures at depth,
and drilling is difficult because of high pressures. *Alborz* is an oilfield
and *Sarajeh* is a "wet" gasfield, whose reserves have been calculated[14,15]
at about 1·5 Tcf of gas and 57·5 MM brl of condensate. About 250,000 brl
of condensate was produced from here in 1970. *Gorgan* is a high-pressure
gasfield, from which gas may be transmitted to Astara in the USSR.

The Offshore Fields

The offshore operations on the Iranian side of the Gulf have been
carried out since 1958 by companies jointly owned by the National
Iranian Oil Co. and foreign interests.

There are two structural trends in this area, the younger running
northwest-southeast and paralleling the axes of the Iranian Tertiary
onshore fields, the older trend running north-south and paralleling the
axes of the Arabian Mesozoic oilfields.

The first (and most northerly) offshore field to be discovered by
SIRIP[16] was *Bahregansar*, in 1960, with two oil reservoirs—dolomites
of the Oligo-Miocene Asmari and also a Middle Cretaceous Sarvak
limestone[17]. This field produced 7·2 MM brl in 1971. A more southerly
test on the same trend at *Norouz* (*Palinurus*) discovered Middle Cretaceous oil during 1967.

Darius (IPAC[18]) near Kharg Island had an output of 36·5 MM brl in
1971 from a porous calcarenitic limestone of Yamama (Lower
Cretaceous) age. *Cyrus* (*Kuroush*) (IPAC), some 50 miles southwest of
Kharg Island, which began production in 1967, had an output in 1971 of
9·1 MM brl from unconsolidated Burgan sands. *Esfandiar* (IPAC) is a
further potential producing structure in the same group awaiting
development.

To the south, northeast of Das Island, the LAPCO[19] *Sassan* field went
into production in 1968 from a dolomitic Arab Zone (Darb) limestone
reservoir. Production in this field, which totalled 49 MM brl in 1971, is
via an 88-mile long submarine pipeline to Lavan Island off the coast of
Iran. *Sassan* is probably in a structural and stratigraphic setting quite
different from the more northerly Iranian offshore fields, being geologically associated with the offshore Qatar and Abu Dhabi accumulations.

---

[14] P. Swain, *Oil Gas J.*, 18.12.61, 59 (51) 60.
[15] Badakshan *et al.*, 6 WPC Frankfurt, II, 13-PD 3, 1964.
[16] SIRIP = Société Irano-Italienne des Pétroles.
[17] P. Mina *et al.*, 7 WPC Mexico, PN9 (10).
[18] IPAC = Iran Pan-American Oil Co.
[19] LAPCO = Lavan Petroleum Co.

Other offshore discoveries, believed to be commercial, were made in 1969 at *Sirri* (Middle Cretaceous limestone reservoir) and *Hendijan*, four miles from *Bahregansar*. The *Rostam* (IMINOCO[20]) field went on production in mid-1969 and produced 17 MM brl in 1971 from Shuaiba, Mishrif and Arab Zone limestones; *Raksh* or *Rostam Northeast* is a further accumulation with similar reservoirs, a small distance to the northeast of the main field.

TABLE 49

IRAN—PRODUCTION OTHER THAN CONSORTIUM, 1971

	MM brl
IPAC	45·6
SIRIP	21·0
LAPCO	49·1
IMINOCO	24·2
NIOC	4·4
	144·3

KUWAIT

Although discovered before the 1939–45 war, oil was not produced commercially in Kuwait until it ended. However, output thereafter rose so rapidly, that by 1966 Kuwait ranked fourth among world oil-producing countries, and was second only to Venezuela as an oil exporter.

In 1971, output reached nearly 1,068 MM brl, and during 1972 it was announced that future production would be stabilized at about this rate. With a land area of only about 6,000 sq miles, Kuwait is certainly far ahead of any other country in the world in terms of oil production per unit area.

By far the largest oil accumulation is *Burgan*, a subsurface dome, with oil-impregnated Miocene surface sandstones and sulphur deposits. Two wells in this field had each produced more than 70 MM brl of oil before the end of 1966—an indication of the great productivity of the 374 wells which are currently oil producers. By mid-1969, the cumulative total production had increased to 10 B brl, making Burgan the first and only oilfield in the world to have produced such as enormous volume of oil[21].

The structure here is a gentle north-south elliptical dome with an area of approximately 135 sq miles. Average dips are 3° and the closure is at least 12,000ft. The gentleness of the folding is probably indicative of

---

[20] IMINOCO = Iran Marine International Oil Co.

[21] J. Davis, *Oil Gas Int.*, Oct. 1970, 100.

vertical movement at depth. Production comes from a number of highly permeable Middle Cretaceous sandstones, often almost uncemented, which are interbedded with shales. The total thickness of reservoir sandstones is over 1,300ft, and they range in age from Lower Cenomanian (Wara) to the Albian (Mauddud) and Burgan formations, the latter being equivalent to the Nahr Umr of Iraq. There is also a deeper oil reservoir in the top Jurassic "Minagish oolite" zone. The *Magwa* and *Ahmadi* accumulations, a few miles to the north and northeast of Burgan respectively, are culminations on the same major structure, with interconnected oil bodies; collectively, these three fields form the "Greater Burgan" field, about 30 miles long by 15 miles wide, which may well turn out to be the largest oil accumulation in the world. Production from this group is over 800 MM brl/a and original recoverable reserves have been estimated at 72 B brl.

The most northerly field in Kuwait is *Raudhatain*, which was found in 1955. The structure here is a dome, slightly elongated along a north-south axis, with flank dips of about 3°. Structural growth began in Cenomanian times and all formations deposited since then show thickening on the flanks, so that it is likely that the uplift, which may still be growing, is due to the movement of salt at depth[22]. Oil occurs here in sands of the Mauddud, Burgan, and Zubair formations (Middle and Lower Cretaceous), of which the Lower Burgan is the principal reservoir.

Smaller accumulations are at *Sabiriyah-Bahrah* (with oil in the Mauddud, Burgan and Zubair formations); *Umm Gudair* (Minagish oolites); and *Minagish* (Wara, Burgan and Minagish oolites).

Offshore exploration in Kuwait territorial waters has so far proved disappointing, but exploration is being carried out on Bubiyan Island.

The recoverable oil reserves of Kuwait are probably not less than 75 B brl.

A large volume of associated gas is inevitably produced with the oil —nearly 500 Bcf/a, of which a large proportion has been flared in the past. However, reinjection of associated natural gas is now practised on a large scale as a method of reservoir pressure maintenance, and gas is utilized in electricity generation and in the fast-developing petrochemical industry. About 48 % of the available gas is now being usefully used in these ways, and before the end of the decade there is likely to be no further flaring of gas in Kuwait. The total gas reserves are about 40 Tcf.

NEUTRAL ZONE

In the diamond-shaped jointly-administered neutral area lying between Kuwait and Saudi Arabia, an accumulation was found at *Wafra*, close to the Kuwait frontier, after several unsuccessful explora-

---

[22] D. Milton & C. Davies, *J.Inst.Pet.*, 51, 493, Jan. 1960, 17.

tory wells had been drilled in the 1950's. Heavy oil occurs here in Lower Eocene and Middle Cretaceous beds, but the main reservoir is the Ratawi (Lower Cretaceous) sandstone. The field produced 47·1 MM brl in 1971. Reserves are very large, but the oil is of low API gravity (18°–24°) and high sulphur content (4·7% by wt.), so outlets for it are restricted.

A second accumulation was discovered in 1962 at *South Fuwaris*, which in addition to Eocene and Middle Cretaceous reservoirs has oil in a deeper oolitic limestone of Ratawi (Lower Cretaceous) age. A third discovery, in 1967, at *South Umm Gudair,* near the Kuwait frontier, is probably a southward extension of the *Umm Gudair* field in Kuwait. The main reservoir here is also of Ratawi age. Production from these fields is now decreasing; it was 45·5 MM brl in 1969 but only 20·3 MM brl in 1971, most of which came from *South Umm Gudair.*

Several important offshore oilfields have more recently been found by Japanese interests. *Khafji*[23] lies on a structure which seems to be a north-ward extension of the Saudi Arabian *Safaniya* offshore field, with oil in Middle Cretaceous sandstones; there is also a deeper Ratawi reservoir. *Al Hout,* further to the north, has an oil reservoir of Ratawi age and deeper Jurassic gas. Another offshore discovery (1967) at *Lulu,* on the same uplift as the Iranian *Esfandiar* field, near the centre of the Gulf has a reservoir which is also of Ratawi age. The newest discovery, *Dorra,* 30 miles from *Khafji,* contains oil in a similar reservoir. The latter fields await development.

Neutral Zone offshore production was 130·6 MM brl in 1971 and is likely to increase rapidly as the newer accumulations are brought into production. The total output of the Neutral Zone (onshore and offshore) was about 198 MM brl in 1971.

OMAN

The first exploration well was drilled in the Sultanate of Oman and Muscat in 1956, the first commercial accumulation was found in 1962, and the first oil was exported in mid-1967. In 1971, the output was about 107·4 MM brl.

The Oman Mountains are composed of massive limestones, ranging in age from Permian to late Tertiary, lying transgressively on a Pre-Cambrian "basement", and ophiolites which have probably been extruded along Late Cretaceous tension faults[24]. To the southeast of the mountains, gentler structures occur in the foothill belt. The first sub-surface structure to be successfully drilled was *Yibal*, in 1962. Gas was

---

[23] M. Minato *et al.,* 7 WPC Mexico, PD3–11.
[24] R. Tschopp, 7 WPC Mexico, 1967, PD2–16.

found here in the Middle Cretaceous (Wasia) limestone, and oil in the underlying Lower Cretaceous Thamama (Shuaiba) limestone. The overlying Upper Cretaceous Aruma shales are an excellent sealing medium. The field was not put on production until late in 1969; it produced about 8·9 MM brl in 1970.

Two large surface folds have since been found to contain major oil accumulations. *Natih* and *Fahud*, lying about 20 miles apart, can be considered[25] as together forming an uplifted "horst," with deep sedimentary troughs both to the north and south from which oil may have migrated into them. Both structures are large, asymmetrical anticlines (*Fahud* is 27 miles long by 5 miles wide) with flank dips of 15°–20°, cut by major northwest-southeast axial faults. The principal fault movements took place in Upper Cretaceous time, and there was subsequent later Tertiary folding and uplift. These are thought to be essentially growth structures related to the movements of deep-seated Cambrian salt. The first well to be drilled at *Fahud* intersected the fault-plane where the productive Wasia formation was cut out; but a second well, drilled early in 1964, was successful. *Natih* and *Fahud* have since been shown to be major oilfields, with good-quality, low-sulphur oil contained in several members of the thick Wasia limestone sequence. *Fahud* is the larger producer (75·8 MM brl in 1970); *Natih* produced 36·5 MM brl in the same year.

A further commercial oilfield was found in 1970 at *Al Huwaisah* in the same central area of Oman, about 15 miles south of Yibal. This is a flattish anticlinal accumulation with a Thamama (Shuaiba) reservoir.

The oil reserves of Oman have been estimated at 4·75 B brl.[26]

Offshore exploration is also being carried out on the Continental Shelf off the eastern (Batinah) coast of Oman.

In **Dhofar** a considerable exploration effort between 1953 and 1966 failed to discover commercial accumulations of hydrocarbons, although deposits of heavy oil were found associated with several diapiric salt uplifts. A later phase of exploration is being directed towards the discovery of commercial accumulations.

QATAR

Although the peninsula of Qatar has a sedimentary area of about 8,000 sq miles, only one oilfield has so far been discovered on land. This is the *Dukhan* field on the west of the peninsula, a low, symmetrical north-south fold with a productive area about 30 miles long and three miles wide. Production commenced here in 1950 from three Upper Jurassic reservoirs termed the "No 3", "No. 4" and "Uwainat" lime-

25 H. Wilson, *Bull.AAPG*, March 1969, 53(3), 626.
26 *World Oil*, August 15, 1972, 60.

stones. Of these, the 185ft-thick "No. 4" limestone, stratigraphically equivalent to the Arab Zone D, is the most important. The "No. 3" limestone reservoir shows stylolite growths (p. 150) which can be dated as Late Lower Cretaceous, a time when the reservoir rock was already largely oil-permeated. As a result of their development, much of the oil in place was isolated in porous lenses which are now surrounded by thin, water-saturated limestones. The reservoir is consequently a difficult one to develop.

The remainder of the peninsula is occupied by a large, north-south trending, gentle uplift which exposes Eocene beds at the surface. Although a number of local culminations have been explored by drilling, further commercial oil has not been discovered.

In the offshore areas to the east and northeast, two anticlinal oil accumulations have been discovered. The reservoirs at *Idd-el-Shargi*, found in 1960, are in Arab Zone A–D equivalents, and there is oil also in the deeper "Uwainat"; while at *Maydan Mahzan* (1963) the "No. 4" limestone (Arab Zone D) is the producing bed. *Bul Hanine* is another offshore accumulation, discovered in 1970, which was producing oil at the rate of 120,000 b/d by end-1972.

The output from Qatar was 156·9 MM brl in 1971—*Dukhan* 81·1 MM brl, *Maydan Mahzan* 61·1 MM brl, and *Id-el-Shargi* 14·7 MM brl. By 1972, production had risen to 163·3 MM brl.

Production is rising as a result of the inception of water-flood secondary recovery operations in the *Dukhan* and *Maydan Mahzan* fields. Oil reserves are approximately 4·8 B brl. Gas from *Dukhan* will be converted into ammonia and urea in a fertiliser plant under construction at Umm Said. A natural gas liquids extraction plant is also being set up.

SAUDI ARABIA

Exploration for oil began in Saudi Arabia in 1936 in the area nearest to Bahrein Island, where a commercial accumulation had already been found. The first discovery (in 1938) was the *Dammam* field, a dome which subsequent exploration showed to be only one of a series of large anticlinal oil accumulations stretching for some 300 miles in several roughly parallel trends along a broad north-south uplift. This is the Hasa Arch, which separates the Dahana basin on the west from the Gulf basin on the east. (Curiously enough, the *Dammam* structure, probably because of its association with a diapiric salt mass, is an exception to the general north-south alignment). This region is already one of the world's most important oil-producing areas, with an output in 1971 of nearly 1,642 MM brl—more than 27 times the production of 60 MM brl in 1946, the year in which post-war development began. Further development is planned to raise Saudi Arabian production

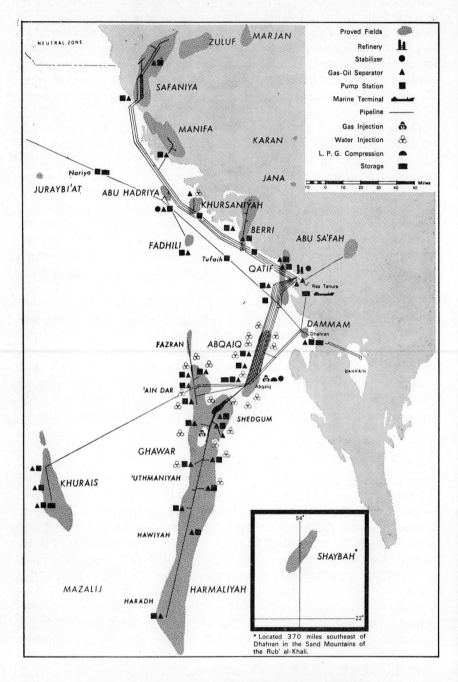

Fig. 30. SAUDI ARABIA—OILFIELDS AND FACILITIES

Source: Arabian American Oil Co.

very considerably over the next decade, perhaps to several times its present level.

The principal reservoirs in nearly all the fields are the thick calcarenitic limestone members of the Upper Jurassic Arab Zone, which are separated and sealed by dense anhydrite beds. These Arab Zone sediments were in fact deposited in four major cycles, each of which included a normal, shallow marine carbonate and an evaporite phase; the overlying Hith anhydrite represents the final deposition of evaporites. The oldest (D) limestone is the most prolific reservoir and may be more than 200ft thick. A few fields, such as *Manifa* and *Fadhili* have stratigraphically higher Thamama oil, while several fields (notably *Abu Hadhriya, Fadhili, Abqaiq, Khurais* and *Ghawar*) also have deeper Middle Jurassic reservoirs (the Darb, Diyab or Araej limestones).

The *Ghawar* field is one of the largest in the world; it is about 135 miles long, and has six culminations, each of which contains a major oil pool; their combined output accounts for about 50% of the overall production of Saudi Arabia. *Abqaiq*, on a trend to the northeast of Ghawar, produces a further 21% of the total, so that these two huge fields between them currently account for more than 70% of the national oil production.

Several of the fields listed, notably *Khurais,* which lies on another trend considerably to the west, await further production facilities which will result in considerably higher future outputs.

A number of partly or wholly offshore oilfields have also been developed. Of these, *Safaniya*, the most northerly Saudi field, is one of the world's largest offshore oilfields, with an output of nearly 281 MM brl/a from two Cretaceous Wasia sandstone reservoirs[27]. Other important offshore (or partly offshore) fields are the eastern group of *Manifa, Berri, Qatif* and *Abu Safah*, the profits from the last of which are divided with Bahrein. Further out in the Gulf, another offshore field, *Zuluf,* has recently been discovered, and agreement has been reached with Iran for joint ownership of the *Marjan-Fereidoon* accumulation.

The *Fereidoon* part of the field, with Farsi island, goes to Iran, and the *Marjan* part, with el-Arabi island goes to Saudi Arabia, each country having rights over 50% of the oil resources of the whole field.

The Saudi Arabian oilfields are characterised by their high per-well productivity (averaging more than 7,000 b/d/well) and hence the small number of wells needed (a total of only about 500—all flowing). Less happily, they also inevitably produce with the oil an enormous output of associated gas (more than 500 B cf/a), of which a very large proportion is still flared, in spite of the introduction of extensive gas reinjection and petrochemical manufacturing schemes.

In contrast to the other Gulf oil-producing states, the land area of

---

[27] Q. Lowman, *Oil Gas Int.,* Nov. 1965, 47.

Saudi Arabia is large, and sedimentary basins probably comprise not less than 70% of it; however, no commercial oil accumulation has yet been announced outside the relatively narrow producing belt near the east coast, except for two recent discoveries at *Shayba*, near the Abu Dhabi frontier, and at *Mazalij*, 150 km south-southeast of Riyadh.

The proved reserves of Saudi Arabia are estimated to be at least 137 B brl*, more than those of any other Middle Eastern state, and gas reserves are probably at least 55 Tcf. In addition, exploratory drilling in the northern Red Sea offshore area in 1969 discovered some gas-condensate in Miocene sandstones which await further testing.

The oil in the Saudi Arabian fields is generally believed to be indigenous to the pellety, oolitic Upper Jurassic limestones in which it is now found. These reservoir limestones grade into anhydrites to the west and into denser limestones to the east, so it is thought likely that the oil originated in the limestones themselves under the cover of the overlying Hith anhydrite, and then accumulated into the anticlinal traps in which it is found, as a result of Upper Cretaceous and later folding[28].

TABLE 50

SAUDI ARABIA—OUTPUT OF ARAMCO† OILFIELDS, 1971

MM brl

Ghawar		
Shedgum	272·1	
Ain Dar	237·0	
Uthmaniyah	192·3	
Hawiyah	63·8	
Haradh	21·0	
Fazran	1·7	
		787·9
Abqaiq		328·7
Safaniya		280·9
Berri		67·7
Qatif		38·5
Abu Hadriya		38·4
Abu Safah		31·6
Khursaniyah		31·3
Fadhili		18·9
Dammam		7·7
Khurais		5·9
Manifa		4·1
Total		1,641·6

Source: Saudi Arabian Ministry of Petroleum.

(In addition to the Aramco output listed above, Saudi Arabia is entitled to one-half of the Neutral Zone production of two of the operating companies.)

In October, 1970, it was announced in Jeddah that Saudi Arabian oil reserves amounted to 147 B brl.

† *Arabian American Oil Co.*

28 M. Steineke *et al.*, *AAPG*, "Habitat of Oil", 1958, 1294.

SYRIA

There are three minor oilfields in the northeast corner of Syria— *Soueide, Rumelan* (*Hamseh*) and *Karatchok*—small anticlines with Upper and Middle Cretaceous reservoirs similar to those in northwestern Iraq; deeper Triassic oil has also been found. Together, these fields hold an estimated 1·3 B brl of recoverable reserves.

Production began in mid-1968, most of the output coming from *Soueide;* during 1971, 40·2 MM brl of oil were produced.

Other hydrocarbon accumulations have been found at Sukhne in east-central Syria, and Jibissa in the northeast, but it is not known whether these are in fact commercial.

TURKEY

Oil was first found in Turkey in 1940 in an asymmetrical anticline at *Raman*, on the east of the Tigris River. The reservoir here is an Upper Cretaceous limestone, and the oil is relatively heavy, as in most of the Turkish oilfields which have since been found. A second anticlinal accumulation, with a similar reservoir, was found in 1951 at *Garzan* in the same area. Intensive exploration in the Mardin Basin has subsequently discovered a number of small oilfields, commencing with *Germik*, on the same trend as *Garzan*, in 1958. This sedimentary basin covers an area of about 100,000 sq miles on the northern edge of the Arabian "Shield" platform, and is bounded on the north by the arc of the Taurus Mountains. Of the score or so producing oilfields, *Beykan* and *Kurkan* are the largest, (with outputs of 4·5 and 4·1 MM brl respectively in 1971), but growth is slow and the overall oil output was still only 24·0 MM brl in 1971. All the reservoirs are Middle or Upper Cretaceous limestones.

In the narrow Adana Basin of southwestern Turkey, a small oilfield has been found at *Bulgurdag*. This accumulation has a Miocene limestone reservoir which overlies a Palaeozoic uplift.

While exploration both in Anatolian Turkey and Thrace still continues, large hydrocarbon accumulations are unlikely to be found, principally because of the absence of the Upper Tertiary evaporites which act as seals in Iraq and Iran, and the general lack of basinal-type source rocks. The frequent occurrences of fresh water with traces of asphalt and heavy oils in the reservoir beds suggest furthermore that water flushing has taken place[29].

Total Turkish oil reserves are now about 200 MM brl, with about 300 Bcf of associated gas.

---

[29] P. Temple & L. Perry, *Bull.AAPG*, 46(9) Sept. 1962, 1596.

OTHER AREAS

A major north-south anticline (the "Hebron arch") of Mesozoic limestone forms the principal structural feature of **Israel**. It plunges southwestwards into a syncline before rising again to form further structures in the northern Negev[30]. In two of the latter folds, minor gas accumulations have been discovered in Jurassic reservoir limestones at *Zohair-Kidad* and *Har-Hakanaim*.

The only oilfield so far discovered in Israel is at *Heletz*, east of Gaza, with its extensions of *Brur* and *Kokhav*, which together produced 0·2 MM brl of oil in 1971 from Jurassic limestone reservoirs. Reserves here are about 2 MM brl. The Palaeozoic and Triassic beds of the Negev area, and the Mediterranean Continental Shelf are future exploration prospects.

In **Jordan** a number of dry wells have been drilled, including one which reached a depth of 14,000ft without success. There are interesting seepages, however, in the Dead Sea area[31].

In the **Lebanon**, test wells have been drilled without success to explore possible Jurassic or Upper Cretaceous reservoirs.

In **Southeast Arabia** and the offshore islands, a few test wells have been drilled unsuccessfully in recent years. Prospects here appear poor in view of the lack of sedimentary thickness likely on the edge of the Arabo-Nubian "Shield". In another fringe area lying to the south of this massif, one unsuccessful test has been drilled in the Hadhramaut near Thamud. The sediments here are thicker, dipping and thickening northwards.

Onshore and offshore exploration has been carried out without any announced success in several of the other Gulf States—including **Ajman, Umm al Qaiwain** and **Fujairah**. In **Ras al Khaima**, an exploratory well discovered a deep accumulation (17,800ft) in offshore water early in 1972, the commercial viability of which is not yet certain, while during 1973 oil indications were reported from a well drilled near Abu Musa Island, between **Sharjah** and Iran.

---

[30] J. Coates *et al.,* 6 WPC Frankfurt, 1963, 1, 26, 21.
[31] F. Bender, *Erdöl u. Kohle*, Oct. 1969, 10, 597.

CHAPTER 6

East Asia

COMPARED with the Middle East, East Asia is much larger in area, but very much smaller in its petroleum production and known reserves. Nearly all the oilfields so far found are relatively shallow, Tertiary sandstone accumulations, as compared with the deeper, largely Mesozoic and preponderantly limestone reservoirs of the Middle East. They are generally structurally related to the most easterly continuations of the Alpine orogenic belts, one branch of which passes northeast-wards through Burma and the Philippines to Japan, while another branch curves southeastwards through Sumatra, Java and New Guinea to New Zealand.

Although there is a great thickness (up to 50,000ft) of Tertiary sediments in the basins of Pakistan, India and Burma, the principal hydrocarbon accumulations so far found in these countries are all related to one geological feature—the unconformity between the Upper and Lower Tertiary: either below it (in Eocene beds as in the Punjab fields) or above it (in the Miocene oilfields of Assam and Burma). The absence of competent cap rocks may be the reason why much of the oil generated in the rest of the Tertiary section has disappeared. Anhydritic shales provide adequate cover in some areas, but elsewhere the potential reservoirs are exposed to the surface.

In the countries further to the east (Indonesia, Borneo, Japan), nearly all the hydrocarbon accumulations so far found occur in shallow Miocene and Pliocene sandstones. Throughout East Asia, only relatively few major oilfields have so far been discovered. One such field is in Burma (*Yenangyaung*), two in Sumatra (*Rantau* and *Minas*) and one in Borneo (*Seria*). Perhaps the violence of the Tertiary orogenies fractured the shallow reservoir beds to such an extent that accumulations on the Middle East scale cannot now be expected to remain. However, little is known of the potentialities of China, and in general it is true to say that the intensity of exploration has so far been very low in proportion to the available sedimentary areas of the whole huge region. War, political upheavals, impenetrable jungles, deserts and marshes have hindered development, as well as the immense distances which separate many of the potential oil-producing areas from likely markets.

The total production of the countries of East Asia amounted to about 607 MM brl in 1971, of which more than half came from Indonesia; but these statistics can only be regarded as approximate, in view of the uncertainty surrounding activities in China. Similarly, reserve figures can be little more than intelligent guesses. Intensive offshore exploration operations are now taking place in the Gulf of Thailand, the Java Sea and the South China Sea, so that new discoveries can be expected in these areas. Several offshore oilfields have already been discovered off Java and Kalimantan.

TABLE 51

EAST ASIA—1971

	oil production† MM brl	proved reserves † B brl
Brunei-Malaysia	70·1	0·82
Burma	6·9	0·04
China	145·0	12·50
India	50·6	0·74
Indonesia	325·7	10·67
Japan	5·5	0·03
Pakistan	3·0	0·03
Taiwan	0·2	0·02
Thailand	0·0*	0·01
Total	607·0	24·86

*40,000 brl.   †estimated

TABLE 52

EAST ASIA—OIL PRODUCTION 1890–1970
(thousands of brl)

Year	Brunei[3]	Burma[1]	China*	India[1]	Indo-nesia[2]	Japan	Paki-stan[1]	Taiwan	Thai-land	Total
1890	—	—	—	118	—	52	—	—	—	170
1900	—	—	—	1,079	4,014	871	—	—	—	5,964
1910	—	—	—	6,138	12,173	1,829	—	—	—	20,140
1920	1,020	—	—	8,375	17,529	2,221	—	—	—	29,145
1930	4,907	—	—	8,292	41,729	1,950	—	—	—	56,878
1940	7,047	7,731	10	2,302	62,011	2,639	—	—	—	81,740
1950	30,995	532	800	1,867	50,148	2,048	1,281	23	—	87,694
1960	34,005	4,078	24,959	3,370	154,526	3,678	2,636	14	—	227,266
1970	58,400	6,570	145,000	49,275	321,629	5,656	2,555	182	40	589,307

* estimated

Notes: 1. Because of the difficulty of separating earlier statistics, the figures for Burma and Pakistan have been allotted to India in the above Table, prior to 1940 and 1950 respectively.
2. The heading Indonesia includes New Guinea for similar reasons.
3. The heading Brunei includes Sarawak (Malaysia).

BRUNEI AND MALAYSIA

Northwest Borneo forms a marginal trough of the Sunda Shelf, in which a great thickness of Tertiary sediments was deposited in a basin whose axis shifted northwards from Eocene to Miocene times[1]. (The basins in the Indonesian part of Borneo are discussed on p. 183). Orogenic activity took place in the Upper Eocene and again in the Pliocene. The two largest oilfields, *Miri* in Sarawak (Malaysia), and *Seria* in Brunei (discovered in 1910 and 1929 respectively) are accumulations in crestal culminations along the same narrow, asymmetrical anticlinal uplift, which has been greatly affected by complex normal faults. The reservoirs in both these fields are Upper Miocene—Lower Pliocene sandstones, and the oil is thought to have originated in the local Oligo-Miocene marine shales.

In **Sarawak**, *Miri* had an annual output of several million brl in the 1920's, but was approaching exhaustion during 1972, with a production of only 500 b/d.

Two offshore discoveries (*Baram* and *West Lutong*) now contribute the bulk of the Sarawak production of about 90,000 b/d; *Bakau* and *Baronia* are smaller offshore fields. *TuKau* is a 1973 discovery.

In **Brunei**, the *Seria* field, which was extensively war-damaged but subsequently rehabilitated and extended offshore, reached a peak output of 115,000 b/d in 1956, but this has since declined to about 44,000 b/d. Most of the Brunei output of about 200,000 b/d comes from the offshore accumulation of *South-West Ampa*, discovered in 1963. This field also has several Tertiary sandstone reservoirs containing large reserves of non-associated natural gas, some of which is now being liquefied for export as LNG to Japan. The gas reserves here have been estimated as about 1·4 Tcf.

Another offshore hydrocarbon accumulation (*Fairley*) was discovered in 1969, 18 miles north of *South-West Ampa*, and extensive offshore exploration is currently being carried out off Brunei and off northwest Sabah, as well as off the east coast of Malaya. The newest discovery, *Champion*, is expected to be put into production at the rate of 18,000 b/d during 1973.

BURMA

Oil was obtained in Burma for several centuries from shallow hand-dug pits before the first wells were drilled with percussion tools in 1885. In 1913, and for many years thereafter, about seven million brl of oil were produced annually; this output rose to a maximum of 7·7 MM brl

---

[1] H. Schaub & A. Jackson, *AAPG* "Habitat of Oil", 1958, 1330.

in 1941, then fell to very little during the war years, before slowly recovering to about its previous level. Production in 1971 is estimated to have been again nearly 7 MM brl of oil, with some 8 Bcf of mainly associated gas.

The principal oilfields have all been in production for many years and are consequently declining. Exploration ceased for a considerable time, but has now been resumed, with reported discoveries of gas near Chauk, oil and gas near Ayadaw, and some oil in the Prome Hills. The *Myananaung* field, found in 1964, now accounts for about one-third of the total oil output. A promising new oilfield (*Mann*) was found in 1968 at Shwepyitha, and was put into production in 1970 at the rate of about 19,000 b/d. Offshore exploration is also under way off the west coast and in the Gulf of Martaban.

The oilfields of Burma are all Tertiary anticlinal accumulations in a north-south trending basin lying between the Arakan Yomas on the west, and the Pegu Yomas and Shan Plateau massif on the east. In this basin, which coincides roughly with the present-day Irrawaddy and Chindwin valleys, three sub-basins are separated by transverse uplifts[2].

Oil seepages are common, together with "mud volcanoes", gas "shows" and saline springs. In the Chindwin area, the "shows" are in Upper Eocene rocks, in the central basin in Eocene or Oligocene beds, and in the south they are usually in Miocene strata.

The reservoir beds are Oligocene and Miocene sands and sandstones of the Pegu system, and the oil is believed to have originated in shales stratigraphically close to the present reservoirs. The rapid alternation of permeable and impermeable beds renders vertical migration improbable.

Yenangyaung, discovered in 1889, is an elongated and slightly asymmetrical dome, with a steeper west flank, and good closure. Numerous dip-faults cross the structure. Production has been from a number of lenticular Upper Pegu sands extending down to about 5,000ft.

Singu (*Chauk*), discovered in 1902, is an elongated faulted dome, with dips on the east flank reaching 80°. Production has come from about 15 sandstone reservoir beds.

The continuation of the Singu fold across the Irrawaddy appears on the right bank as the *Lanywa* structure. This field was first drilled in 1927 on a sandbank reclaimed from the river. The structure is regular, pitching gently, and is little affected by cross-faults; there are several productive sandstone reservoirs.

The *Yenangyat* field lies on the same fold trend as *Singu* and *Lanywa*, but is *en échelon** with them. It is a long, narrow accumulation, with a small northern extension at *Sabe*. *Minbu, Palanyon* and *Yethaya* are three small culminations on an anticlinal north-south fold lying south of Yenangyaung, in an area remarkable for the number and size of its

---

[2] H. Tainsh, *Sci.Pet.,* 6, 1, 113, 1953, and *Bull.AAPG,* 34, 823, 1950.

* *i.e. parallel but not aligned.*

"mud volcanoes". Both flanks of the fold are very steep, the east side being actually overturned in places, and there is a strong pitch both to the north and to the south. Very large "mud volcanoes" occur along a crestal strike-fault, some being 50ft high.

Indaw is a solitary field in the Chindwin basin, which was first discovered in 1918. The structure is an elongated, asymmetrical dome with a gently-dipping western flank and a steeply-dipping eastern flank. Limited production has come from shallow Lower Pegu sands.

In the Thayetmyo district, south of Yenangyaung, *Padaukpin* is a steep anticline with flank dips of 60°, whose output has never exceeded 50 b/d. *Pyaye* is a slightly asymmetrical dome containing a large volume of high-pressure gas. At *Yenanma*, west of the Irrawaddy, a homocline with gentle eastward dip contains several Oligocene sandstone reservoir lenses.

CHINA

Small quantities of oil and gas were obtained for many centuries from shallow pits and hand-dug wells in different parts of China. There are records of one such ancient "commercial" development in 211 B.C. in the Szechuan area. In modern times, minor oilfields were discovered in Sinkiang (*Tu-shan-tze*) in 1897 and Shensi (*Yenchang*) in 1907. A wartime peak output of 2·3 MM brl/a from these and other minor accumulations had fallen to only 0·5 MM brl/a in 1949, the year in which scientific techniques were first introduced into the Chinese petroleum industry. A remarkable period of growth resulted, so that oil output in 1971 is believed to have reached 145 MM brl, with gas output not less than 800 Bcf. In addition, some 15 MM brl/a of shale oil is obtained from the extensive Manchurian deposits. As a consequence of this growth, petroleum imports from the USSR terminated several years ago, and some 2–3 MM brl/a of oil has been regularly exported by pipeline to North Vietnam.

Although the geology of China is complex and not thoroughly explored, it is known that its huge land area includes at least 20 large interior and coastal sedimentary basins, some of which are composite, i.e. they are modern topographic basins overlying older and deeper structural basins.

The interior basins are usually filled with Permian terrestrial strata; marine beds are confined to the Szechuan and a few other basins.

Although hydrocarbons are known to exist in most of these basins, the major fields are confined to only a few of them:

(a) *Dzungaria Basin*

The area of this basin, which lies in Sinkiang province in northwest China, is nearly 50,000 sq miles, and it contains a thick Late Permian to Tertiary continental sequence overlying strongly folded Palaeozoic rocks. The sediments thin northwards from a trough paralleling the Tien Shan mountains to a platform area with gentle folds and block-fault structures.

On this platform, several large oil accumulations have been found—notably *Urho*, which has Jurassic and Triassic continental-facies reservoirs, of which the latter holds most of the estimated reserves of 100 MM brl of oil; and *Karamai*, a broad dome which also has Jurassic and Triassic terrestrial sandstone reservoirs containing about 640 MM brl of oil.

(b) *Pre-Nan Shan Basin*

There are at least four fold fields in the Yumen area, with Eocene-Miocene continental-facies reservoir rocks. Total reserves here are between 0·8 and 1·2 B brl, and *Ya-erh-hsia* is the largest field.

(c) *Tsaidam Basin*

At least 18 anticlinal oilfields have been found in this area, with Tertiary terrestrial sandstone reservoirs. *Leng-hu* has three productive culminations, and the reservoir age is late Miocene-Oligocene. The estimated recoverable reserves here are 550 MM brl. *Nan-shan* has probable recoverable reserves of 200 MM brl from a Miocene reservoir. *Yen-hu* is a large gas accumulation in the centre of the basin, with a Pliocene continental-lacustrine sandstone reservoir and reserves of at least 1·5 Tcf of gas. The overall gas reserves of the Tsaidam basin are thought to total 2·8 Tcf.

(d) *Szechuan Basin*

This appears to be a halokinetic area in which the movement of Middle Triassic salt affected the tectonic history. The thick Permian-Cretaceous sedimentary sequence of the south and southeast thins towards a platform area in the north and northwest. The structures on the platform are gentle domes, whereas in the south and southeast of the basin there are many long parallel folds related to salt movements.

Several large hydrocarbon accumulations have been discovered in the platform area: *Lung-nu-ssu, Yinshan, Nan-Chung* and *Peng-lai-chen* are broad domes with oil in Late Jurassic reservoirs. The *Shih-you-kon—Tung-hsi* (and other) gas accumulations lie in the southern fold belt and have between them gas reserves totalling about 18·5 Tcf in fractured Middle Triassic limestones.

(e) *Sunglaio Basin*

This is a platform area in which broad, taphrogenic structures occur. There are at least four oilfields, of which the largest is the *Taching* (*Daqing*) anticline, discovered in 1963, 80 miles northwest of Harbin. The productive reservoirs are a series of more than 20 Lower Cretaceous sandstones. The structure has maximum dimensions of 30 miles by 12 miles and this is probably the largest oil accumulation so far discovered in China, with reserves of 0·5 B brl.

(f) *North China Basins*

These cover an area of about 150,000 sq miles of complex sedimentation. The *Shengli North* and *South* fields are Mesozoic oil accumulations producing together about 5·7 MM brl/a.

Reserves

The total of proved recoverable oil reserves is thought to amount to a minimum of 12·5 B brl, but this figure will almost certainly be greatly increased as the country's petroleum industry is developed. Its 1970 state of development has been compared to the United States petroleum industry in the year 1900[3]. The ultimate oil production likely to be obtained from oil-shale distillation has been estimated at 5·1 B brl.

INDIA

The subcontinent of India is made up of a buttress of crystalline rocks —the Gondwana massif or Indian "Shield", on the circumference of which lie a number of sedimentary basins, covering an area of some 400,000 sq miles. These basins, which are filled with marine and freshwater sediments ranging from Jurassic to Recent in age, have so far been only lightly explored for oil[4].

Assam Fields

On the northern and eastern borders of the "Shield", a deep geosynclinal belt separates it from the Himalayan foothills, and includes four separate basins: from west to east, the Rajasthan, Punjab, Ganga and Assam basins.

The Tertiary Assam basin, from which much of the oil production of India has come, lies on the flank of the Arakan Yoma-Naga-Chin mountain chain[5]. Seepages of gas and oil are common, and many "mud volcanoes" and "paungs" (mud wallows) occur along the Arakan coast.

The Tertiary strata have been strongly folded into long, narrow anticlines paralleling the eastern hill boundary zone, with many strikefaults and major thrusts. Oil was obtained in Assam from shallow wells and seepages for many years before drilling commenced. Gas pressures are often very high, and drilling has consequently been generally difficult.

The *Digboi* field[6] is an asymmetrical, sharply folded anticline with its steeper (northern) flank faulted. It forms a local culmination at the northeastern end of the Jaipur uplift. The field was first drilled in 1889, but the yield of oil obtained was very small until 1922, when deeper

[3] A. Meyerhoff, *Bull.AAPG*, 54(8), Aug. 1970, 1567–1580
(from which reference most of the information contained in this section has been drawn).

[4] A. Ghosh, *Unesco Min. Res. Dev. Series* 10, 1959.
[5] G. Wilson & W. Metre, *Sci.Pet.*, 6, 1, 119, 1953.
[6] *Inst. Pet. Rev.*, Nov. 1965, 389.

drilling resulted in increased production from several Miocene sandstones, which reached a wartime maximum of about $2\frac{1}{2}$ MMbrl/a. At the present time, production has declined to about 700,000 brl/a.

The *Bardarpur* field was discovered in 1915, but was exhausted by 1933 after comparatively small production. It is a small, sharply-folded dome with its eastern flank overthrust.

Nahorkatiya and *Moran* are more gently folded but markedly faulted anticlines with roughly east-west axes. They were discovered in 1953 and 1956 respectively, and now together produce about 14 MM brl/a of oil from sandstones in the Upper Oligocene Barail series. Oil has also been found at *Rudrasagar, Lakwa* and *Kusijan*, in the same area.

Gujarat Basin

In the west of India, several accumulations of hydrocarbons have been found in Tertiary reservoirs in the Gujarat Basin. This is bounded by faults and has been described as a "graben" or "half-graben", with block-faulting as the dominant tectonic influence[7].

Natural gas seepages had been known for some time on both sides of the Gulf of Cambay in the broad, flat Cambay sub-basin, before a test well drilled on a north-south fold at Lunej in 1958 found non-commercial oil in Paleocene-Eocene beds above the Deccan lava flows. The *Cambay* field, which was subsequently discovered, is divided into several blocks by cross-faults, and produces gas with condensate and some viscous oil. Other oil accumulations which have since been found include *Kalol* in the narrowest, northern part of the basin, and *Ankleshwar*, the largest of this group of oilfields, with an annual output of some 21 MM brl. Smaller fields are *Wavel, Navagam, Dholka, Bereja* and *Kosamba*. Offshore exploration is also being carried out in this area.

The total production from the Indian oilfields is estimated to have been nearly 50 MM brl in 1971. Intensive exploration may add to the proved oil reserves of only 740 MM brl of oil and about 1·4 Tcf of gas. It has been suggested that, when all onshore basins have been adequately explored, the ultimate oil resources of India could be as much as 19 B brl of oil[8].

INDONESIA

Submarine surveys have shown that a narrow belt of strongly negative gravity values, nearly 2,500 miles long, lies to the south of and parallel to the "island arc" of the Indonesian Archipelago. On either side of this negative belt is a strongly positive zone. This is clearly therefore an area of great isostatic disequilibrium, likely to be associated with earthquakes

---

[7] A. Raju, *Bull.AAPG,* 52(12), Dec. 1968, 2422.
[8] N. Kalinin, 3rd ESCAFE Symposium, Tokyo, Nov. 1965.

and vulcanism. The Indonesian "island arc" is actually composed of a double arc—an inner, volcanic arc and an outer Tertiary island arc. The complex structure of the region may be explained[9] in terms of "plate tectonics" if it is assumed that the Outer Banda Arc (extending from Timor through Tanimbar, Kai and Seram to Buru) formed part of the Australian continental margin from at least the Permian, becoming in part detached after the break-up of Gondwanaland, then undergoing large rotations as the continent drifted northwards.

It seems that in Eocene times, the Sunda Shelf of southeast Asia was fringed by a number of basins which separated it from the Sahul Shelf of Australia. A great thickness of Tertiary sediments was deposited in these basins and subsequently affected by Alpine folding movements. In Sumatra, Java, and Southern Borneo, the folding was Plio-Pleistocene, and in Northern Borneo it was Plio-Miocene.

The largest of the many islands of Indonesia, whose total area exceeds 1·9 million sq km, are therefore basically long anticlines with exposed cores of pre-Tertiary rocks, on either flank of which lie Tertiary basins.

Oil accumulations mainly occur in the area between the axial geanticlinal uplifts and the Sunda Shelf, in structures which are usually gentle asymmetric folds, usually associated with some degree of thrusting. Reservoir beds are fairly shallow Miocene sandstones, but in West Irian they are limestones. In Ceram and Tarakan, the reservoirs are Pliocene in age, with some additional Upper Triassic oil in Ceram.

The petroleum development of these islands began in about 1890 and output increased rapidly to a pre-war peak of 62 MM brl in 1940, but subsequently fell to as little as 12 MM brl in 1945. A period of political difficulty followed, until new legislation in the late 1960's stimulated exploration, both onshore and offshore, by joint State-owned and foreign companies.

As a result, oil production reached a peak of 380 MM brl in 1972, and will probably be over 400 MM brl in 1973. Intensive offshore exploration is currently under way, and the first oil discovery was made in 1970 in the Java Sea, about 80 miles east-northeast of Djakarta. The fields since discovered off the coasts of Java and Kalimantan are expected to add considerably to the Indonesian output of crude oil. Most of the oil produced, which is mainly light, paraffinic and of low sulphur content, comes from Sumatra. The overall reserves of Indonesia have been put as high as 10·7 B brl of oil and 5·3 Tcf of gas.

Sumatra

Sumatra currently produces about 95 % of the total Indonesian output from about 40 fields, its crudes being of excellent quality, with low sulphur contents, medium gravities and paraffinic bases.

---

[9] M. Audley-Charles *et al., Nature Phys. Science,* 239, 35, 1972.

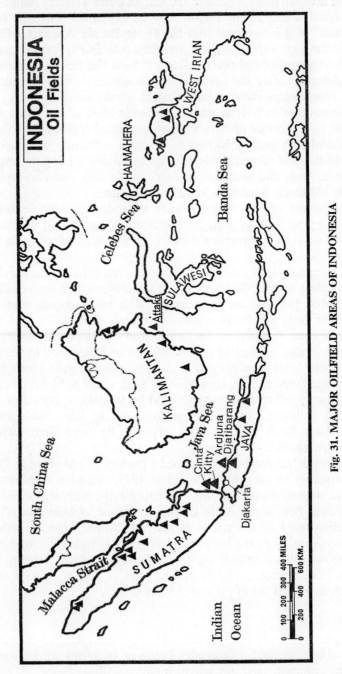

Fig. 31. MAJOR OILFIELD AREAS OF INDONESIA

Source: Petroleum Press Service.

An elongate sedimentary basin with an area of some 91,000 sq miles extends the length of Sumatra and is divided into three subsidiary basins, of which the northernmost is the deepest, with some 25,000ft of sediments[10,11]. These basins contain a large number of fairly shallow oil accumulations whose total original reserves were probably about 11 B brl; of these about $2\frac{1}{2}$ B brl have been produced. The fields include at least two "giants" (*Duri* and *Minas*).

(a) North Sumatra Basin

About a dozen oilfields have been developed in a narrow strip along the northern coast of Sumatra near Pangkalam Brandan, all of which are faulted anticlines with Middle or Upper Miocene sandstone reservoirs.

The most important of these fields is *Rantau*, discovered in 1929, which is a cross-faulted anticline, with a number of shallow Upper Miocene sandstone reservoirs, from which 11·84 MM brl of oil was produced in 1968. A similar accumulation, found in 1965 at *Guedongdong* north of Rantau, produced 1·64 MM brl in the same year, and is thought to hold a further 10 MM brl of reserves. Among the Perlak group of fields is *Telaga Said*, of interest as the oldest Sumatran field, from which oil was exported to Singapore as long ago as 1893.

(b) Central Sumatra Basin

Much of this basin is covered by swamps which have made exploration very difficult. The Tertiary sedimentary thickness here is less than in the north, being not more than 8,000ft at the most. The two main groups of oilfields around *Duri* and *Lirik* are mainly anticlines with reservoirs in Lower Miocene sands. *Duri* itself is a very large anticlinal accumulation, with reserves of some 2 B brl. *Minas*, discovered in 1944, is the largest oilfield in the Far East, with estimated recoverable reserves of not less than 7 B brl, and an output of about 300,000 b/d. Other fields in this area are *Rangau, Petani, Pematang* and *Banko*; three newer discoveries (1970–72) are *Petapahan, Sintong* and *Suram*. Another large producer is *Kotabatak*, southwest of *Minas*.

(c) South Sumatra Basin

The pre-Tertiary Tiga Pulu mountains separate the Central basin from the South Sumatra basin, in which another great thickness (15,000ft+) of Tertiary sediments is found. In this area, there are several sub-basins separated from each other by three Plio-Pleistocene uplifts.

All the accumulations are anticlinal structures with axes paralleling the Barisan Mountain range, which limits the basin to the southwest. Oil is produced from Lower or Upper Miocene sandstones, and there are surface seepages. The location of the accumulations in this area

---

10 E. Kissling, *Sci.Pet.*, 6, 1, 109, 1953.
11 R. P. Koesoemadinata, *Bull.AAPG*, Nov. 1969, 53(11), 2368.

illustrates the principle (p. 61) that in a sedimentary basin oil tends to occur in older beds near the stable "Shield" and in younger beds near the central, deeper part of the basin.

Only a few of these fields are now commercially important—notably *Talang Akar-Pendopo* (discovered in 1917), *Talang Djirak* (1930), *Tempino* (1931), *Kenali-Asam* (1931), and *Talang Djimar* (1937). The *Belimbing* field is a newer discovery, found in 1964. Some of the associated gas flared from the *Prabumulih Bapat* field is to be piped to Palembang for industrial use. Offshore, the *Cinta* field was the first discovery off the east coast of South Sumatra, followed by *Kitty, Zelda, Gita* and *Nora*. Several of these fields were being put into production during 1972; they are thought capable of a production rate of at least 100,000 brl/day.

Java

Drilling near seepages began in East Java as early as 1871, and commercial production was obtained for the first time near Surabaya in 1888. Extensive exploration of the east-west trending East Java Tertiary basin since then has resulted in the discovery of some 26 oilfields around Surabaya and Rembang (Tjepu)[12,13]. Of these, the Tjepu fields are the more important—with *Kawengan* (discovered in 1926), and the older *Ledok* (1896) easily the most prolific. In these fields, oil is trapped in the porous Upper Miocene Ngrajong sands which lie directly above a thick sequence of shales. The structures are anticlines, nearly all with pronounced asymmetry, with their steeper flanks often associated with thrust faults. In the *Lidah-Kruka* fields of the Wonokromo (Surabaya) area, the reservoir sands are younger—Upper Pliocene to Lower Pleistocene in age. Some oil has also been found at *Lepok* and *Kertegen* on Madura. The oilfields occur in a relatively small area in the northeast of Java, and exploration in the west, although continuing, has so far found few commercial accumulations. The largest of these is *Djatibarang*, 125 miles east of Jakarta, which is expected to produce 60,000 b/d by the end of 1973.

The basin is limited to the south by a line of volcanoes which coincides with a Quaternary geanticlinal axis paralleling the island's long axis.

The oil output of the onshore Java fields is now only about 0·6 MM brl/a, but important offshore discoveries have recently been made which will greatly expand this output in the near future. The first of these offshore fields to be found off the north coast of West Java was *Ardjuna,* and there are a number of other accumulations known in this area and off Madura Island. The fields developed in the Java Sea produced oil at the rate of 160,000 b/d in 1972, and their output will probably reach 270,000 b/d by the end of 1973.

[12] R. van Bemmelen, "Geology of Indonesia", 1949.
[13] J. Weeda, *AAPG*, "Habitat of Oil", 1958, 1359.

Borneo (Kalimantan)

The island of Borneo or Kalimantan has an area of 850,000 sq km. It lies inside and apart from the outer "island arc" system of Indonesia. Three Tertiary sedimentary basins surround a central core of Cretaceous and older metamorphic and igneous rocks. In late Pliocene and early Pleistocene times, complex folding movements produced a series of anticlines paralleling the coasts. Oil has been found in some of these structures, in reservoir sandstones overlying middle Tertiary shales, which were probably the source rocks[14]. The three basins are: on the northwest coast, Brunei-Sarawak (see p. 173); in the northeast, Tarakan; and in the southeast the so-called Borneo geosyncline, which is divided by the Meratus Mountains into the Koeti and Barito sub-basins[15]. All the basins contain a thick Tertiary sandstone and shale section.

In the Tarakan basin, oilfields have been found on the offshore islands of *Tarakan* and *Bunju* (*Boenjoe*)*. The latter is a gentle structure with several Pliocene sandstone reservoirs.

The *Pamoesian* field on Tarakan Island, which is itself a large anticlinal uplift, comprises several culminations, and has produced oil since 1905 from a number of shallow Pliocene sands and gravels.

Some 300 miles to the south, in the Koetei basin near Balikpapan, the oilfields of *Sanga-Sanga* (found in 1897) and *Sambodja* (1910) lie on the same long fold parallel to the coast. This anticline is thrust-faulted and asymmetric, with a steeper west flank; the producing horizons are various shallow sands of Upper Miocene age. These fields have together produced about 300 MM brl of oil in the course of their long history.

Offshore exploration in the same basin has now resulted in the discovery of two accumulations (*Attaka* and *Badak*) which are thought to be capable of producing at least 100,000 b/d, with reserves of some 315 MM brl. Total output of Indonesian Kalimantan is now about 8 MM brl/a.

Tandjung, discovered in 1938, is the only inland field, the only accumulation so far found in the Barito basin, and the only Eocene reservoir in Indonesia. Because of lack of outlets, oil was not produced here until 1961. The field is situated on a fold paralleling the Meratus Mountains, and is markedly asymmetrical. It also produces oil from Upper Miocene sands.

Ceram

On the northeast coast of this island, which is one of the most easterly of the Indonesian Archipelago, oil has been produced since 1897 in the *Boela* field, with its extensions of *Nif* and *Lemum*. Although it has never accounted for more than about 1 % of the total Indonesian output, the

---

[14] J. Weeda, *AAPG*, "Habitat of Oil", 1958, 1337.
[15] N. Haile, paper to *Geol. Soc. London,* 21 June 1967.
*N.B. Indonesian place names are alternatively spelled with "oe" or "u".

accumulation is of geologic interest in that oil is found in strata of Upper Triassic age, from which it has migrated along a gently-dipping unconformity into very shallow Pliocene sands. Part of this field extends offshore.

TABLE 53

INDONESIA: Major Oilfields

	Output 1971 MM brl	
Sumatra		
Minas	249·2	
South Sumatra fields	22·5	
Rantau	15·7	
Duri	13·6	
Lirik	10·2	
Cinta (offshore)	2·5	
Djambi	1·6	
		315·3
Java		
Ardjuna (offshore)	1·5	
Onshore fields	0·6	
		2·1
Kalimantan		
Tanjung	. 4·7	
Bunju	2·0	
Tarakan	0·5	
		7·2
Irian Barat	0·5	0·5
Bula (Maluku)	0·5	0·5
Total		325·6

West New Guinea (Irian Barat)

In the western part of New Guinea, there are two major lines of uplift, the northern one forming the Van Rees mountains and the southern one the Central Mountains. These ranges stretch from Australian Papua on the east to the Vogelkop peninsula on the extreme west of the island. A thick Tertiary sedimentary sequence occurs in three basins which lie respectively to the north, in between, and to the south, of the mountains, composed themselves of igneous and sedimentary strata of Palaeozoic to Tertiary ages.

The inaccessible nature of much of the interior has hampered oil exploration, but it is known that there are large areas of sedimentary rock affected by Alpine folding and therefore of potential interest for petroleum accumulation.

Four small oilfields have been found on the Vogelkop peninsula in the west of the country—*Klamono, Wasian, Mogoi* and *Kasim*.

Klamono, although discovered in 1936, did not produce oil until 1948. The structure is a small dome, with several cross-faults. The reservoir formation is a shallow Upper Miocene reef limestone, from which about 2 MM brl/a is currently produced.

The *Wasian* and *Mogoi* anticlines in the eastern Vogelkop are gentle structures with surface closures in Pliocene strata and oil contained in Upper Miocene limestones. During 1972, a Miocene reef accumulation was discovered at *Kasim* in the Salawati Basin.

JAPAN

The Japanese islands form a typical tectonic "island arc", with all the concomitant phenomena of crustal instability—earthquakes, vulcanism, etc. Pre-Tertiary formations are therefore unlikely to be of interest for hydrocarbon accumulations, which are in fact confined to the relatively narrow belts of Tertiary sediments.

Oil was first found in commercial quantities in 1892, and annual production usually was between 2 and 3 MM brl until the last few years, when an accelerated exploration effort has resulted in an increase of output, which reached a peak of 5·5 MM brl in 1971.

Most of this oil comes from a number of small fields in two sedimentary basins on the island of Honshu, in Niigata and Akita Provinces; some oil and much gas is also obtained from the Kanto Plain, east of Tokyo. There are also a few small accumulations in the northern island of Hokkaido. Most hydrocarbon production comes from anticlines in which faulting is the dominant trapping feature; there are also some stratigraphic traps along the northwestern margin of the Niigata basin[16]. Reservoirs are generally in Miocene and Pliocene sandstones, with some fractured shales and volcanic tuffs also providing locally permeable reservoirs[17].

Yabase (discovered in 1933 near the centre of the Akita basin), one of the larger oilfields, is typical of Japanese accumulations. Production comes from a number of Mio-Pliocene sandstone reservoirs in a narrow anticlinal fold.

In addition to oil, a total of nearly 72 Bcf of natural gas was produced in 1969, from several Mio-Pliocene sandstone and gravel accumulations; of these, the most important is *Niigata*, in northwestern Honshu.

The recoverable reserves of crude oil are about 26·9 MM brl, with probably 172 MM brl of natural gas liquids and about 700 Bcf of natural gas.

Domestic hydrocarbon production can satisfy only a very small proportion of the high and rapidly increasing consumption of petroleum required by the expanding Japanese economy. The demand for oil products grows by about 15% p.a.; it amounted to some 1500 MM brl in 1972, nearly all of which was imported from the Middle East and Indonesia (in which areas Japanese-controlled companies have important interests, particularly offshore).

[16] UNESCO *Min.Res.Dev.Series* 10, 1959.
[17] C. Pollock, *Sci.Pet.*, 6, 1, 129, 1953, and K. Magara, *Bull.AAPG*, 52(12) Dec. 1968, 2466.

o

Japan stands third as a world consumer of petroleum products, behind only the USA and USSR. It is estimated that petroleum provided 67·2% of all primary energy needs in 1970, a proportion which will rise to 72·8% in 1975[18].

PAKISTAN AND BANGLADESH

The two major sedimentary basins of **Pakistan** are the "foreland" Indus basin and the "hinterland" Baluchistan basin, separated by a geanticlinal axial belt. In the north of the Indus basin there are numerous oil seepages, along the foothills of the Margala and Kala Chitta ranges, the northern flank of the Salt Range, and at Khaur, 45 miles southwest of Rawalpindi, where there is a large area of oil-soaked sandstone. Most of the seepages are associated with gypsiferous marls overlying nummulitic shales and limestones, and are probably related to fault-planes. Oil shows also occur in a Tertiary basin in Baluchistan, and many large "mud volcanoes" are known in the southern part of this basin; but drilling in this area of salt diapirs has been unsuccessful[19].

Several small, anticlinal oilfields have been found on the Potwar plateau area of the Indus basin, between the Salt Range and the outer Himalayas[20]. Their total output was only 2·6 MM brl in 1971.

Although drilling for oil commenced in the Punjab as early as 1869, commercial production was not obtained until 1915, when the shallow oil-sands of the *Khaur* dome were discovered. By 1935, some 3 MM brl of oil had been recovered from this field, but thereafter production declined to a negligible level. The reservoir formation here was the Miocene Murree Sandstone series, and there has also been deeper production from an Eocene limestone.

Another productive anticline discovered in this area in 1935 is *Dhulian,* southwest of Khaur. The structure here is a broader fold, containing in basal beds of the Murree series, and also in deeper Eocene limestones. At the present time, *Dhulian* is the most productive Pakistan oilfield, with an annual output of some 1·8 MM brl.

A few others among the many anticlinal folds in this area have also been found to be oil-bearing. Thus, *Joya Mair* produces viscous oil from Eocene limestone in a narrow, steeply-folded anticline; and *Balkassar* yields better-quality oil from a similar limestone reservoir in a more open fold. The *Tut* field, discovered a few miles west of Dhulian in 1968, produces from a Jurassic sandstone, and the *Meyal* structure, also discovered in 1968, has found oil in Eocene beds, notably the

[18] K. Morimoto, 7 WPC Moscow, 1968, A1 113,11.
[19] S. S. Ahmed, *Bull.AAPG,* 53(7), 1480, 1969.
[20] E. Pinfold, *Sci.Pet.,* 6, 1, 1953, 125.

Sakesar limestone. Oil is also reported from the *Kot Sarang* structure, 70 miles southwest of Rawalpindi. Drilling in these fields is complicated by very high pressures, which may owe their origin to the hydraulic flow resulting from the outcropping of reservoir formations in the nearby Himalaya Mountains.

The oil in these accumulations is thought to have been formed in the Eocene limestones (probably Hill Limestones and Lower Chharat Stage) and to have later migrated upwards into the porous Murree sands, probably along fault-planes. The Upper Tertiary rocks are generally of fresh-water facies, while the Eocene beds below the unconformity are marine.

In the southern part of West Pakistan (Sulaiman and Kirthar provinces), there are a number of large gasfields. The most important of these is *Sui*, a gently folded dome on the regional Jacobabad "high" with an east-west axis 34 miles long. The Eocene limestone reservoir is estimated to hold reserves of about 6·3 Tcf. In 1970, output from Sui was 93·4 Bcf, with only about 30,000 brl of condensate, i.e. the gas is remarkably "dry". When the further pipelines currently being constructed have been completed, the offtake from Sui may reach about 110 Bcf/a or more.

Several other gasfields have been discovered in this area, associated with culminations on the same regional "high". They are at *Mazarani, Zin, Uch, Kandhkot, Kairpur* and *Mari*, with total estimated recoverable reserves of some 7 Tcf of gas.

It is not clear why such large volumes of gas have accumulated here, apparently with no associated oil. The reservoirs are porous limestones of Paleocene, Lower Eocene and, in some cases, Middle Eocene ages. The shales intercalated in these limestone sequences may have been the source rocks in which the gas was formed, although there are also older bituminous shales. On the other hand, the limestones, which are probably of biohermal origin, may themselves have been the source of the gas, while at the same time providing a suitable reservoir formation. Another possibility is that the gas was formed in the Cretaceous and Lower Eocene shales, and that no oil was formed at the time because of the relative shallowness of the seas and unfavourable conditions of sedimentation.

In **Bangladesh**, gas has been discovered in Tertiary folds on the periphery of the Bengal Basin at *Sylhet, Chhatak, Titas, Kailas Tila, Rasidpur* and *Habibganj*. The reservoirs are the Boka Bil or Bhuban Sandstones of Miocene age. Of these fields, only *Sylhet* and *Chhatak* are as yet connected by pipeline to industrial outlets, while *Titas* was brought into production in 1967 to supply the city of Dacca.

Exploration drilling for oil has been concentrated in the Jaldi area south of Chittagong, but so far without commercial success; this also applies to the offshore exploration in the Bay of Bengal.

PAPUA—NEW GUINEA

The Australian territory of Papua—New Guinea lies on the eastern side of the 141st parallel, which divides it from the Indonesian territory of West Irian. The continuation of the uplifts of the two mountain chains of West Irian forms a mountainous backbone which runs the length of the country. These Central Ranges are composed of igneous and meta-morphic rocks with considerable areas of unmetamorphized Jurassic, Cretaceous and Tertiary strata[21].

Oil seepages have been known for many years at Matapau on the northern side of the Central Ranges, where there is a narrow zone of folded and faulted Lower Miocene and Pliocene beds. Other seepages were also discovered in 1911 at Kariava, south of the mountains, near the mouth of the Vailala River. These, and other "shows" in this area, are in early Tertiary strata in a steeply folded sedimentary zone forming the southwestern foothills of the Central Ranges.

A number of exploratory wells have been drilled, the earliest being in 1913, without as yet finding commercial oil accumulations. They have generally been located on the complex anticlines of the southwestern Papuan Basin. However, several natural gas accumulations have been found in the Aure Trough, notably at *Kuru, Bwata* and *Puri*, where large faulted anticlines with Upper Miocene limestone reservoirs some-times contain a little oil as well as gas. Other accumulations of gas have been detected in deeper Mesozoic sandstones at *Iehi, Barikewa* and *Omati* in the western shelf area[22,23]. Total onshore gas reserves so far found have been estimated to be 500 Bcf. Offshore gas discoveries have also been made in the Gulf of Papua (*Uramu* and *Posca*).

TAIWAN (FORMOSA)

Several minor accumulations of oil and gas have been discovered in recent years in the Neogene sedimentary belt on the west side of the island of Taiwan[24].

Production of oil amounted to about 224,000 brl in 1971, coming mainly from three Plio-Miocene fields in the Miaoli area, where a locally important petrochemical industry has been developed. The oldest of these fields, *Chuhuangkeng*, discovered in 1904, currently produces only very small amounts of oil and gas, and the bulk of the current output comes from *Tiehchenshan* (1965), and *Chinshui* (1938).

Oil reserves are thought to be about 18·9 MM brl, and gas reserves are about 1·0 Tcf. Offshore exploration is currently under way.

21 K. W. Gray, *Sci.Pet.*, 6, 1, 139, 1953.
22 J. Casey & M. Konecki, 7 WPC Mexico, 1967, PD 2–19, p. 30.
23 C. Tallard, *Oil and Gas Int.*, July 1970, 50.
24 S. Chang, *Bull.AAPG*, 52(12), Dec. 1968, 2438.

OTHER AREAS

A small sedimentary basin has been found in the **Central Philippines**, bounded approximately by latitude 10°–12° N. and longitude 123°–125° E. The *Maya* (*Libertad*) anticline, roughly in the centre of the basin, produced some oil and gas after it was drilled in 1961[25]. The structure is about three miles long, with a northeast-southwest trend and a plunge to the southwest. The petroliferous reservoir was a very shallow (282ft) sandstone of Middle to Upper Miocene age.

There are oil prospects in **Afghanistan**[26] in the geosynclinal area north of the Himalaya and Hindukush Mountains, where there is a very thick sedimentary section. Only small amounts of oil have been found south of Keleft, but Russian-aided exploration has resulted in the discovery of several large gasfields in the north of the country at *Khwaja-Gogerdak, Sultan Kot, Khwaja-Borhan* and *Yatim Dagh. Khwaja-Gogerdak*, in the Oxus plains, is the largest of these, with reserves which have been estimated at 1·3 Tcf. A pipeline has been constructed to transport gas to the Soviet city of Tashkent. The total gas reserves of Afghanistan have been estimated to be 5 Tcf.

In **Portuguese Timor** some Triassic oil has been found at *Ossulari*.

In **Thailand**, the small *Boh Ton Kam* and *Mae Suhn Lang* oilfields have been developed in the intermontane Mae Fang basin, near the northwest frontier with Burma. This basin contains a thickness of mainly Tertiary sediments of probable Mio-Pliocene age in which a zone of surface seepages occurs. Production is about 40,000 brl/a. There are extensive deposits of oil-shale in the west of the country.

Considerable offshore exploration activity is currently under way in the Gulf of Thailand, in the South China Sea and off the coasts of Japan, in all of which areas thick sedimentary sequences occur.

---

[25] F. Mamaclay and R. Grey, 6 WPC Frankfurt, 1, 32, 1963.
[26] V. O. Ganss, *Erdöl u. Kohle,* July 1968, 7, 381.

CHAPTER 7

Africa and Australasia

AFRICA—NORTH OF THE SAHARA

MUCH of the 30 million sq km of the land area of Africa is made up of a "Shield" of igneous and metamorphic rocks. However, geochronological work has shown[1] that the continent was subjected to basement reactivation and orogenesis during the pre-Cambrian and early Palaeozoic, as a result of which a distinction is possible between three stable cratons (West Africa, Congo and Kalahari) and the intervening areas of mobility, tectonism and granitization. The same "Pan-African thermo-tectonic episode" may have produced a number of downwarpings round the edges of the continent which were subsequently filled with sediments as a result of Mesozoic and Tertiary marine transgressions.

The prospects for oil and gas accumulations in Africa are therefore largely confined to these peripheral sedimentary basins, most of which contain a fairly thick sedimentary sequence, which includes salt deposits. The age of the salt varies—thus, in Morocco it is Permo-Triassic, but in Gabon it is Cretaceous. In some cases, diapiric salt domes are known, and most of these coastal basins extend out to sea onto the Continental Shelf, being possibly related in their origin to the rifting of the Atlantic. They are therefore of interest as possible areas of oil accumulation, and exploration has been concentrated in them, both onshore and offshore. On the West Coast of Africa the Mesozoic basins are, from North to South, those of: Spanish Sahara, Senegal-Mauritania, Ivory Coast, Dahomey, Nigeria, Gabon, Congo and Angola. There are also some inland basins, such as those of Chad-Niger, to be explored for oil.

In the east, a Miocene evaporite basin fills the Red Sea, "graben" and is productive of oil in Egypt and under exploration in Sudan; while Mesozoic-Tertiary basins occur in Somalia, Abyssinia, Tanzania and Mozambique. None of these so have far been shown to contain commercial oil accumulations.

---

[1] W. Kennedy, "Salt Basins Around Africa", I.P. 1965, 9.

Although seepages had been observed for many years in a number of areas, and oil had been produced on a modest scale in Egypt since 1908, the prospects for the discovery of large-scale accumulations of oil in Africa were never considered high until exploration was eventually successful in the Algerian Sahara in 1956.

The Atlas Mountains of North Africa are formed by a folded geosynclinal belt of Mesozoic and Cenozoic sediments with small areas of igneous rocks and Palaeozoic strata[2]. To the south, there is a wide marine embayment area, divided into a number of important basins stretching across Algeria and Libya into Egypt. The thickness of the sediments in these basins is usually no greater than about 12,000ft and their age ranges from Tertiary to Palaeozoic. There is a major unconformity between the continental Carboniferous formations and overlapping Cretaceous strata.

A new chapter in African oil history commenced with the discovery in 1956 of the *Hassi Messaoud* oilfield in Algeria, and one of the largest gasfields in the world at *Hassi R'Mel*. Algerian oil output has since multiplied remarkably—24 *thousand* brl in 1950 to more than 370 MM brl in 1970, in which year the industry was largely nationalised.

Remarkable also has been the initiation of an export trade in liquefied natural gas, which was begun in 1963 between Algeria and Britain and France. This was the first project of this type ever to be carried out on a large-scale international basis anywhere in the world. Initial sales to Britain and France were at the rate of 17·6 Bcf/a of gas apiece, but in 1968 the export rate to France was increased to 70 Bcf/a, and this was again increased in 1972 to 124 Bcf/a with the completion of the regasification plant at Fos. In addition, projects are far advanced for the export of large volumes of LNG (from liquefaction plants at Arzew and Skikda) to the east coast of the USA, while further exports may be made to various countries of Europe in addition to Britain and France.

The discoveries in Algeria drew attention to the potentialities of the neighbouring sedimentary basins, and further spectacular success was soon achieved in Libya, where within a remarkably short time a country with no known oil resources had been transformed into one of the world's major oil producers and exporters, with a production which reached 1212 MM brl in 1970. (Governmental restrictions have since limited further growth in Libyan oil output.)

New discoveries in the Western Desert and offshore in the Gulf of Suez have given a fresh impetus to the long-established Egyptian oil industry, making up for the loss of the Sinai fields.

In West Africa, the renewed investigation of the Niger River delta

---

[2] G. Brognon & G. Verrier, *Bull.AAPG,* 1966, 1, 108.

area from 1955 onwards produced remarkably rapid results, large oil accumulations being discovered in Tertiary sandstone reservoirs, both onshore and offshore. In spite of the disruption caused by the Biafran war, Nigerian output reached 396 MM brl in 1970, and 556 MM brl in 1971.

Oil production from the African continent increased from less than 7 MM brl/a to more than 2200 MM brl/a between 1950 and 1970, with still larger increases only awaiting the provision of transportation facilities and governmental sanction. Three countries (Algeria, Libya and Nigeria) produced about 90% of African output in 1971, so that the output from the other countries of the continent is still relatively small. Apart from Egypt, only Gabon, Tunisia, Angola and Morocco have yet achieved commercial oil production, although some oil has also been found in the Congo (Brazzaville) and Senegal. However, intensive exploration in the peripheral sedimentary basins, both onshore and offshore, is likely to produce additional hydrocarbon discoveries before very long.

TABLE 54

AFRICA—OIL PRODUCTION 1920–1970
(thousands of brl)

Year	Algeria[1]	Angola[2]	Congo (Br.)[3]	Egypt	Gabon	Libya	Morocco	Nigeria	Tunisia	Total
1920	4	—	—	1,042	—	—	—	—	—	1,046
1930	16	—	—	1,996	—	—	—	—	—	2,012
1940	—	—	—	6,505	—	—	27	—	—	6,532
1950	24	—	—	16,373	—	—	305	—	—	16,702
1960	67,613	477	—	23,698	5,991	—	695	6,552	—	105,296
1966	255,123	4,387	530	43,298	10,970	548,924	730	151,340	4,814	1,020,116
1968	332,485	5,384	309	62,567	33,560	949,305	681	51,906	24,686	1,460,882
1970	373,090	35,770	146	119,229	39,729	1,212,048	323	395,841	32,226	2,208,402

[1] Includes condensate.
[2] with Cabinda.
[3] Congo (Brazzaville).

TABLE 55

AFRICA—OIL AND CONDENSATE OUTPUT 1971

1. *North of the Sahara*

	MM brl
Algeria	289·6
Egypt	107·3
Libya	996·4
Morocco	0·3
Tunisia	31·9
Total	1,425·5

2. *South of the Sahara*

	MM brl
Angola (and Cabinda)	41·3
Congo (Brazzaville)	0·3
Gabon	41·8
Nigeria	556·1
Total	639·5

Total Africa 2,065·0

TABLE 56

AFRICA—ESTIMATED OIL RESERVES

	B brl
Algeria	9·84
Angola-Cabinda	1·00
Congo (Br.)	0·00*
Egypt	1·00
Gabon	0·60
Libya	28·00
Morocco	0·00**
Nigeria	10·00
Tunisia	0·41
Total	50·85

*8·0 million
**7·8 million

MOROCCO

Most of the small and declining Moroccan crude oil output (totalling only about 311,000 brl in 1971) comes from the *Sidi Rhalem* field in the Essaouira basin, where the reservoir rock is made up of fractured dolomitic limestones of Argovian-Callovian (Jurassic) age. *Kechoula* and *Jeer* are gasfields in the same basin. Some oil also comes from small accumulations in the Meknes region, of which the most important is *Haricha*, where the reservoir is a Dogger (Jurassic) sandstone. A belt of similar but smaller fields runs southwest from Sidi-Kacem (*Bled Khatara, Bled-Eddoum, Sidi-Fili,* etc.). The older discoveries of the *Djebel Tselfat* and *Bou Draa* anticlines have Liassic reservoirs, and *Ain Hamra* has some oil in Miocene sandstone lenses.

ALGERIA

Algeria has an area of about 2,300,000 sq km, of which more than 90% comprises the Sahara. In the north, the coastal Atlas and Sahara Atlas mountain ranges, separated by a plateau, are composed of Tertiary, Cretaceous and Jurassic limestones and marls resting on "basement" rocks. The folds here are faulted, with steeper flanks to the south. Two small anticlinal oilfields have been found in this area, at *Oued Gueterini* (1949) and *Djebel Onk* (1960), with Eocene and Upper Cretaceous reservoirs respectively; but neither has produced more than a small fraction of the Algerian output. By 1964, *Djebel Onk* had ceased to operate, leaving *Oued Gueterini* as the only non-Saharan field, currently producing less than 20,000 brl/a.

The Sahara incorporates part of the African "Shield", in the form of a series of pre-Cambrian platforms of much-metamorphosed beds

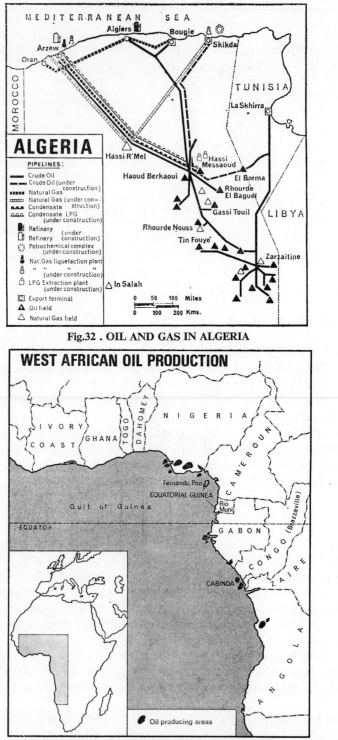

Fig.32 . OIL AND GAS IN ALGERIA

ALGERIA

PIPELINES:
— Crude Oil
- - Crude Oil (under construction)
▪▪▪ Natural Gas
▭▭▭ Natural Gas (under construction)
▬▬ Condensate
▬▬ Condensate LPG (under construction)
🏭 Refinery
🏭 Refinery (under construction)
⬡ Petrochemical complex (under construction)
🛢 Nat. Gas liquefaction plant
🛢 " " " " (under construction)
🛢 LPG Extraction plant (under construction)
▣ Export terminal
▲ Oil field
△ Natural Gas field

MEDITERRANEAN SEA
Algiers · Bougie · Skikda · Arzew · Oran
MOROCCO · TUNISIA · LA Skhirra · LIBYA
Hassi R'Mel · Hassi Messaoud · El Borma
Haoud Berkaoui · Rhourde El Baguel · Gassi Touil
Rhourde Nouss · Tin Fouye · Zarzaitine
In Salah

0   50   100 Miles
0   100   200 Kms.

WEST AFRICAN OIL PRODUCTION

IVORY COAST · GHANA · TOGO · DAHOMEY · NIGERIA · CAMEROUN
Fernando Poo · EQUATORIAL GUINEA · Rio Muni
Gulf of Guinea · EQUATOR · GABON · CONGO (Brazzaville) · ZAIRE
CABINDA · ANGOLA

● Oil producing areas

Fig. 33. OIL-PRODUCING AREAS OF WEST AFRICA
Source: Petroleum Press Service.

which stretch across North Africa, from the Reguibat massif of Algeria to the Arabo-Nubian "Shield" area of Egypt and the Sudan.

Between the massifs, the pre-Cambrian beds form the "basement" beneath a great variety of sediments of different ages, which are sometimes raised as regional arches or depressed as downwarps, depending on the "basement" configuration.

There are several important sedimentary basins: adjoining the frontier with Morocco are the Tindouf and Bechar basins; roughly in the centre of the country are the Reggane, Hassi Homeur and Ahnet basins; and on the east, contiguous with Tunisia and Libya, are the Triassic and Illizi (or Polignac) basins, forming together the so-called Saharan "shelf" area[3] from which most Algerian oil and gas production has so far come. In the latter region, a thick Palaeozoic section is unconformably overlain by Mesozoic strata; to the north it is separated from the Cretaceous-Jurassic Atlas basin by the Saharan flexure zone. The Hoggar Mountains in the southern Sahara are made up of folded and faulted Palaeozoic beds and mark the southern edge of the zone of sedimentary basins.

Only rare oil and gas shows had been recorded in Algeria, prior to the commencement of systematic exploration in 1947. A few traces of oil had been found in the Kenadsa coal basin, and of gas in water wells in the remote In Salah area. The absence of surface indications and the generally widespread cover of Quaternary and Tertiary strata necessitated the use of extensive seismic surveys before exploration drilling began in the Sahara in 1954. Gas was soon found in a test well at Ahnet, and the first major oil discovery was made in 1956, at *Edjele* in the Illizi Basin. In the same year, another oilfield was found in the same area at *Tiguentourine,* while a major oil accumulation was discovered at *Hassi Messaoud*, and one of the world's largest gasfields at *Hassi R Mel*.

Between 1960 and 1965, political and economic difficulties caused a falling-off in exploration activity in Algeria. However, after the signature of the Franco-Algerian agreement and the setting up of "Sonatrach", the state oil company, progress was once more made.

There are some 35 producing oilfields in Algeria, whose output in the peak year of 1970 totalled 373·1 MM brl of crude oil and condensate; several other fields awaited connection to a pipeline system. All the reservoirs are sandstones of Palaeozoic age—mainly Cambrian to Devonian, with some Triassic reservoirs also. The total remaining recoverable oil reserves are estimated to be about 9·8 B brl.

Table 57 shows the principal oilfields, the age of the main producing horizon and the approximate production of each in 1971.

Algeria is also an important producer and exporter of natural gas, with a current output of some 120 Bcf/a of non-associated gas, which is produced mainly from *Hassi R'Mel*.

---

[3] A. Megatelli *et al.,* "Exploration for Petroleum in Europe & N. Africa", IP, 1969, 271.

TABLE 57

ALGERIA—PRINCIPAL OILFIELDS, 1971

	Crude oil production 1971 1000 b/d	Age of principal reservoir
Hassi Messaoud S.	239·1	Cambrian
Hassi Messaoud N.	185·8	Cambrian
Rhourde Baguel	55·1	Cambrian
Gassi Touil	48·7	Triassic
Zarzaitine	41·8	Carboniferous, Devonian
Tin Fouyé N.	16·2	Devonian
El Gassi-El Agreb	35·1	Cambrian
Edjele	18·7	Carboniferous, Devonian
Haoud Ber Kaoui	28·2	Triassic
Tin Fouyé Tabankort	27·5	Devonian, Ordovician
Hassi R'Mel	14·7*	Triassic
Tiguentourine	4·1	Carboniferous, Devonian
Ohanet S.	9·1	Ordovician
El Borma	21·9	Triassic
Other fields	47·5	
Total	793·5	

*Condensate

THE OILFIELDS

1. Illizi (Polignac) Basin

The first Saharan commercial oilfield found was at *Edjele*, close to the Libyan frontier. This is an asymmetrical faulted surface anticline, with oil in Carboniferous (Visean), Devonian, and Cambro-Ordovician reservoirs. The structure is divided by transverse faults with northern and southern producing sectors. About 40 miles west of Edjele is *Tiguentourine*, a field in which vertical faults cut a large, gentle fold. *Zarzaitine*[4], a faulted homoclinal accumulation, also produces from reservoir formations of these ages. Another group of oilfields was discovered on the northwest edge of this basin in 1960/61, of which the most productive is *Ohanet*, an anticline with a faulted west flank and several Devonian reservoirs.

2. Central (Triassic) Basin

Hassi Messaoud is a major oilfield which structurally comprises a gentle dome with a north-northeast—south-southwest axis and two culminations separated by a fault zone. The reservoir is composed of a number of Cambrian-Ordovician sandstones at depths of 10–11,000ft, overlain unconformably by Triassic sandstones and evaporites.

---

[4] J. Groult *et al., J.Pet.Tech.,* July 1968, 18, 7, 883. For a full description of the geology and geography of the Algerian oilfields (up to 1963) see C. Verlaque, *Mem.Sect.Geogr.* Paris, 1, 1964, 11. (Le Sahara Pétrolier).

The two sectors of this field together produce nearly half the Algerian oil output; production from *Hassi Messaoud* totalled 0·5 MM b/d in 1970. Recoverable reserves here are estimated to be some 2·0 B brl. The accumulation is due to the existence of a Palaeozoic uplift stretching northwards from the Hoggar (the "Amguid trend"), the top of which was truncated by pre-Triassic erosion. There is a widespread development of the Cambrian reservoir sandstone, and the anhydritic Trias acts as an excellent sealing medium.

El Gassi-El Agreb is about 35 miles south-southwest of *Hassi Messaoud*, with a similar structure on the same "high". This is also a double structure with two culminations separated by faults. Pre-Triassic erosion is more marked than at *Hassi Messaoud* and the Ordovician and part of the Cambrian are absent. The northeastern sector is the most productive part of the field.

El Adeb Larache is another Palaeozoic field in this basin, with a structure sharply divided by faulting into two sectors whose axes are nearly at right angles to each other.

The Amguid-El Biod plateau links the fields of the Illizi and Central Basins. *Rhourde Baguel* produces oil and gas from Cambrian sandstones; while *Rhourde Adra* and *Rhourde Nouss* are Triassic and Ordovician gas producers.

One of the largest gas and condensate accumulations in the world is at *Hassi R'Mel*. The first well drilled here (in 1956) found abundant high-pressure natural gas at a depth of about 7,000ft in a series of very porous and permeable Triassic sandstones, underlying a salt series. The structure is an elongated ellipsoidal anticline, some 80 km long by 40 km wide, with very gentle dips. The probable reserves of recoverable gas are about 61 Tcf, with about 1·0 B brl of condensate. About 105 Bcf/a of marketable natural gas is obtained from this field after piping to the Mediterranean coast; there it is converted into LNG for export to Europe.

Exploration drilling carried out in the other main basins of Algeria has so far proved unsuccessful, but non-commercial shows of oil and gas have been reported from tests in the Tindouf (Palaeozoic), Bechar, Reggane, Hassi Homeur and Ohanet basins, in each of which there is a considerable thickness of sediments.

The origin of the oil and gas in Algeria is uncertain. While large volumes have been found in Cambrian strata, these are unfossiliferous, continental deposits, and are very unlikely source rocks. It is true that the hydrocarbons could have migrated *downwards* into the present reservoirs from overlying Silurian or Devonian shales. However, the Cambrian sandstones themselves were subject to prolonged uplift and erosion in Upper Palaeozoic and Lower Mesozoic times, so that any oil held in them would presumably have then been dissipated. There is evidence of uplift during deposition of the reservoir beds, so an alternative explanation is that they were charged with hydrocarbons derived from the unconformably overlapping Triassic shales[5].

In the Edjele area, where Carboniferous shale is overlapping, this may have been the local source rock.

TUNISIA

Seepages have been known for many years in a zone of isoclinal folding in Tunisia, at *Ain Rhelal* and *Sloughia* in Helvetian (Miocene) strata, and at *Kef bou Debbous* in the Cretaceous. Natural gas has been produced from a small field (*Djebel Abderrahmane*) at Cap Bon. (A well drilled in this field in 1965 reached the record depth for Africa of 18,801ft before being completed as a Triassic gas producer.) Other hydrocarbon discoveries have been made in recent years in the north at *Ali ben Khalifa*, west of Sfax (gas), and at *Douleb-Semmama* (Cretaceous oil).

The first commercial oilfield was found in 1964 at *El Borma*, in southwestern Tunisia, close to the frontiers with Libya and Algeria. The reservoir here is a Triassic sandstone, and recoverable reserves have been estimated as 300 MM brl of oil. Late in 1968, another oil discovery was made at *Kasserine*. The output from the Tunisian fields was 31·9 MM brl in 1971.

Offshore exploration is being carried out in the Gulf of Gabes, and was successful in finding two accumulations in 1971—the coastal field of *Sidi el Atayem* and the larger offshore *Ashtart* accumulation, 80 km southeast of Sfax. Both these fields are expected to enter production in 1973. The onshore oil reserves of Tunisia are estimated at 410 MM brl.

LIBYA

Libya has an area of about 1·76 MM sq km, of which 62% (1·10 MM sq km) is occupied by sedimentary basins. Traces of oil and gas had been found in occasional water wells for many years, although there have been no reports of surface "shows".

Systematic exploration for oil began for the first time in 1954, and within three years commercial accumulations were discovered. The first exports of Libyan oil were made before the end of 1961, and development thereafter was remarkably rapid, so that by 1966 output amounted to 549 MM brl, surpassing that of Iraq and making Libya the world's seventh ranking oil producer within a period of only 5½ years.

The proximity of Libyan ports to European markets, and the lightness

---

5 F. E. Wellings, *Inst.Pet.Rev.*, July 1960, 213.

and low sulphur content of most Libyan crudes are factors which have encouraged an extraordinarily rapid expansion of oil output, unequalled anywhere in the world, particularly when it is remembered that in the initial stages many months had to be devoted to the dangerous work of clearing the thousands of buried mines left over from the 1939–45 war. Furthermore, since most of the oilfields lie 80–300 miles inland, pipelines have had to be built to carry the crude to five Mediterranean terminals. By mid-1969, six pipelines totalling 1,500 miles in length had been built.

In 1970, production reached 1212 MM brl, and it seemed probable that Libyan oil output would soon exceed that of Venezuela; but partial nationalisation and economic and political restrictions resulted in a reduction of activity, so that production in 1971 actually decreased to 996 MM brl.

The surface of Libya consists of a belt of "sabkha" (evaporite crust) along the coast, with an interior stony desert of flat Miocene limestones interspersed with sand dunes. The major tectonic elements are two large Palaeozoic-Mesozoic basins in the west and east of the country, which are separated from each other by a major uplift, and separated internally from a large subsidiary basin.

In the west, the Murzuk basin is about 600 km long and 1,000 km wide, and contains predominantly marine Palaeozoic and continental Mesozoic sediments. It is bounded on the west by the Hoggar Mountains and on the east by the Tertiary basalts of the Djebel Soda-Djebel Haroudj and the massif of the Tibesti Mountains. The latter cover an area of more than 100,000 sq km and are the highest mountains of the Sahara, with a central volcanic mass penetrating Cambrian and Ordovician sandstones and Precambrian basement rocks.

The Gargaf uplift, trending west-southwest, separates the main Murzuk basin from the Homra platform. Cambrian sandstone is exposed on its northern flank, and there are outcrops of Precambrian "basement" rocks in the centre of the uplift, which is cut by many faults, mainly trending east-northeast. On its southern edge, Devonian and Carboniferous rocks are exposed.

The Homra platform itself is a stable shelf area which has been formed over a westerly-trending Palaeozoic basin.

The uplift separating the Murzuk basin from the Kufra basin to the east forms the plain of the Serir Tibesti, sloping gently northeastwards and covering an area of some 50,000 sq km.

The Kufra basin in the east of Libya is about 500 km long and 900 km wide. It has been formed by structural elements of varying ages and trends, and contains Palaeozoic strata which are partly marine, and Mesozoic strata which are entirely continental. In the north, block-faulting along older northwesterly trend lines during the Cretaceous has resulted in the formation of the Sirte basin or embayment, which is

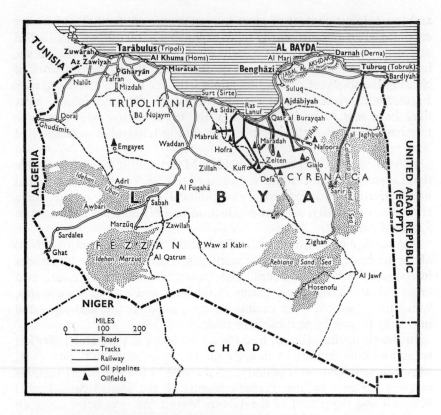

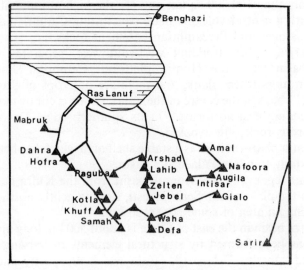

**Fig. 34. MAJOR OILFIELDS AND PIPELINE SYSTEMS OF LIBYA
(with details of Sirte Basin.)**

limited on the north by the stable shelf of the Cyrenaica platform and on the west by the north-northwest-trending Hon "graben", Cretaceous to Tertiary in age[6]. The Calanscio-Auenat uplift is an important subsidiary trend, Hercynian to Mesozoic in the north and Caledonian in the south, which runs north-northwest from the east of the Sirte basin towards Djebel Auenat.

Most of the important oilfields have been found in the Sirte basin, where the reservoirs are usually of Paleocene, Eocene or Cretaceous age. Smaller accumulations have also been found in Triassic and Palaeozoic strata in the west of the country, near the borders of Algeria and Tunisia.

One of the earliest and also the largest oilfields to be discovered in the Sirte basin was *Zelten*, some 200 miles south of Benghazi, which was found as a result of the remarkably accurate seismic profiling of a broad gentle surface Miocene anticline, which is repeated in the subsurface. The discovery well was drilled in 1959, and the field produced nearly 193 MM brl of oil in 1970 from Lower Eocene limestones. The Zelten structure was influenced by an underlying pre-Cretaceous fold during Cretaceous and early Tertiary times[7]. It is bounded on the west by a normal fault system, with downthrow to the west, and periodic rejuvenation of the structure has been the major cause of accumulation. The field has three culminations separated by intervening "lows".

Prolific production is obtained from a shelly limestone with pockets of reefal material which occurs in a carbonate sequence; a dense, compact limestone and overlying shale unit provide an efficient cap rock for this reservoir, which has an average thickness of 350ft and is at about 5,400–6,000ft below the surface.

Zelten lies centrally in a belt of oilfields from *Arshad* in the north to *Waha* and *Defa* in the south. To the west are *Dahra, Hofra, Raguba* and *Samah*; while to the east lie *Amal, Nafoora, Augila* and *Gialo* (Table 58).

In the Calanscio "sand sea" in southern Cyrenaica, near the southeastern margin of the Sirte Basin, a new productive area was opened-up by the discovery of the *Sarir* field in 1961, followed by a second similar accumulation at *Sarir North*, and a smaller field about 12 miles to the northwest at *L–65*.

The main *Sarir* accumulation is essentially a combination trap— partly structural with closure to the north dependent on faulting, and partly stratigraphic, due to a local unconformity with the reservoir capped by unconformable shales. *Sarir North* and *L–65* are separate traps with individual oil/water contacts. The reservoir formation is a series of thick, Upper Cretaceous sandstones. The oil produced is a very

---

[6] L. Conant and A. Goudarzi, *Bull.AAPG*, 51, 5, May 1967, 719.

[7] W. Fraser, 7 WPC, Mexico, 1967, PD 2–18.

P

waxy, high pour-point crude (with a low gas/oil ratio) which requires heating to enable it to be pumped along the pipeline to the coast 320 miles away at Hariga.

During 1967/68, an important group of oilfields was discovered west of *Nafoora*—the *Idris* (renamed *Intisar*) fields. The *Intisar A* bioherm is oval in plan and measures about 2½ by 3 miles at its greatest dimensions. The Paleocene reef limestone reservoir is about 1,200ft thick here and contained over 2·0 B brl of oil-in-place. A test well in *Intisar D* produced the record Libyan output from a single well of 74,867 b/d from a highly permeable Paleocene reef limestone. The total producible reserves of four of these reefal fields have been estimated as over 3·0 B brl, assuming only a 20% recovery factor. They are part, in fact, of a very prolific zone of "pinnacle" reefs or isolated bioherms developed on a broad flat anticline[8]. These bioherms began as foraminiferal banks built up on local mounds on a shallow sea floor, and were added to by algae and coral growths.

Considerable volumes of associated gas are produced with the oil in Libya (about 700 Bcf in 1970) and largely flared. Schemes for the liquefaction and export of part of this gas have been developed, and one of the largest liquefaction plants in the world has been built at Mersa el Brega. Exports of LNG from here to Spain and Italy were due to commence in 1972.

TABLE 58

LIBYA—PRINCIPAL OILFIELDS, 1971

	Production MM brl	Age of reservoir
Intisar A, C and *D* (formerly *Idris*)	182·1	Paleocene
Zelten	114·8	Eocene
Sarir with Sari N. and *L-65*	148·1	Upper Cretaceous
Gialo	124·5	Oligocene, Eocene, Paleocene
Nafoora	86·2	Eocene, Paleocene, Cretaceous, Cambrian
Amal	54·9	Cambrian, Ordovician
Defa	58·5	Paleocene
Waha	45·5	Eocene, Cretaceous
Raguba	35·9	Cretaceous, Cambrian, Ordovician
Augila	25·2	Cretaceous, Cambrian, Ordovician
Samah	18·7	Cretaceous, Cambrian, Ordovician
Dahra	13·0	Paleocene
Other fields	89·0	
Total	996·4	

[8] C. Terry & J. Williams, "Exploration for Petroleum in Europe & N. Africa", IP, 31, 1969.

TABLE 59

LIBYA—OIL EXPORTS, 1970

Destination	1000 b/d
Italy	806·3
W. Germany	686·2
Britain	493·2
France	421·4
Holland	302·4
Spain	161·8
Belgium	128·5
USA	94·0
Switzerland	49·0
Trinidad	48·1
Bahamas	32·8
Denmark	23·9
Others	64·7
Total	3,312·3

EGYPT

A considerable thickness of Palaeozoic sandstones (mainly the so-called Devonian-Carboniferous "Nubian" series[9]) was laid down on a pre-existing peneplained surface of folded metamorphic and igneous rocks on the western edge of the Arabian "Shield". From early Mesozoic to Cenomanian times, only the northern part of Egypt and Sinai was covered by the sea, and there are therefore marked variations of facies in the Cretaceous deposits. Tertiary sedimentation was more uniform, and the Eocene and Miocene beds contain both suitable source and reservoir rocks and an excellent plastic seal in the Miocene anhydrite shales.

The tectonic history of the eastern part of Egypt is complex. It seems likely that epeirogenic movements probably associated with the slow bulging of the Arabian "Shield" have produced tension faults, which resulted in the depressions of the Red Sea and Gulf of Suez. As a result of step-faulting and the subsidence of the "basement", the overlying Mesozoic and Tertiary beds have been split up into fault-blocks with much ancillary folding and distortions in the more plastic strata[10].

Although hydrocarbon "shows" are rare, a number of important oilfields have been discovered on both sides of the Gulf of Suez.

At *Gemsah*, on the west, oil was found as long ago as 1908. The total output from this field only amounted to some 1·3 MM brl up to 1927, when production virtually ceased. The structure is a steep anticline lying over a granite ridge, and the reservoir bed is a dolomitic reef limestone of Miocene age.

---

[9] R. Pomeyrol, *Bull.AAPG*, 52(4), April 1968, 587.
[10] M. Youssef, *Bull.AAPG*, 52(4), April 1968, 601.

At *Abu Durba*, north of Gemsah, minor production has been obtained from two inclined fault-blocks. The field is of geologic interest as an example of an "open" structure in which the reservoir horizon outcrops at the surface in the form of a homocline of oil-saturated Nubian sandstones, but inspissation of the paraffinous crude has formed a natural seal which has prevented total dispersal of the oil.

Hurghada was the first major oilfield to be discovered in Egypt, in 1913, and has produced some 48 MM brl during its life, which is nearly at an end. Two asymmetrical fault-blocks form local uplifts, the oil being produced from Miocene lagoonal dolomites and the Nubian (mainly Carboniferous) sandstone.

Ras Gharib was found in 1938, after a gravity survey which showed that previous wells drilled here in 1921–3 on surface evidence had been incorrectly sited. The structure here is an eastward-dipping homocline of faulted Cretaceous and Carboniferous rocks, which is unconformably overlain by a Miocene fold. Production of about 2·5 MM brl/a is obtained from Miocene limestones, Cretaceous limestones and sandstones, and Carboniferous sands, trapping being due to the unconformity and the faults.

The *Sudr* field was discovered in Sinai in 1946, about 30 miles southeast of Suez. Production here comes from a Miocene sandstone and the fissured and cavernous Eocene limestones. The field consists of two upthrown blocks, unconformably overlain by the Miocene anhydrite shales.

On the west side of the Gulf of Suez, several oilfields have been discovered since 1958—*Bakr, Kareem, Umm al Yusr*, and the offshore *Amer* accumulation; these have reservoirs in Miocene, Eocene and Cretaceous sandstones and limestones. *Bakr* is by far the largest of these fields, with an output of 4·2 MM brl in 1970.

Offshore, in the Gulf of Suez, a number of important accumulations have been found. The largest is *Morgan Marine*, which is a major anticlinal accumulation with a Miocene sandstone reservoir. This field only started to produce oil in 1967, but its output was 98·5 MM brl of oil in 1969, or 82% of the total Egyptian output.

In the sedimentary area west of the Nile, about 20 miles south of *El Alamein*, a new oilfield has been discovered, from which 2·4 MM brl was obtained in the last four months of 1968, and 13·3 MM brl during 1969. The reservoir here is a Lower Cretaceous dolomite[11]. A gas/condensate discovery has been reported from a structure lying 70 miles southwest of Alamein, in the Qattara Depression at *Abu Gharadiq*. Gas has also been found at *Abu Madi* and *Al Wastani* in the Nile Delta, and offshore at *Abu Qir*. Perhaps potentially more important than any of these, was the discovery in 1968 of oil by a test well drilled at *Imbarka*, 200 km west of any previous hydrocarbon discovery (Alamein) and only 400 km east of

---

11 H. el Banbi, *World Oil*, Jan. 1970, 67.

the nearest Libyan accumulation. Here, too, the reservoir is a Lower Cretaceous sandstone. Other new oil discoveries southwest of *Alamein* are *Yidma* and *Al-Razzaq.*

The prospects for the Egyptian oil industry now look considerably brighter than they did a very few years ago.

The source rock for the Gulf of Suez oil accumulations is considered[12] to have been either the Miocene *Globigerina* marls or the Eocene chalky limestones, both of which formations are capped by anhydritic shales, so that lateral migration took place into whatever porous reservoir strata were structurally contiguous.

TABLE 60

EGYPT—PRINCIPAL OILFIELDS AND PRODUCTION, 1968–1971

MM brl

	1968	1971
Ras Gharib	3·33	2·55
Bakr	3·67	3·22
Kareem	0·67	0·64
Morgan	51·64	89·63
Alamein	2·41	8·64
Others	0·85	3·19
Total	62·57	107·27

It is interesting to see how, thanks to the rapid increases of output from *Morgan* and *Alamein*, the overall Egyptian output has increased in spite of the occupation of the Sinai oilfields, which produced about 13 MM brl in 1967. *Belayim* (both onshore and offshore) was the largest producer, with Middle Miocene sandstone and Cretaceous limestone reservoirs.

Other fields along this coast are, to the north, *Asl* and *Matarma*, near Sudr. These fields are essentially westward-dipping upthrown blocks, broken and bounded by faults. Further south, *Rudeis, Ekma, Sidri* and *Feiran* are other accumulations of the same type. Most of the reservoirs are Miocene/Eocene in age.

AFRICA—SOUTH OF THE SAHARA

ANGOLA

Oil was discovered in Angola in 1955 at *Benfica*[13] in the Cuanza basin. The reservoir here is a Lower Cretaceous dolomitic limestone. Several other small fields were subsequently found at *Tobias* in Quicama (1961), at *Catete,* southwest of Luanda, in 1966, and at *Quenguele Norte* (1968), where there are Miocene and Eocene reservoirs.

[12] P. Van de Ploeg, *Sci. Petroleum*, 6, 1, 1952, 151.
[13] G. Brognon & G. Verrier, *Bull.AAPG*, 1966, 1, 108.

Offshore oil was found in 1968 on the continental shelf of the coastal enclave of Cabinda, which lies between the north and south forks of the Congo river. The accumulation had several sandstone reservoirs of Tertiary and Cretaceous age.

During 1969, the first full year of production operations, the output from Cabinda district was 12·2 MM brl, making an overall total for Angola and Cabinda of 17·5 MM brl. By 1971, the combined output had risen to 41·3 MM brl. Offshore exploration is being carried out near the mouth of the Congo, and a new onshore accumulation has recently been discovered at *Cabeca da Cobra*.

GABON

The *Ozouri* oilfield at the base of the Port Gentil peninsula in Gabon was discovered in 1956, followed by a cluster of other onshore fields— *Lopez, Port Gentil, Tchengué* and *M'Bega*, and offshore fields, *Anguille*[14] *Torpille, Tchengué-Ocean* and *Bonite*. These accumulations all occur in a sedimentary basin containing diapiric Lower Cretaceous evaporite structures, and the reservoirs are mainly Senonian (Pointe Clairette) sandstones.

Another onshore field, *Gamba*, also on the coast but southeast of the other fields, came into production early in 1967 at the rate of about 14 MM brl/a.

Oil production rose by 35% in 1968 compared with the previous year, and in 1971 reached the peak of 41·8 MM brl. It is expected that output may increase to the rate of 140,000 b/d (about 51 MM brl/a) by the end of 1973, as the important *Grondin* offshore oilfield comes into production.

NIGERIA

Exploration for oil in Nigeria began as long ago as 1908, although surveys using modern techniques were not carried out until 1937, and there was a cessation of operations during the 1939–45 war. The initial reservoir target was the Cretaceous, since there are extensive heavy oil seepages north and east of the main Niger delta in rocks of this age. The seepages east of Lagos in fact form one of the largest known oil "shows", the oil escaping updip along beds dipping out to sea and reaching the surface at the contact between Cretaceous sandstones and metamorphic "basement" rocks.

The first commercial accumulation of light oil was discovered at *Oloibiri* in 1955. The age of the reservoir here was Tertiary, and this at once directed attention to the reservoir rock potentialities of the

[14] Y. Belmonte, *Rév. Franc. de l'Energie*, 168, March 1965.

Tertiary sandstones in the deltas of the Niger and Benue rivers. The coastal sedimentary basin in this area has been the scene of three distinct depositional cycles between mid-Cretaceous and Recent times, the third of which, starting in the Eocene, marked the continuing growth of the main Niger delta[15].

The output from *Oloibiri* was in fact somewhat disappointing, but its discovery was soon followed by a number of others in the Port Harcourt and Afam areas, of which *Imo River, Bomu, Umuechem* and *Koro Koro* proved to be the most prolific. Other onshore discoveries were made in the Mid-West Region, e.g. at *Olomoro* and *Uzera*. A number of important offshore fields were also discovered, e.g. *Okan* and *Obagi*. In this part of West Africa, the slope of the Continental Shelf is very gentle, so that a large area of shallow water is accessible for exploration.

Production from most of these Nigerian fields comes from very porous Tertiary sandstones of the Agbada series, which sometimes contain oil and associated gas, and sometimes only non-associated gas. The Agbada beds in which hydrocarbons have been found range in age from Eocene to Pliocene.

Most of the traps are associated with growth faults and their fronting "rollover" structures[16]. Deeper offshore drilling into the Cretaceous may in the future be undertaken to investigate the origin of the seepages mentioned above, but there are indications that high pressures may be encountered.

During 1966, the last full year before the outbreak of the Biafran war, 84% of the Nigerian oil output of 152 MM brl came from 13 onshore fields, and the balance from two offshore fields. A number of other accumulations had been discovered, both onshore and offshore, but had not been connected to points of outlet.

As a result of hostilities, output dropped to only about 52 MM brl by 1968, but increased rapidly again after the end of the war. By 1971, it had attained the peak of 556 MM brl, the oil coming from about 42 onshore and 12 offshore accumulations in roughly equal proportions, and being mainly exported through Bonny. In 1972, production was 652 MM brl.

The newer onshore fields include *Jones Creek*, northwest of Warri, and *Forcados* in the Niger estuary; and offshore *Delta, Delta South* and *Delta Meren*. The very rapid growth of Nigerian output in the last few years has largely been due to the development of fields such as these, which had been discovered earlier but had been awaiting pipeline and terminal facilities. Thus, with the taking into service of the new Qua Iboe export terminal in 1971, six new offshore fields could be put into production. A further terminal at Brass was due to be completed early in 1973. Overall reserves in Nigeria are thought to be at least 10 B brl. The ac-

---

15 K. Short & A. Stauble, *Bull.AAFG*, 51(5), May 1967, 761.
16 E. Frankel & E. Cordry, 7 WPC Mexico, 1968, PD2 (12).

cumulations are numerous but relatively small, with several sandstone reservoirs in each. Development therefore is more complex than is the case with the large oil accumulations in areas such as the Middle East; however, there are now some 60-70 known accumulations awaiting production facilities, so it is probable that Nigerian output will exceed 2 MM b/d in 1973, i.e. a rate of about 750 MM brl/a.

Nigerian crude oils have a low sulphur and a high but variable paraffin content. It has been suggested[17] that some at least of the source material was of terrestrial vegetable origin. Their low sulphur content makes these crudes particularly sought-after in pollution-conscious countries. Furthermore, their closeness to seaboard and relative proximity to the markets of Britain, Europe, Africa and South America, are additional reasons to explain the speed with which the oilfields of Nigeria have been developed.

SOUTH AFRICA

Efforts have been made for many years without success to find commercial petroleum accumulations in South Africa, and a number of unsuccessful exploration wells have been drilled in different localities, some in the neighbourhood of local seepages. The State-owned Southern Oil Exploration Corporation (SOEKOR) was set up in 1964 to organise and develop the search for petroleum. Stimulated by the offshore discoveries made in the neighbouring territories of Angola and Mozambique, a number of concessions were granted covering the sediments (up to 15,000ft thick) on the Continental Shelf all round the coast. The first success was the discovery of "wet" gas in a well drilled early in 1969 at Pleftenburg Bay, about 150 miles west of Port Elizabeth. This well showed on test a production potential of about 25 MM cf/day of gas with some 100 b/d of condensate[18].

OTHER AREAS

On the west coast of Africa, between the fields of Gabon and Angola, the *Pointe Indienne* oilfield of **Congo (Brazzaville)** began production in 1960, but its output was small and declined to only 146,000 brl in 1970. In that year, offshore drilling discovered the *Emeraude* accumulation with an Upper Cretaceous reservoir which was expected to produce oil at the rate of 10,000 b/d in 1972, rising perhaps to twice this output in 1973–4.

The *Diam-Biada* field in **Senegal** produced some 33,400 brl of oil between 1960 and 1962 and was then closed down.

---

[17] R. K. Dickie, *J.Inst.Pet.*, Feb. 1966, 38.
[18] W. Harris, *IP Review*, Nov. 1969, 304.

In **Mozambique**, there are oil seepages in the Inhambane area, and Cretaceous gas discoveries have been reported from two onshore test wells near the coast at *Pande* and *Temane*.

Considerable offshore exploration activity is under way on the Continental Shelf round much of the West African coast—off Spanish Sahara, Mauritania, Senegal, Portugese Guinea, Equatorial Guinea, Ghana, Dahomey and Togoland, Ivory Coast, Cameroon and South-West Africa. (The first success reported so far was an exploration well drilled off Saltpond in **Ghana**, which found some oil in Cretaceous and Devonian beds.) In the interior, the award of concessions and the start of exploration activity have been announced in Chad and the Central African Republic in the west, and Tanzania, Somali, Kenya and Ethiopia in the east.

In **Madagascar,** there are widespread occurrences of bituminous sandstones around Morafenobé, on the eastern side of a sedimentary basin on the west of the island. These lie stratigraphically in the Isalo series of Middle Lias age, and small quantities of heavy oil have been obtained from shallow wells drilled into this formation.

AUSTRALASIA

AUSTRALIA

The western and central parts of Australia are largely made up of an ancient Pre-Cambrian "Shield", with which are associated a number of intracratonic and peripheral sedimentary basins[19]. The presence of the rigid "Shield" has protected the sediments from large-scale diastrophism, and in general the continent has been by-passed by the Tertiary orogenies which produced the fold structures of southeast Asia.

In eastern Australia, diastrophism has been more marked and there has also been extensive igneous activity. However, it was the Caledonian and Hercynian (rather than the Alpine) orogenies that affected this region.

The total area of sediments in Australia is about 1·7 million sq miles; they range from Pre-Cambrian to Tertiary in age. No major oil or gas seepages are known, but methane is frequently found in shallow water wells.

The search for petroleum was carried on sporadically for many years, beginning as long ago as 1892, when a well was drilled in the Coorong area of South Australia. Natural gas was found near *Roma* in Queensland in 1900, and a little oil at *Lakes Entrance* in Victoria in 1924; there was no further success until 1953, when non-commercial oil was discovered

---

[19] J. C. Cameron, 7 WPC, Mexico, 1967, PD 2(7).

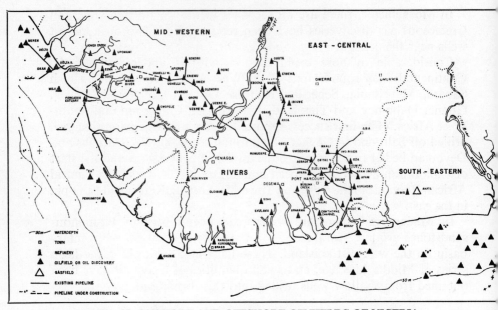

Fig. 35. ONSHORE AND OFFSHORE OILFIELDS OF NIGERIA

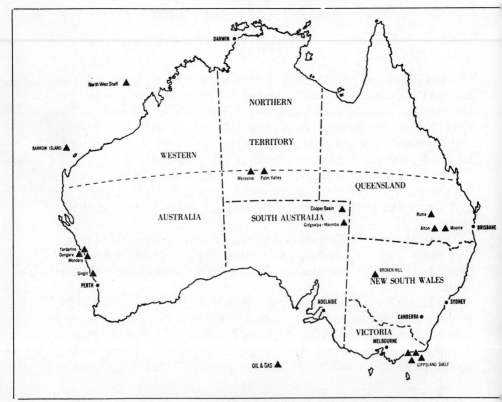

Fig. 36. OIL AND GAS ACCUMULATIONS IN AUSTRALIA

Source: Petroleum Press Service.

at Rough Range in Western Australia. Small quantities of indigenous hydrocarbons were commercially used for the first time in 1961 when gas from Roma provided fuel for local electricity generation. A small oil accumulation was discovered at *Moonie* in Queensland in 1961, which was put into production in 1964, together with a similar small accumulation at *Alton*, also in Queensland. However, in 1969, Australian oil output rose sharply to 15·7 MM brl, as a result of the discovery and putting into production of the *Barrow Island* field off the West Coast and the first of the Bass Strait offshore fields. Production increased to nearly 113 MM brl in 1971, and by the end of that year Australia was producing 55% of its domestic petroleum requirements. An extensive pipeline building programme also brought into use some of the huge reserves of natural gas discovered in recent years in several parts of the country, which are thought to total 13·9 Tcf. It is estimated that by 1980, the consumption of oil products in Australia will have increased to 335 MM brl, compared with the 1970 consumption of 187·4 MM brl.

West Australia

Two elongated peripheral sedimentary basins lie along the west coast of Australia—the Perth Basin to the south and the Carnarvon Basin to the north. In the latter, the non-commercial oil mentioned above was found in a Devonian limestone reservoir, at Rough Range. After considerable further exploration, natural gas accumulations were discovered in Jurassic reservoirs at *Gingin* and in Triassic sandstones at *Yardarino*, and also subsequently at *Dongara* and *Mondarra*. The first commercial oilfield was found in 1964 on *Barrow Island* in the Carnarvon Basin. The principal reservoir here is a Lower Cretaceous (Neocomian) sandstone sequence (the "Windalia Sands"), which is of good porosity but low permeability, so that hydraulic fracturing and water-flooding techniques have been used to stimulate production. Deeper Upper Jurassic reservoirs are also known, but are much less productive, giving only 3% of the total annual output.

The recoverable reserves of the *Barrow Island* field have been estimated at 162 MM brl; production commenced in 1967, at a rate of 9,000 b/d, and this rose to 47,000 b/d during 1970.

Considerable further exploration has been carried out in the Carnarvon Basin, resulting in discoveries of gas in several offshore wells (e.g. *Flinders Shoal* and *Flag No. 1*). A well drilled offshore northeast of Barrow Island, named *Legendre*, was successful in discovering an oil accumulation which awaits evaluation.

What may prove to be important discoveries have also been made on the Continental Shelf of northwest Western Australia: several wells have found gas and condensate (*Rankin, Goodwyn, Angel* and *Scott Reef*) in Triassic reservoirs and gas reserves of 50 Tcf have been mentioned which may involve a very large investment in LNG export facilities.

Central and South Australia

Accumulations of natural gas have been found in two intracratonic basins in Central Australia, parts of which extend into South Australia. In the Amadeus Basin, which covers some 50,000 sq miles, there is a great thickness of 20–30,000ft of Upper Proterozoic to Ordovician strata[20,21]. Gas has been found in a large east-west surface anticlinal fold at *Mereenie*, about 150 miles west of Alice Springs, in a Cambro-Ordovician sandstone reservoir; there are also traces of oil in the under-lying Cambrian beds. *Palm Valley* is another large surface anticline with gas in a similar reservoir.

The Eromanga Basin is an extension of the Great Artesian Basin of eastern Australia and contains mainly Mesozoic rocks. Anticlinal gas accumulations have been found in the Cooper Sub-Basin at *Gidgealpa* and *Moomba* in Permian sandstone horizons. At *Tirra Warra*, 20 miles north of Moonie, an exploration well in mid-1970, found oil in a Lower Permian reservoir which awaits evaluation. At Ooraminna, in the same remote area, the first Pre-Cambrian gas to be discovered outside Siberia was found; although not commercial, this discovery has been of great significance in upgrading unmetamorphosed Pre-Cambrian strata elsewhere. Other gas accumulations have been found at *Daralingie* and *Toolachee*, in South Australia.

Eastern Australia

The Great Artesian Basin covers about 200,000 sq miles of Queensland, New South Wales, and South Australia, and is filled with Palaeozoic to Cretaceous strata. One of its lobes is the Surat Basin, extending over some 40,000 sq miles of southeastern Queensland, which contains Jurassic to Tertiary beds[22], unconformably overlapping the southern extension of the older Palaeozoic Bowen Basin, in which sedimentation was Permian to Triassic. Oil and gas have been found in Jurassic, Lower Triassic and Permian reservoirs in the Surat Basin, and gas in the Lower Permian sandstones of the Bowen Basin.

The *Moonie* field was the first commercial oilfield to be found in Australia; it is a gentle anticline about 5 miles long, 150 miles west of Brisbane on the eastern flank of the Surat Basin. Production began in 1964 from reservoir beds of Lower Jurassic (Liassic) age—the locally named "Precipice" sandstone sequence, which varies in thickness between 130 and 300ft. (Hydrodynamic effects may account for the tilt to the northwest of the oil-water interface in this structure.) Proved primary oil reserves are 27 MM brl, and the rate of production is now about 2·6 MM brl/a of oil and 610 MM cf/a of gas. The oil is thought to

[20] C. Siller & E. Webb, 7 WPC Mexico, 1967, PDZ-10.

[21] A. Froelich & E. Krieg, *Bull.AAPG*, 53(9), Sept. 1969, 1978.

[22] D. Hogetoorn, 7 WPC, Mexico, 1967, PDZ-8.

have been derived from underlying marine Permian rocks or from overlying Lower Jurassic shales[23].

A second oil accumulation was found in 1964 at *Alton*, 55 miles southwest of Moonie, in a fold on the western flank of the Surat Basin. It produces 1,000–2,000 b/d from a Lower Jurassic sandstone reservoir; the oil is sent by truck to Moonie to be introduced into the Moonie-Brisbane pipeline. The total recoverable reserves here are not more than 2 MM brl.

Other minor anticlinal oil accumulations in this area are at *Conloi* and *Bennett*, north of Moonie; some further gas accumulations have also been found. The reservoirs are generally Lower Jurassic sandstones, and the source rock is believed to be the associated Lower Jurassic shales.

Gas was discovered as long ago as 1900 at *Roma* in Queensland, in the northwest of the Surat Basin. The gas reservoir here is a Jurassic sandstone arched over a buried basement "high". Since 1961, this gas has been used to supply the local power station, and a small amount of oil is also produced (about 8,000 brl/a). Several other anticlinal accumulations of natural gas have been found in the Surat Basin in recent years (at *Bony Creek, Leichhardt, Pickanjinnie, Pine Ridge,* etc.)—all with Precipice sandstone reservoirs. (Gas began to move from the Roma area to Brisbane early in 1969 on the completion of a pipeline whose capacity is 30 MM cf/d.)

Further natural gas deposits have been discovered in the Bowen Basin, in an arc stretching from about 175 miles south of Roma to Rolleston, in the north. *Westgrove, Rolleston* and *Arcturus* are buried anticlines with roughly north-south axes, which have Permian sandstone gas reservoirs.

A further gas accumulation has been found at *Gilmore* in the Adavale Basin in Central Queensland, another remnant Palaeozoic basin underlying the Mesozoic strata of the Great Artesian Basin. The reservoir here is a thick Devonian sandstone in a faulted anticline. Huge oil-shale deposits occur in Queensland at Julia Creek, about 400 miles from the coast, and there are plans for developing these as sources of fuel for the Mount Isa mines.

Southeastern Australia

Oil-shale deposits were worked for many years in the Glendavis area of New South Wales, and small quantities of oil and gas were found in 1924 in the *Lakes Entrance* area of eastern Victoria, where there is a narrow strip of Tertiary sediments. The oil reservoir here is a glauconitic sandstone of Eocene age, with some gas in deeper Cretaceous strata. The structure may be the results of deposition on a buried Pre-Mesozoic uplift.

23 R. Mathews *et al., Bull.AAPG*, 54(3), March 1970, 428.

This area lies within the relatively small onshore section of the Gippsland sedimentary basin, which covers some 20,000 sq miles of southeastern Victoria and the adjoining Continental Shelf, more than two-thirds of the basin lying offshore, where a great but variable thickness of Mesozoic and Tertiary strata was deposited in a rapidly sinking "graben" resulting from the east-west faulting of the underlying Palaeozoic basement[24].

Offshore exploration began at the end of 1964, and has resulted in the discovery of several important oil and gas accumulations in the Gippsland Basin and the adjoining Bass Basin, which is separated from it by a buried granite ridge[25]. The offshore gasfields are: *Barracouta* and *Marlin* (gas and condensate) and *Snapper* (gas only). Gas is found in Lower Tertiary (Eocene-Cretaceous) sandstone reservoirs in the Latrobe Valley Coal Measure formation, with an additional deeper Upper Cretaceous reservoir. The gas reserves of *Barracouta* probably amount to about 1·8 Tcf, with perhaps 7 MM brl of condensate, while *Merlin* may hold as much as 3·5 Tcf of gas.

Good-quality oil has also been discovered in the *Kingfish* and *Halibut* structures, about 40 miles out to sea, and in two smaller accumulations, *Mackerel* and *Tuna*.

Halibut is interesting as an example[26] of a combination trap resulting from the truncation of the southwest-dipping Latrobe beds by a late-Eocene erosion surface and their subsequent sealing by the Oligocene Lakes Entrance mudstones. There is also structural closure of about 500ft at the unconformable Latrobe surface.

The total reserves of the major offshore fields mentioned have been

TABLE 61

AUSTRALIA— PRODUCTION AND RESERVES, 1971

	Production (MM brl.)	Reserves (MM brl.)
Queensland		
Roma area, Moonie, Alton, Bennett	1·0	5·2
Western Australia		
Barrow Island	16·2	141·8
South Australia		
Gidgealpa, Moomba	—	8·8
Northern Territory		
Mereenie-Palm Valley	—	75·0
Victoria		
Offshore Gippsland fields	96·3	1,727·4
Total	113·5	1,958·2

Source: National Development Dept.

24 T. Traill, *APEA Jnl.*, 1968, 8, 2, 67.
25 L. Weeks and B. Hopkins, *Bull.AAPG*, May 1967, 742.
26 E. Franklin and B. Clifton, *Bull.AAPG*, Aug. 1971, 1262.

estimated to be about 1·7 B brl of oil (mainly in *Kingfish*) and 8·6 Tcf of gas. These fields are clearly of great economic importance as they are only about 150 miles from Melbourne. During 1972, they produced about 281,000 b/d, and their potential is probably 350,000 b/d.

In **Tasmania**, although there has been sporadic drilling in the small sedimentary coal basins containing Permo-Triassic beds, no commercial hydrocarbons have yet been found.

NEW ZEALAND

Several small sedimentary basins occur round the coasts of both the North and South Islands of New Zealand, which have been sporadically explored for oil over a period of many years, but with little success until recently.

The Taranaki Basin, in the southwest corner of North Island, extends over the offshore area—the South Taranaki Bight—between the islands. Onshore, it contains the *Kapuni* anticline, in which natural gas and condensate were discovered in 1959 in a thick Upper Eocene reservoir sequence, which also contains coal measures. The structure here is probably the result of the "draping" of Tertiary beds over an underlying "high". Production began in 1970: potential output is estimated to be 4,500 b/d of condensate and 67 MM cf/day of gas. Reserves in this field are thought to be 26 MM brl of condensate with 460 Bcf of gas.

In the west of the same basin, at Alpha near New Plymouth, three shallow wells which were drilled near seepages have produced a small quantity of oil (only 150,000 *gallons* p.a.) for several years at Moturoa, together with some 4·6 MM cf/a of gas[27,28].

A number of oil seepages occur in the east coast basins of North Island and the northern part of South Island, which contain a thick sequence of marine Cretaceous and Eocene strata, but tectonic deformation has been severe.

Unsuccessful wells have been drilled on anticlines in the Westland basin, in the Southland basin of South Island, and in the Waikato basin of North Island.

Considerable exploration activity is taking place in various offshore areas: off the west coast of North Island, off the Bank Peninsula of South Island, and in the vicinity of Stewart Island. The first success achieved (in 1969) was the discovery of the *Maui* offshore hydrocarbon accumulation in the continuation of the Taranaki Basin, which awaits development. This is a very large gasfield, with a potential productive capacity of 300–500 MM cf/d of gas, and considerable volumes of condensate, from what is probably an Eocene reservoir[29,30].

[27] D. Watt, 8th Commonwealth Min. Met. Cong., 7, 222, 1965.
[28] E. Lehner, 8th Commonwealth Min. Met. Cong., 4, 331, 1965.
[29] R. Sprigg, *et al., Bull.AAPG*, 53(9), 1956, Sept. 1969.
[30] *Oil and Gas Int.*, Oct. 1970, 71.

CHAPTER 8

The United States (1)

THE traditional predominance of the United States in the world oil industry is shown by the fact that it produced more than 40% (90·5 B brl) of all the crude oil produced anywhere in the world in the century between 1859 and the end of 1969. Following the early discoveries in Pennsylvania, the north-eastern states were initially the dominant oil producers; but subsequent developments in other areas—California, Texas, Oklahoma, Kansas, the Rocky Mountains and finally the Gulf Coast—steadily shifted the centre of gravity of the oil industry southwards and westwards, until in 1971, only seven of the 31 states which produced oil commercially (Texas, Louisiana, California, Kansas, New Mexico, Oklahoma and Wyoming) together accounted for 89% of the total national output. In fact three states only—Texas, Louisiana and California—together produced as much as 72% of the oil output of the whole United States.

The total number of oil and gas wells that have been drilled in the United States is more than 1,500,000, and between 25,000 and 30,000 new wells are added every year. There are about 512,000 oil-wells still functioning, although about 75% of this number together produce only 13% of the total oil output, averaging only 3·7 b/d per well. Because of the large number of these "stripper" wells, the average daily production per well in the United States is also very low—about 18·6 brl. For the same reason, more than 90% of the oil-wells are currently on some form of artificial lift; less than 10% are flowing under natural pressures.

The fact that the United States has produced more oil (and gas) than any other country has been due partly to its favourable geological circumstances, partly to an early start, and partly to the intensity with which the search for hydrocarbons has been carried on. For some years prior to 1946, the USA regularly produced 60–65% of the world's annual crude oil output; however, since 1946 this proportion has steadily decreased due to the increasing importance of other sources (Table 5), until in 1972 it was only 19·2%. In other words, although US oil production has been slowly rising, it has become a smaller fraction of the world total with each succeeding year; and the 1972 production (3,469 MM brl) was actually less than that of 1970 (3,517 MM brl).

TABLE 62

NORTH AMERICA—OIL PRODUCTION
(thousands of brl)

Year	USA	Canada	Mexico	Total
1860	500	—	—	500
1870	5,261	250	—	5,511
1880	26,286	350	—	26,636
1890	45,824	745	—	46,569
1900	63,621	913	—	64,534
1910	209,557	316	3,634	213,507
1920	442,929	196	157,069	600,194
1930	898,011	1,522	39,530	939,063
1940	1,353,214	8,591	44,036	1,405,841
1950	1,973,574	29,044	72,443	2,075,061
1960	2,574,933	189,534	99,049	2,863,516
1970	3,517,505	438,023	156,530	4,112,058

TABLE 63

THE DECLINING PROPORTION OF US CRUDE OIL PRODUCTION RELATIVE
TO WORLD OUTPUT

Year	%
1946	63·2
1950	52·0
1956	42·7
1960	33·3
1966	25·4
1969	21·1
1971	19·9
1972	19·2

The prosperity of the United States has largely been based on the availability of plentiful supplies of liquid and gaseous hydrocarbons. Since 1900, total energy consumption has risen by an average of 3·2% p.a. against a population increase of 1·4% p.a., but in recent years the rate of energy consumption growth has been nearer 5% p.a. The per capita consumption of oil rose from 6·3 brl/head in 1925 to over 25 brl/head in 1969, paralleling the growth of the Gross National Product. During the latter year, liquid petroleum accounted for 44·5% of primary energy consumption, and natural gas a further 32·4%—i.e. more than three-quarters of the national energy requirement was provided by petroleum, although this requirement had doubled from 1,093 MM tons c.e. in 1950 to 2,189 MM tons c.e. in 1969.

TABLE 64

USA—CRUDE OIL PRODUCTION, 1972

State	MM brl
Alabama	10·0
Alaska	73·5
Arizona	1·0
Arkansas	18·5
California	348·0
Colorado	31·9
Florida	15·8
Illinois	35·4
Indiana	6·3
Kansas	74·4
Kentucky	9·8
Louisiana	908·7
Michigan	13·1
Mississippi	61·4
Missouri	0·1
Montana	33·8
Nebraska	9·2
Nevada	0·1
New Mexico	110·6
New York	0·9
North Dakota	20·5
Ohio	7·9
Oklahoma	206·7
Pennsylvania	3·7
South Dakota	0·2
Tennessee	0·4
Texas	1,296·4
Utah	25·9
West Virginia	2·7
Wyoming	141·9
Total	3,468·8

Source: Bureau of Mines and API.

With such a large production of liquid oil, the United States inevitably also produces an enormous volume of natural gas. The total output (associated and non-associated) reached a peak of 22·1 Tcf in 1971, derived from some 113,000 gas-wells, of which about 15,000 also produced condensate. However, the *potential* gas output is probably as much as 40 Tcf/a, if limitations of markets or transportation facilities are not considered.

The production of associated gas, of course, closely parallels the production of oil and its geographic distribution; while five states (Texas, Louisiana, Kansas, New Mexico and Oklahoma) account between them for more than 90% of the non-associated gas production.

In recent years, there has been a notable and continuing falling-off of exploratory drilling activity in the United States, from a peak of about 16,000 wells in 1956 to only 6,327 wells in 1972, reflecting the increasing difficulty experienced in finding economically viable hydrocarbon accumulations. The proportion of exploratory wells which produced some oil or gas was 20·73% in 1955 but only 15·72% in 1971;

or, put another way, the ratio of "dry holes" drilled for each producing well increased steadily over the period, from 3·82 to 5·36.

About six years are needed for the size of a newly discovered oilfield to be established, so that the volume of hydrocarbons discovered by exploratory drilling in any one year can only be approximately calculated. However, there is no doubt that the United States is consuming considerably more hydrocarbons than are being replaced by new discoveries. This depletion of known hydrocarbon reserves is a trend that has been evident for some time. The proved liquid petroleum reserves (i.e. crude oil plus natural gas liquids) of the United States amounted to 37·77 B brl at end-1969; this was actually appreciably lower than the equivalent figure for end-1965. The ratio of reserves to production has fallen to only about 10 years' supply at current rates, if the Alaskan North Slope reserves are not included. Even allowing for 10·1 B brl from Alaska, the total at end—1972 was only 43·1 B brl.

TABLE 65

USA—PROVED LIQUID PETROLEUM RESERVES AT YEAR-END (B brl)

	Crude oil	Nat. gas liquids	Total
1965	31·35	8·02	39·37
1969	29·63	8·14	37·77
1971*	38·06	7·30	45·36
1972*	36·30	6·80	43·10

* Includes (for the first time) Alaska reserves.

(It is true, however, that a further 5·1 B brl of "indicated additional reserves" were estimated to exist at the end of 1972. This is a rather loose term which describes reserves not considered to be "fully proved" or which are "expected to be produced from known reservoirs in response to the future application of known improved recovery techniques".)

The US proved oil reserves are decreasing even more notably when expressed as a proportion of world reserves:

TABLE 66

US PROPORTION OF WORLD PROVED OIL RESERVES

1945	1965	1969	1971*	1972*
35·3%	11·0%	7·0%	6·7%	6·2%

* including Alaska

The distribution of the known reserves in the United States is given in Table 67, from which it will be seen that three states only (Texas, Louisiana and California) together contain 78% of the total outside Alaska; Texas alone holds more than 45% of the proved liquid oil reserves of the "contiguous USA".

TABLE 67

USA—STATES WITH LARGEST PROVED OIL RESERVES, end-1971

	Crude oil B brl	Natural gas liquids B brl	Total liquid petroleum B brl
Texas	13·0	3·1	16·1
Louisiana	5·4	2·5	7·9
California	3·7	0·1	3·8
Oklahoma	1·4	0·4	1·8
New Mexico	0·7	0·6	1·3
Wyoming	1·0	0·1	1·1
Kansas	0·5	0·3	0·8
Other States	2·3	0·2	2·5
Alaska	10·1	0·0	10·1
Total	38·1	7·3	45·4

Major petroleum accumulations are certainly becoming progressively harder to find in the United States, necessitating deeper drilling, development in remote areas (e.g. Alaska), or exploration on the Continental Shelf. Up to the present time, exploration in the USA (as elsewhere) has mainly been concerned with the discovery of structural accumulations, simply because these have always been the easiest traps to find. However, the analysis of the "giant" oilfields so far discovered (defined on p. 223) has shown that as many as 28 % of these are effectively stratigraphic traps, although discovered more or less accidentally in the course of exploration for a structural trap. It is likely, therefore, that further stratigraphic accumulations remain to be found by deliberate search.

As a result of the "prorationing" arrangements that were set up (principally in Texas, Louisiana and, to a lesser degree, in California) to keep crude oil prices stable, US oil output has generally been less than the productive capacity of the existing wells. Thus, while output in 1971 was 3,478 MM brl (i.e. 9·5 MM b/d) it was estimated that the *productive capacity* of the existing installations was such that a production rate of 10·6 MM b/d (i.e. 3,869 MM brl/a) could have been available within 90 days of the end of that year if required (p. 336).

However, for economic and refining reasons it has always been necessary to fill the ever-increasing gap between requirements and domestic production by the import of crude oil and oil products— mainly residual fuel oils. Thus, in 1970, about 1·44 MM b/d of crude oil and 1·88 MM b/d of residual fuel oils were imported, principally from Canada, Venezuela, the Middle East and Africa. These oil imports grow year by year at the rate of between 8–11%, and in 1971, oil imports into the USA reached their highest-ever level as a result of the relaxation of the long-standing restriction which since 1959 had limited imports into PAW Districts I–IV, i.e. East of the Rockies, to 12·2% of domestic production. Altogether, imported oils of all kinds accounted for some 30% of total US consumption in 1971.

The relative importance of the different sources of imported oil varies according to fiscal and economic factors. For example, air pollution considerations are already affecting the pattern of imports. The East Coast, which accounts for some 66% of the total US fuel oil consumption, imports most of its residual oil requirements, mainly from Venezuela; but Venezuelan fuel oil is particularly high in sulphur, and processes for reducing the sulphur would increase the cost of this oil by up to a third. It is therefore being increasingly displaced by low-sulphur crudes from other sources.

During 1972, the demand for liquid oils in the United States was approximately 16·53 MM b/d, of which exports of crude oil and refined products accounted for only about $1\frac{1}{2}\%$. The demand was satisfied by the production of 9·48 MM b/d of crude oil and 1·73 MM b/d of natural gas liquids, and the import of the balance of about 4·63 MM b/d of crude oil and oil products. (The balance of about 0·4 MM b/d is accounted for by "refinery overage".) Imports in 1973 are expected to be 6 MM b/d.

TABLE 68

US OIL IMPORTS

| | | 1000 b/d | |
Source	1959	1969	1970
Canada	92·0	548·0	658·3
Venezuela	455·0	428·5	401·3
Other W. Hemisphere	36·0	65·9	26·2
Africa	—	183·6	95·1
Middle East	317·0	211·1	189·0
Far East	65·0	88·7	70·6
Total crude oil imports	965·0	1,525·8	1,440·5
Fuel oil and products	814·0	1,557·4	1,875·3
	1,779·0	3,083·2	3,315·8

Future Demand and Supply

Oil consumption in the United States is expected to be at least 30% higher in 1980 than in 1970—i.e. to reach about 7 B brl/a (say 19 MM b/d). There are considerable doubts as to how this increased demand is to be met. Prior to the end of 1971, some 428·6 B brl of crude oil-in-place had been discovered, of which 96·5 B brl had been produced.

The remaining recoverable reserves were then estimated as 38·1 B brl, which, when added to the oil already produced, would give an ultimate crude oil recovery of 134·6 B brl and an average overall recovery factor of only 31%. Of course, this is likely to be improved in the future, so that the ultimate production from known resources may be at least 141 B brl. Furthermore, in estimating total ultimate petroleum re-

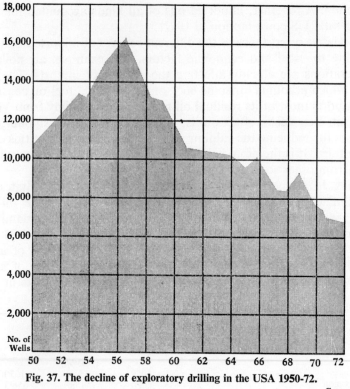

Fig. 37. The decline of exploratory drilling in the USA 1950-72.

Source: API.

sources, due weight must be given to likely future discoveries. However, it seems from all the evidence that these will be on a declining scale.[1]

Looking further ahead, to 1985, it seems likely that in that year the demand for petroleum liquids will be not less than 25·9 MM b/d[2]. In that year, total liquid oil production (including an estimated 2 MM b/d from Alaska) may not be more than 11·1 MM b/d, i.e. rather less than in 1970. In 1985, therefore, some 14·8 MM b/d of oils will need to be imported, representing not less than 57% of total petroleum consumption. (The resultant economic problems are discussed on p. 361.)

Intensive offshore exploration is likely to be carried out off the coasts of the United States in the next few years in an endeavour to bridge the gap between domestic production and consumption. Although at present limited to only three areas (off the Gulf Coast, Southern California and Southern Alaska) there are further possibilities, notably off the Atlantic Coast, where the Georges Bank Trough, the Baltimore Canyon Trough and the Blake Plateau Trough are potential target areas.

The US Geological Survey has indicated that about 208 B brl of oil

[1] M. Hubbert, *Bull.AAPG*, 4(10), Oct. 1965, 1720; see also p. 350.
[2] *A 1972 Chase Manhattan Bank report suggests a demand of 30·2 MM b/d in 1985.*

may eventually be recoverable from the Continental Shelf. However, this can only be considered as indicative of potential reserves, the development of which would entail considerable environmental opposition as well as necessitating huge capital outlays.

In this context it is of interest to note that the USA has very large proved deposits of oil-shales, in particular those in the Eocene Green River formation underlying large areas of Colorado, Wyoming and Utah. The first pilot-scale plant designed to obtain shale oil by the distillation of these shales has been in operation since 1970. Future success with nuclear explosion fracturing and heating *in situ,* and the subsequent production of resultant shale oil by drilling operations, may perhaps make available large volumes of future oil at an acceptable cost.

Distribution of Hydrocarbons

The North American continent is made up of a number of geosynclines surrounding a central stable cratonic region; local sedimentary basins and marginal coastal plains have developed upon the geosynclines.

Within the approximately 3 MM sq miles of sediments lying within the boundaries of the present United States, a wide variety of geological conditions occur which have been favourable for the accumulation of hydrocarbons. In this respect the USA is comparable with the USSR, and there are many similarities between the petroleum occurrences of both countries.

It is obviously possible to consider here only a few of the very large number of oil and gas accumulations—exceeding 23,000—that have been found in the United States in the course of the past century or so*. However, most of the important fields lie within a comparatively few "petroliferous provinces", within each of which the structure and stratigraphy of the principal accumulations are similar (Table 69).

They include 310 "giant" accumulations (defined here as containing at least 100 MM brl of recoverable oil or 1 Tcf of gas) which produce together more than half the current oil output and contain more than 57% of the remaining proved reserves of the United States. The states of Texas, Louisiana and California have most of the "giants"—121, 53 and 43 respectively. Oklahoma has 27, and the Rocky Mountain states 18 between them. New "giants" are still occasionally found, but the rate of discovery has tailed off in recent years.

---

* *The literature describing American oilfields is as prolific as the fields themselves. Readers are recommended to the publications of the American Association of Petroleum Geologists, Dallas, and in particular to their exhaustive Memoirs and Special Publications—e.g. "Geology of Giant Petroleum Fields", "Stratigraphic Oil and Gas Fields", "Geophysical Case Histories", etc. An invaluable reference work covering the whole of the United States in a single volume is "Petroleum Geology of the United States" by K. K. Landes (Wiley Interscience).*

TABLE 69

USA: PRINCIPAL PETROLIFEROUS PROVINCES

I. **The Northeastern and East-Central States**
   1. Appalachia
   2. Cincinnati Arch
   3. Michigan Basin
   4. East Interior Coal Basin (Illinois)

II. **The Mid-Continent Area (West Interior Coal Basin)**
   1. Kansas
   2. North and Central Oklahoma
   3. South Oklahoma
   4. Arkansas

III. **Rocky Mountains Province**
   1. Colorado
   2. Wyoming
   3. Montana
   4. East Flank (Dakotas and Nebraska)
   5. West Flank (Utah and Idaho)
   6. "Four Corners" Intermontane Area

IV. **Texas**
   1. North Texas
   2. The Panhandle
   3. West Texas
   4. Balcones and Mexia Fault Zones
   5. Sabine Uplift and East Texas Basin
   6. Gulf Coast Embayment in Texas
      (a) SE Texas—Upper Gulf Coast
      (b) SW Texas—Lower Gulf Coast

V. **Louisiana and the S.E. States**
   1. North Louisiana
   2. South Louisiana
   3. Mississippi and the S.E. States

VI. **California**

VII. **Alaska**

I. THE NORTHEASTERN AND EAST-CENTRAL STATES

1. Appalachia

A Palaeozoic geosyncline extended from the present New York State southwestwards as far as Alabama, and a great thickness of sediments was deposited within this depression. Oil subsequently accumulated in gentle anticlinal folds paralleling the edge of the Hercynian mass of the Appalachian Mountains. To the west and northwest, the basin was bounded by the arch of the Cincinnati uplift. Although most of the oilfields occur in structural traps, local variations in reservoir rock porosity have also been a major factor in accumulation. The reservoirs consist of Palaeozoic (Pennsylvanian-Cambrian) sandstones and limestones.

In this area, there appears to be a general relationship between the types of hydrocarbons found and the degree of metamorphism, which can be estimated by the changes brought about in the local coal deposits. The resultant Carbon Ratio Theory sought to show[3] that where the percentage of fixed carbon exceeded 70 no oil or gas pools would be found; when it was between 70 and 60 gasfields only would occur; and only when it was below 60 would oilfields be found. An explanation for the distribution of oil in the western part and gas in the eastern part of Appalachia based on this theory has indeed been qualitatively successful, but there are sufficient exceptions to make quantitative interpretation doubtful.

The oil production of the region has been steadily declining since the last century. The large number of wells drilled is related to the fact that in these areas hydrocarbons have been longest sought, and, since the fields discovered have been close to centres of population and industry, markets for relatively small producers have always been available.

The states of New York, Pennsylvania, West Virginia and Ohio were the first of the United States to be explored for petroleum, starting with the Drake discovery well in Pennsylvania in 1859; and they have been intensively explored for many years. By comparison, there has been relatively little exploration activity in the New England states or in New Jersey, Delaware, Maryland and North Carolina. In Virginia, about 2·8 B cf/a of natural gas is produced, and only a very small volume of oil, mainly from Berea sandstone and Trenton limestone reservoirs.

In **Pennsylvania**, a number of small oilfields occur in the central part of the Appalachian geosyncline, near the western limits of the sandstones which then pinch out into shales. Accumulations have been found in locally porous portions of widely deposited sand masses, where lithological factors outweigh structural conditions.

The largest and most famous oilfield in Appalachia is *Bradford*, which extends from Pennsylvania into New York State. This field was discovered in 1871, reached a peak production of 23 MM brl in 1881, and has been producing continuously ever since. The present output, stimulated by water flooding, is about 3·6 MM brl/a from about 25,000 producing wells spread over an area of 85,000 acres. The structure consists of two broad anticlines with gentle dips, plunging southward and converging northward.

The principal reservoirs are the Chemung (Devonian) sandstones, the thickest of which is the Bradford Sand. The oil was probably formed close to its present position—possibly in local carbonaceous shales—and is of high quality, being rich in lubricating fractions.

During 1972, 3·7 MM brl of oil were produced in Pennsylvania. There

---

[3] D. White, *J. Wash, Acad. Sci.,* 5, 189, 1915.   *See Fig. 5, p. 38.*

are nowadays two major reservoir zones—the shallow Bradford, Glade, Clarendon and Red Valley sandstones, and the deeper Lower Devonian Oriskany sandstones. In addition, gas is produced from the Lower Silurian Medina sandstone. The most important of the deeper zone gasfields are *Punxsutawney-Driftwood* (10 pools), *Leidy, Ellisburg* and *Summit*.

In the southwestern corner of **New York State** a number of minor oilfields have been found with anticlinal structures and Devonian sandstone reservoirs. The *Allegany* field has for many years been the largest producer, although it is now nearing exhaustion and has been supplanted by *Cattaraugus*. *Busti* is the most important of the other small pools.

Although about 5,000 oil-wells are still productive in the state, out of more than 53,000 wells that have been drilled there, the overall production is relatively small (about 0·9 MM brl in 1972).

The oil-producing area of **West Virginia** is a continuation of that of Pennsylvania. The main structural feature is the Warfield-Chestnut Ridge anticline which runs from northern Pennsylvania through West Virginia and into east Kentucky. Associated with it are the secondary features of the Parkersburg-Lorain syncline, the Volcano-Burning Springs uplift and the Cambridge anticline.

The principal reservoirs are the Big Injun (Lower Mississippian), Riley-Benson (Upper Devonian), Oriskany (Lower Devonian), and Newburg (Upper Silurian) sandstones.

There are some interesting examples in this area of the synclinal accumulation of oil in sands which carry no water. Thus, in such fields as *Copley* and *Cabin Creek,* oil is found in the vicinity of a plunging synclinal axis, while gas occurs above, extending across neighbouring anticlines. This seems to be the result of the local absence of water— perhaps due to the dehydrating effect of clay minerals in the reservoir beds, or to pressure reduction.

The oil output of West Virginia was about 2·7 MM brl in 1972, but gas production was more than 230 Bcf, largely from the *Kanawha-Jackson* gasfield, which has been producing gas since 1880 from a number of local crestal accumulations in an area covering nearly 200,000 acres. The original reserves here may have been as much as 11 Tcf.

In eastern **Ohio** there is a further extension of the Pennsylvanian oilfield area within the Appalachian geosyncline. Oil and gas are found in various Permian and Mississippian sandstones on the flanks of weak folds. In parts of eastern Ohio there is also deeper-seated production, from the Lower Devonian (Oriskany) sandstone, and from the "Clinton" sand, which is actually the Medina sandstone of Lower Silurian age. Accumulations are the result of shoreline deposition of this sand along an unconformity and in the channels of ancient deltas[4], and are therefore

---

[4] W. Knight, *Bull.AAPG*, July 1969, 53(7), 1421.

examples of stratigraphic trapping. The main targets for hydrocarbon exploration are the "Clinton" sandstones, the Lower Mississippian Berea sandstone, and the Copper Ridge Cambrian limestone. *Moreland* and *Denmark* are the principal producing oilfields. (The oil accumulations of northwest Ohio are described in Section 2.)

In eastern **Kentucky**, the Pennsylvanian trend continues in the eastern part of the state. The major structural feature is the Chestnut Ridge uplift from West Virginia, which in this state has been modified by structural movements along two parallel fault zones.

The hydrocarbon accumulations in East Kentucky are related to this structural trend. (The West Kentucky fields which lie in the East Interior Coal Basin are described on p. 231.) *Irvine,* on the southwest edge of the Appalachian basin, has oil chiefly in the Lower Devonian Corniferous limestone. *Big Sinking* is a stratigraphic trap discovered in 1917, which still produces over 2 MM brl/a of oil from a Silurian dolomite reservoir. *Big Sandy* is a huge, combination-type gas accumulation with several trapping features.

The total production from the East Kentucky fields was 3·2 MM brl of oil in 1971.

In western **Maryland** several small oil and gas accumulations (e.g. *Accident* and *Negro Mountain*) have been found, with reservoirs of Lower Devonian (Oriskany) age.

2. The Cincinnati Arch

The Appalachian geosyncline is bounded on the west by the uplift of the Cincinnati Arch. This runs northward from Alabama and Mississippi across Tennessee and Kentucky, bifurcating north of Cincinnati, with one branch passing through western Ohio and into Ontario and the other through northern Indiana. Hydrocarbon accumulations are found on crestal undulations of this great regional uplift.

The *Lima-Indiana* field, in northwest Ohio and eastern Indiana, was discovered between 1884 and 1889, and reached its maximum production of 25 MM brl/a in 1904. The field is really made up of a number of separate, narrow, elongated pools that extend from near the crest of the Arch in Delaware County, Indiana, northeastwards to Toledo in Ohio. Nearly 100,000 wells have been drilled in this field, which is approaching exhaustion. The principal reservoir has been the dolomitic upper 25–50ft of the Ordovician Trenton limestone, which was probably also the source rock; oil has also come from the Silurian Clinton limestone.

The Cincinnati Arch continues into Kentucky and then Tennessee. In **Tennessee**, a relatively small volume of oil is produced (351,000 brl in 1972) mainly from Middle Ordovician (Hermitage) beds. *Oneida West* is the largest field, one of several shallow accumulations near the crest of the Arch.

3. The Michigan Basin

The Michigan Basin was probably related in its origin to the Keeweenawan (Pre-Cambrian) movement which produced the Lake Superior geosyncline. Tangential pressure from the north and northeast, acting against the Wisconsin land mass, is believed to have caused the downwarping, and later stresses then modified the old depression into its modern form. There is a close relationship between the tectonic origin of the Cincinnati Arch and that of the Michigan depression.

Although some drilling had been carried out with moderate success in earlier years, the development of the oilfields of this area really began in 1925. Oil accumulations have been found in local culminations along many of the narrow, anticlinal folds which traverse the southern part of the peninsula in a series of parallel northwest-southeast trends, probably related to deep-seated faults.

Many of the traps are also locally affected by unconformities. The *Saginaw* field, which was the first to be discovered has a shallow, Mississippian (Berea) sandstone reservoir; however, the principal reservoirs are Mid-Devonian and Lower Devonian limestones and dolomites. Oil production in Michigan is declining (it was 13·1 MM brl in 1972) but may increase again as a result of the discovery of a number of Ordovician and Silurian pools along the axis of the *Albion-Scipio* trend. The fields along this 35-mile long zone of faulting and fracturing (which probably overlies a "basement" fracture) now account for nearly 40% of Michigan's oil production. Hydrocarbon accumulation here is largely dependent on the local dolomitisation of the Trenton limestone, while denser dolomite acts as a cap rock. Carbonate rock reservoirs now account for some 99% of the oil produced in Michigan.

4. The East Interior (Illinois) Coal Basin

This basin includes the states of Illinois, southwestern Indiana and northwestern Kentucky. It is limited on the east and northeast by the Indiana branch of the Cincinnati Arch and on the southwest by the Ozark uplift of Missouri.

The sediments are similar in character to those laid down in the Appalachian and Michigan basins. Under a mantle of glacial drift, the Palaeozoic strata range in age from Pennsylvanian to Ordovician. In the extreme south of Illinois, Cretaceous and Tertiary rocks occur, due to the encroachment of the Missouri embayment.

The principal structural feature is the La Salle anticline system which extends for some 240 miles from the northwest corner of Illinois, to the Wabash River. This is not a continuous arch, but a series of elongated domes whose axes are sometimes *en échelon*, with oil and gas pools occurring on local structural crests towards the southeast.

Around the southern end of the La Salle uplift lies the Illinois basin, the deeper parts of which contained deep water and were consequently sediment-starved during early Middle Mississippian time[5].

In **Illinois**, oil accumulations occur along the La Salle and DuQuoin uplifts, and also in the basin between them. *Old Illinois* is the largest producer, a "giant" with several pools. The reservoir beds are Pennsylvanian-Lower Mississippian sandstones which are sometimes found on the crests of local domes and sometimes forming "sand lens" type stratigraphic traps. Parts of this field have been in production since 1905, and output still exceeds 10 MM brl/a. Other "giant" anticlinal accumulations are *Salem, Louden, Clay City, New Harmony,* and *Dale*. Local productivity is largely controlled by permeability variations.

In western Illinois there are a number of scattered fields on the west flank of the basin, on a structural "high" extending north from the Ozark uplift.

Most of these fields are associated with anticlinal structures, although some, like the *Jacksonville* gasfield, are essentially sand-lens traps. Oil comes from porous zones in the Spergen (Mississipian) limestone, and gas has been found in shattered Pennsylvanian black shales.

The main structural features of **Indiana** are parts of the Cincinnati Arch and the Illinois Basin. Oil is produced in this State in relatively minor volumes (6·3 MM brl in 1972) mainly from Mississippian and Pennsylvanian beds, but also from Devonian and Ordovician (Trenton Limestone) reservoirs. The first Ordovician oil discovery was in the old *Trenton* field (1886), and the very porous Trenton reservoir here produced some 105 MM brl of oil before its virtual exhaustion. An important discovery in 1966 was *Urbana*, with a dolomitized Trenton Limestone reservoir.

In southwestern Indiana, a number of small oil- and gasfields occur on local wrinkles on a gently-dipping monoclinal slope. Examples are the *Francisco* and *Tri-county fields*. Reservoir beds here are the Pennsylvanian and Chester sandstones.

The oilfields of western **Kentucky** account for the greatest part (6·9 MM brl) of the state's total oil output of 10·7 MM brl in 1971. The principal structural feature in this area is the 100-mile long *Rough Creek* system of east-west faults and folds. A number of small associated pools have Chester and Pennsylvanian sandstone reservoirs, and also some Devonian and Silurian limestone production. The oil is believed to have been formed in each case in the nearest shale section. *Hitesville* is the largest productive anticline, with several oil reservoirs of Mississippian and Pennsylvanian ages.

---

[5] J. Lineback, *Bull.AAPG*, Jan. 1969, 53(1), 112.

II. THE MID-CONTINENT AREA

West Interior Coal Basin

This basin covers a large part of the states of Kansas and Oklahoma, (and also parts of Texas, Arkansas, Missouri and Nebraska). In the south, the east-west line of the Arbuckle-Wichita-Amarillo Mountains gives rise to a separate petroleum province (described on p. 252). In the northeast, the Ozark Mountains extend into the corner of Oklahoma nearest to Missouri along a northeast-southwest axis. On their western flank, a wide homocline dips to the west and is intersected by the north-south uplift of the Nemaha buried "granite ridge". This is a narrow, Pre-Cambrian crystalline mass, which continues as the Cushing-Seminole ridge into northern Oklahoma.

There are three principal petroleum-producing areas: (i) a trough running across central Oklahoma and east Kansas which separates the buried ridge from the Ozark uplift; (ii) the central Kansas basin, which lies west of the ridge and is bounded by the Amarillo uplift in the south and the central Kansas uplift in the north; (iii) the Arkansas Valley, which runs from east Oklahoma into west Arkansas. Most of the oil-fields lie in a wide belt along the western front of the Ozark uplift, or are closely associated with the Nemaha ridge. Smaller accumulations also occur in local culminations on the gentle northwest-southeast arch of the central Kansas uplift.

The most important reservoirs in this area are of Pennsylvanian, Mississippian, Silurian and Ordovician age. Many of the fields are stratigraphic accumulations resulting from the local deposition of porous Pennsylvanian sands which wedge out in unconformities on the edges of the uplifts. Another type of stratigraphic trap has resulted from the deposition of "shoestring" sands filling the channels of ancient streams or formed as sand bars in shallow marine waters. Many other Pennsylvanian reservoir fields, and the great majority of the Ordovician producers, contain oil in anticlinal structures, whose axes may parallel the Kansas, Ozark or Nemaha trends.

1. Kansas

Oil production in Kansas reached a peak of 124·5 MM brl in 1956, but has since declined to 74·4 MM brl in 1972. One of the oldest and largest fields is *El Dorado*, which has been producing oil since 1916 and still has an output of about 2·5 MM brl/a. The structure is anticlinal, the sediments which were laid down over the buried Nemaha ridge being subjected to later folding and faulting. The surface and subsurface structures are roughly parallel, surface domes being underlain at depth by three Ordovician domes which form the major reservoirs.

Oil and gas are also found in gently-dipping Pennsylvanian sandstones and in the Mississippi Lime. Because of the erosion that took place after the major uplift, the Ordovician beds are in contact with the Pennsylvanian Cherokee formation, and oil has accumulated along the unconformity in porous zones in the Viola, St. Peter and Siliceous limestones. The oil may be indigenous to these limestones, or it may have migrated into them from the highly organic Chattanooga shales.

An important oilfield on the central Kansas uplift is *Kraft-Prusa*. Here, the youngest reservoirs are Pennsylvanian limestones in local uplifts which reflect "highs" in the pre-Cambrian "basement". The main reservoir is the eroded surface of the Cambro-Ordovician Arbuckle dolomite underlying the unconformity and truncated against the central Pre-Cambrian "bald head".

A number of other fields have been found in anticlinal structures on the flanks of the Nemaha ridge. The *Bemis Shutts* field, discovered in 1928, is a domal structure which still produces about 3·3 MM brl/a. *Chase-Silica* and *Trapp* are other anticlinal accumulations.

In cast Kansas, a different variety of oil accumulation is common—palaeogeomorphic traps (p. 52) in the form of long, narrow sand bodies which are known as "shoestrings" on account of their sinuous shape. They may be several miles in length and are generally about 50ft thick and 1,000–1,500ft wide. In cross-section they have flat to convex tops and bottoms, with nearly vertical sides. The porosity of the sand in these deposits is very variable.

Oil pools occur wherever the porosity of the sand has been sufficient for it to act as a lithologic trap. Production typically comes from long, narrow bodies of Bartlesville sand, surrounded by the shale beds which are the probable source rocks. *Greenwood* is a stratigraphic trap which produces important volumes of gas.

The *Hugoton* field in southwest Kansas (together with the nearby *Panhandle* field in Texas) forms one of the largest gas accumulations in the world, with original reserves (1927) of some 70 Tcf. The productive area covers nearly 5 million acres, extending over a length of some 250 miles. About 7,200 wells currently produce about 1·4 Bcf/a of gas, from a thickness of about 250ft of dolomite and limestone beds with shale breaks in the lower (Big Blue) series of the Permian. The porosity is thought to be due to chemical solution by ground-waters and to some sub-aerial weathering. The structure is an eastward-dipping homocline, the gas accumulation being controlled by variations of lithology. The proportion of clastics increases westwards, with a consequent reduction of the limestone and dolomite reservoir porosity, so that the field is therefore an excellent example of a stratigraphic trap.

2. North and Central Oklahoma

North of *El Dorado,* the granite uplift is not productive in its central areas, since erosion has removed most of the Mississippian and Ordovician beds. To the south, however, nearly every culmination along the axis

Fig. 38. PRINCIPAL STRUCTURAL

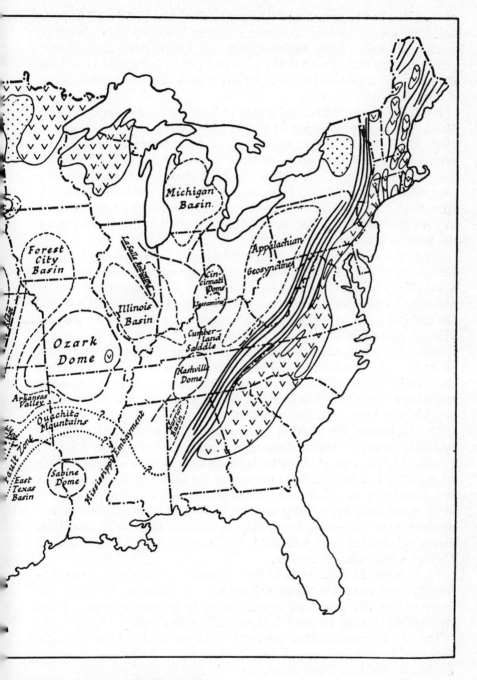

FEATURES OF THE UNITED STATES

R

of the ridge has produced an oil accumulation. Most of the oil in these fields is found around the unconformity between the upturned, weathered edges of the Mississippian and Ordovician series and the gently dipping Pennsylvanian beds. Oil also occurs in Pennsylvanian and Permian sands. Only a few of the many "giant" accumulations can be described here.

Glenn Pool was the first major field to be discovered in Oklahoma (in 1906). The peak production of 80,000 b/d was reached as long ago as 1908, since when the output has steadily declined; however, the field still produces about 4 MM brl/a. *Burbank* is another example of a stratigraphic trap which is affected by the local structure. Production here comes from an undulating monocline of Burbank sandstone, which is stratigraphically equivalent to the Bartlesville formation. Accumulation in both these fields has been controlled by variations in the porosity of the sand, which wedges out up-dip.

Cushing was discovered in 1912, and has since produced nearly half a billion brl of oil. It lies rather to the east of the main Nemaha trend, on a subsidiary parallel uplift. The structure is an asymmetrical, gentle anticline 20 miles long, with a north-south axis and four culminations. Oil and gas occur in various Pennsylvanian sands down to the Bartlesville, as well as in Wilcox and Arbuckle limestones below the unconformity. There seems to be a regional segregation of oil in the west of the structure and of gas in the east, which is explicable on the hypothesis that the oil migrated from the west or southwest. The reservoir formation is remarkable in that it has inclined oil-gas and oil-water levels—both being higher in the east than the west.

The *Oklahoma City* field is structurally a buried, faulted anticline, probably underlain by a deep-seated continuation of the Nemaha granite ridge, intersected by a buried extension of the nearby Arbuckle Mountains. The large fault on the east of the field is post-Mississippian but since nearly the full Pennsylvanian sequence is unfaulted, the movement must have been sharp and short-lived. The accumulation has resulted from the sealing by overlap of younger (Pennsylvanian) beds across an eroded and truncated ("bald-headed") central dome of Cambro-Ordovician limestone.

Production of oil comes from Pennsylvanian sands above the unconformity, and truncated Simpson (Ordovician) sandstone at the unconformity; but the major oil accumulation is in porous zones of the Arbuckle limestone, to which it is probably indigenous.

The uplifted north-central area of Oklahoma, north of the Arbuckle Mountains, contains several very prolific pools, including *Seminole, Earlsboro, Bowlegs, Little River, Pearson Switch, St. Louis,* etc.

The surface formations are Upper Pennsylvanian rocks. In the pre-Pennsylvanian there is a well-marked uplift, probably related to a buried ridge, and the oil accumulations are mainly located in small folds on the main uplift. Most commercial production comes from the Misener sand (Mississippian), the Hunton formation of the Devonian-Silurian and the Wilcox sand of the Simpson formation (Ordovician).

3. South Oklahoma

The oilfields of this area should correctly be considered as in the same province as those of north Texas and the Texas Panhandle. The dominant structural feature in this region is the Hercynian uplift of the Arbuckle and Wichita Mountains, on the flanks of which many oil- and gasfields are found. The province contains the Ardmore, Marietta and Hollis basins, and is bounded to the north by the Arkoma and Anadarko basins. Each of these basins contains many oil accumulations.

The *Crinerville* field is structurally controlled by the nearby Criner Hills. The latter, which lie about 15 miles to the south of the Arbuckle Mountains, consist of a complex mass of folded and distorted Ordovician rocks, against the west side of which the Pennsylvanian surface rocks of the Crinerville anticline are faulted. Oil is produced from Lower Pennsylvanian sandstones and Ordovician limestones.

Cement, discovered in 1917, is another anticline with similar reservoirs and a production of about 2·7 MM brl/a.

Hewitt (1919) is an anticline overlying a buried uplift. The surface structure is a north-south anticline, which increases in sharpness with depth. The Pennsylvanian beds overlie the Hewitt hills, which are part of the north rim of the buried south-eastern extension of the Wichita Mountains. Oil comes from several Pennsylvanian sandstones; production is 3·8 MM brl/a.

Healdton is similar in structure, directly overlying the Wichita Mountains-Criner Hills axis. Oil production was established here as long ago as 1913 from Pennsylvanian (Hoxbar) sandstones, but a new phase of production from dolomitic zones in the Ordovician Arbuckle limestone began in 1960.

In the "Oklahoma Panhandle" area, northeast of the Amarillo Uplift, hydrocarbons have been found in recent years in a continuation of the Anadarko basin of Arkansas, mainly as a result of deep drilling into the Devonian Hunton dolomite and the Simpson and Arbuckle (Ordovician) formations. The producing structures are usually closures against the upthrown side of faults. Examples are *North Custer City, Carter-Knox* and *Gageby Creek*. A well drilled in the last-named field is believed to have a world record potential output of 1·7 Bcf/d of gas.

4. Arkansas

The Ozark uplift also affects northeastern Arkansas. It is bounded to the south by the Arkoma Basin which separates it from the Ouachita uplift. There is little oil production in this part of the state, but "dry" gas comes from Pennsylvanian (Atoka), Mississippian and Silurian sandstones.

Southeastern Arkansas forms part of the Gulf Coast province and contains the northern end of the Monroe uplift, part of the Desha basin and the Mississippi embayment.

"Dry" gas is produced from Pennsylvanian (Atoka), Mississippian and Silurian sandstones in the Arkoma basin, and oil and gas from Jurassic to Tertiary strata in a number of relatively small fields in the Mississippi embayment. Principal reservoirs are Smackover and Cotton Valley (Jurassic), and the two largest accumulations are *El Dorado* and *Magnolia*, which are both anticlines with additional oil in flank pinch-out traps. The *Smackover* field is a combination trap which has been producing oil since 1922, first from Upper Cretaceous sandstones and, since 1939, from deeper Jurassic limestones.

III. ROCKY MOUNTAINS PROVINCE

A great thickness of Mesozoic sediments was laid down in the Cordilleran geosyncline which covered large areas of the present states of Montana, Wyoming, Utah, Colorado, Arizona and New Mexico, and overlapped into North and South Dakota and Idaho. These sediments were subsequently raised and folded in the late Cretaceous orogeny, and the erosion products were then deposited as Tertiary sediments in several intermontane basins—the most important of which are the Big Horn, Powder River, Green River, Uinta, Julesberg-Denver and San Juan Basins. These are vertically composite sedimentary and topographic basins, and with their peripheral belts of folded rocks form the dominant structural features of Wyoming and Colorado; in Montana there are three major uplifts—the central Big Snowy Anticlinorium, the northern Sweetgrass Arch, and the eastern Cedar Creek anticline.

The Rocky Mountains province contains a large number of strongly developed anticlines, extending in a broad belt from the Canadian border to New Mexico. Many of these folds have steep dips and large closures, and are strikingly exposed around the peripheries of the basins; anticlinal traps also occur beneath the central Tertiary cover.

Hydrocarbons have been found at various stratigraphic levels, in Palaeozoic and Mesozoic strata. Of these, the Cretaceous beds have given the greatest oil production from various sandstones in the Montana, Colorado and Dakota formations.

Oil and gas are also found in the Comanchean Morrison, the Jurassic Sundance and the Triassic Chugwater sands, in the Permian Embar limestone, in the Pennsylvanian Tensleep sandstones and the Mississippian Madison limestone. Oil also occurs, but to a lesser degree, in Ordovician and Cambrian rocks.

The Rocky Mountains states produced about 263 MM brl of oil in 1972, of which 54% came from Wyoming. There are 18 "giant" accumulations in this province.

1. Colorado

The oilfields of Colorado are comparatively small, with a total output of 31·9 MM brl in 1972. One group of simple anticlinal accumulations in the southern parts of the Julesburg-Denver and Green River basins produces oil and gas principally from Cretaceous and Jurassic sandstones and Pennsylvanian limestones. Examples are *Collins, Wellington, Moffat* and *Iles*.

Another group of fields in this area is of special interest in that they produce oil from fractured shales, often independently of local structure. Of these, the best example is the *Florence* field, the second oldest oilfield in the United States, which has produced oil continuously since 1862. The field is actually situated in a synclinal flexure and oil is found in joints and fissures throughout some 3,500ft of Pierre shales, the local basal member of the Montana (Cretaceous) beds.

The oil was probably formed locally in the shales and subsequently accumulated whenever adequate secondary porosity developed as a result of the erosion of overlying strata and the consequent relief of pressure which allowed fissures to open out in the shales.

The *Canyon City* field in a small embayment of the same basin also produces oil from the Pierre shales, but the structure here is a southward-plunging anticlinal "nose". At *Rangely* and *De Beque* in the Uinta (Piceance) basin in northwest Colorado, the equivalent (Mancos) shales are also productive. *Rangely* is the largest oilfield in Colorado. Discovered in 1902 as a surface anticline, some 20 miles long by 8 miles wide in the centre, it produced shallow Mesozoic oil for many years before the discovery of deeper Palaeozoic reservoirs in 1932. This field, which has produced some 425 MM brl of oil in the course of its history, still has an annual output of about 15 MM brl. The principal reservoir is the Weber sandstone.

In northwest Colorado, exploration is being actively carried out for Pennsylvanian and deeper reservoirs; in the southwest, the Paradox basin is also a focus for exploration activity. Several important discoveries have also recently been made in the southeast of the State, where the Las Animas Arch lies on the northwest flank of the Anadarko Basin. The oil reservoirs in the *Golden Spike, Ladder Creek, Smoky Creek* and *Timber Creek* fields are Mississippian in age. Pennsylvanian gas reservoirs have also been recently discovered in this area.

2. Wyoming

This is the most important oil-producing state of the Rocky Mountains province, with an output of nearly 155 MM brl in 1971. There are a large number of fields, usually anticlinal structures, whose main production comes from the Cretaceous Wall Creek (Frontier) sands; production has also been found in Pennsylvanian Embar-Tensleep beds.

In several of the Wyoming intermontane basins, newer stratigraphic-type traps in reservoirs of from Palaeozoic to Tertiary ages have also been found. Accumulations on the flanks of folds which were previously thought to be due to "hydrodynamic tilting" effects (p. 55) have more recently been shown to be partly, at least, due to stratigraphic trapping. The Pennsylvanian (Minnelusa) and Cretaceous (Dakota and Muddy) accumulations in the Powder River basin are probably mainly stratigraphic traps of this type.

(i) Big Horn Basin

Around the periphery of this basin lie numerous productive anticlines. This was in fact the classical area for the study of long-distance migration effects. The compaction forces acting on the sediments in the centre of the basin seem to have driven their contained fluids outwards, so that the basinward flanks of the anticlines round the edges are richest in oil. As a corollary, those anticlines which lie in the "shadows" of prominent folds on the periphery have been relatively starved of oil.

In this area, an extensive permeable bed (the Dakota sandstone) was certainly available to act as a "carrier" for long-distance migration. Many anticlines are indeed richer in oil on the basinward side (e.g. *Greybull*), while others certainly appear to have starved the less favourably placed folds in their "shadows". For example, the Torchlight and Grass Creek domes seem to be in this relationship to the Lamb and Little Grass Creek anticlines respectively.

A later theory has been developed[6] to account for the distribution of the fluid contents of the oilfields here. It is suggested that the hydrocarbons were generated from the organic sediments of the Permian *Phosphoria* beds, and as a result of normal primary migration processes had accumulated by early Jurassic time in various stratigraphic traps in Permian and Pennsylvanian beds. These hydrocarbons were subsequently released by the late Cretaceous orogeny, and migrated into older reservoir rocks. The vertical redistribution of an original common pool into several separate pools was subsequently accomplished by "hydrodynamic tilting" within the Tensleep zone, leakage or redistribution of fluids through fault zones, or escape of hydrocarbons to the surface as a result of the breaching of the original Triassic cap rocks.

Some of the more important oilfields of the Big Horn Basin are the following:

Grass Creek, discovered in 1914, is a structure which coincides with a topographical depression. It is an asymmetric anticline, its basinward or eastern flank dipping at 11° and its mountainward or western flank dipping at 29°. Oil is produced at the rate of about 5 MM brl/a from Cretaceous Frontier sandstones, Permian Embar limestones, and Pennsylvanian Tensleep sands.

---

[6] D. Stone, *Bull.AAPG*, 51(10), Oct. 1967, 2056.

Greybull was first drilled in 1907. It is an anticline with the basinward limb steeper than the mountainward dip, and several faults intersect the northern part of the dome, with the downthrow to the northwest. The principal oil production has come from Cretaceous Dakota sandstones.

Elk Basin is a large, elliptical anticlinal accumulation on the border between Wyoming and Montana. The field was originally discovered as a producer from Cretaceous Frontier sands as long ago as 1915; in 1946, however, prolific Mississippian Madison limestone oil was found in a deeper stratigraphic trap[7]. The anticline, which is 8 miles long and 4 miles wide, with dips of 25° on the west flank and 45° on the east, is cut by three sets of normal faults. It is currently the largest producing oilfield in the Rocky Mountain province, with an output of about 20 MM brl/a.

Cottonwood Creek was developed in 1953 on the east side of the Big Horn basin. It is a stratigraphic accumulation on the southwest flank of the Hidden Dome anticline (itself an oilfield), which is an example of trapping by updip facies change[8].

The *Oregon Basin* field, on the west side of the Big Horn Basin, is an anticline with two culminations. The oil reservoirs are the Phosphoria Permian limestone, the Pennsylvanian Tensleep sandstone and the Mississippian Madison limestone.

Other important anticlinal reservoirs of the Big Horn Basin include the *Byron, Lamb, Torchlight, Black Mountain* and *Kirby Creek* fields.

(ii) Wind River Basin

More than 60 oilfields have been found, mainly in the peripheral anticlines of this composite topographic and structural basin; they fall into two general groups—those on the east side of the basin, with a north-west axial trend, and those on the west, paralleling the Wind River Mountains.

Among the most important fields in the first group are the *Poison Spider, Notches* and *South Casper* anticlines. Among those in the second group are the *Maverick Springs, Pilot Butte, Big Sand Draw* and *Lander* anticlines.

No fewer than 17 different reservoir beds are known in this area[9], of which the most important are the Pennsylvanian Tensleep sandstone, Permian Park City formation, Cretaceous Cloverly, Thermopolis (Muddy Sandstone), Frontier and Lance Formations; and the Paleocene Fort Union bed. Exploratory drilling has not yet tested the older rocks in the central, structurally deepest part of the Wind River basin.

---

[7] J. McCaleb & D. Wayhan, *Bull.AAPG*, Oct. 1969, 53(10), 2094.

[8] J. McCaleb & E. W. Willingham, *Bull.AAPG*, 51(10), Oct. 1967, 2122.

[9] W. Keefer, *Bull.AAPG*, Sept. 1969, 53(9), 1839.

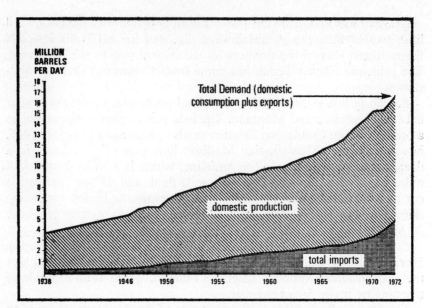

Fig. 39. USA—THE INCREASING GAP BETWEEN DOMESTIC PRODUCTION AND
DEMAND

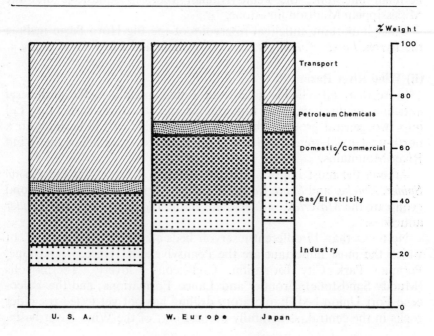

Fig. 40. USA—OIL CONSUMPTION BY SECTORS, 1971 (COMPARED WITH
W. EUROPE AND JAPAN)

Source: BP Statistical Review.

(iii) Laramie Basin

This small basin lies in southeastern Wyoming and is virtually an extension of the Wind River basin. Comanchean and Cretaceous rocks are exposed at the surface, and there are only relatively few oilfields, of which the most important is *Rock River*. This is a simple, prominent anticline trending a little to the west of north, with some 1,600ft of closure. The structure is asymmetrical, the east or basinward flank having dips as high as 70°, while those of the western, mountainward flank do not exceed 35°. Production comes from the Dakota series (Muddy Sands) and the Sundance.

Rex Lake, Medicine Bow and *Simpson Ridge* are similar sharp structures with oil in Colorado and Dakota sands, and some additional Jurassic production.

(iv) Green River Basin

This is the largest of the Wyoming structural and topographic basins, with oil pools along its east and west sides and in the centre. Of these, the eastern group is the most important: *Lost Soldier* is an elliptical dome with 3,000ft of closure, a southwest flank dipping at 45° and a northeast flank dipping at 35°. There are a number of normal dip faults on the northeast flank of the dome which have played an important part in the accumulation of oil. Production from this field comes from several reservoirs, of which the most prolific have been the Cretaceous Dakota and Lakota sandstones, the Jurassic Sundance and the Pennsylvanian Tensleep. Current production is about 4·2 MM brl/a.

Wertz is a dome contiguous to Lost Soldier but with its crest some 2,000ft structurally lower. It is an asymmetrical, elliptical structure with a closure of 1,400ft, and a north-west axial pitch of 13°. Faulting is present but is not so important as on the Lost Soldier structure; production comes from Dakota and Tensleep reservoirs.

Mahoney is a broad flat dome with a sharply dipping northern flank; *Bunker Hill* is a smaller structure. *Ferris* is a sharp elongated anticline with maximum dips on its flanks of 45° to the northeast and 25° to the southwest. Oil in these fields comes chiefly from Dakota sands, with minor Sundance and Tensleep production.

The western fields include *Big Piney*—really a group of several anticlines and stratigraphic traps[10]. These are oil and gas accumulations, and include the *Chimney Butte, Hogsback, Tip Top* and *La Barge* fields. The reservoir is composed of a series of lenticular sandstones ranging in age from Paleocene to Lower Cretaceous. Stratigraphic trapping conditions are dominant in the Tertiary, with some structural control in the lower beds.

The *Baxter Basin* field, in the middle of the basin, is a long, faulted

---

[10] M. L. Krueger, *AAPG*, "Natural Gases of N. America", 1968, 4, 780.

dome, with gas from the Frontier, Dakota and Sundance. It is the largest structure in the area, with a length of 90 miles and a width of 12 miles.

(v) Powder River Basin

This basin covers the northeastern corner of Wyoming and the southwestern corner of Montana; it contains the most important oilfields in Wyoming.

Salt Creek was discovered in 1889 and has proved to be a very large oil producer. It is a well-marked anticline on the eastward slope of the Big Horn Mountains, running roughly north-south and extending for about 20 miles (including the Teapot dome). The average width of the structure is 5 miles and it is slightly asymmetrical, dips on the west flank ranging from 15°–29° and on the east flank from 5°–10°.

Complicated cross-faults are present which have played a significant part in the accumulation of oil and gas. The field is divided by these faults into three main sections:

(i) The northern *Shannon* dome, which is really a faulted segment on the main anticlinal axis. This was the discovery pool and is now virtually exhausted.

(ii) The central *Salt Creek* dome, which is the main producer, with a closure of 1,600ft and 22,000 producing acres.

(iii) The southern extension of the *Teapot* dome, a plunging anticline limited on the north by transverse and reverse faults. This is essentially a fault-block structure rather than a true domal uplift.

Oil is produced from Cretaceous, Jurassic, Pennsylvanian, and Mississippian reservoirs, of which the second Wall Creek sand of the Cretaceous Colorado formation has been the most important producer. Current production is about 12 MM brl/a, and total production from this field is approaching 500 MM brl.

Lance Creek, discovered in 1918, has also been a prolific producer from a large asymmetrical anticline, whose steeper flank dips 15°–32° on the west and northwest, while the opposite flank is very gently inclined at 2°–4°. Production has chiefly come from the Wall Creek, Muddy, Dakota and Lakota sands of the Colorado and Dakota series, but the field is now nearly exhausted.

Mule Creek is about 100 miles east of Salt Creek. Oil comes from a double anticline, trending north-south.

Other important structures in this basin are *Big Muddy, North Casper Creek, Billy Creek, Osage, Upper Thornton.* Oil has been found in Montana, Colorado and Dakota sandstones, with deeper Tensleep oil.

The discovery of the *Bell Creek* field in the north of the Powder River Basin in 1967 (actually in Montana), opened up a new phase of exploration activity in the basin, most of which lies in Wyoming. The accumulation here is a stratigraphic (lithologic) trap in Muddy (Lower Cretaceous) offshore sand-bar reservoirs, some of which are 15 miles long. A number of often similar stratigraphic or structural-stratigraphic traps have since been found in the same area, notably *Hilight, Ute, Kitty, Recluse, Collums, Gas Draw*, etc. Thus, *Hilight* is a local structural "nose" on a very gentle southwest-dipping homocline, with accumulations in a sandbar-type reservoir. *Ute* and *Kitty* are the largest of these fields, with production in 1969 of 4·2 and 3·7 MM brl respectively.

3. Montana

Montana is second to Wyoming as a Rocky Mountains oil-producing state, with a 1972 output of 33·8 MM brl. A large number of oil and gas accumulations have been found at widely separated locations which are related to independent tectonic features.

The most important structural trend is the Sweetgrass Arch, a regional uplift some 70 miles wide, which originates in the Little Belt Mountains and plunges northwest into Alberta. The major gas- and oilfields of the state are associated with this uplift, many of them with accumulations in lenticular sandstone traps of early Cretaceous Kootenai sandstone.

The *Kevin-Sunburst* dome, discovered in 1922, lies only 15 miles south of the Canadian border on the northward-plunging end of the Sweetgrass Arch uplift. Oil is not produced at the culmination of the fold, presumably because of absence of porosity in the reservoir rock at this point, but local structure influences accumulation where porosity is present.

Gas is found in the Sunburst sandstone in Kootenai (Dakota) shales and in the so-called "Ellis sand", a weathered limestone at the unconformable contact of the Jurassic Ellis formation and the Madison (Mississippian) limestone, whose porosity and permeability are very variable. Oil is also found in the Madison limestone proper.

Pondera and *Bannatyne* are comparatively small structures related to the Sweetgrass Arch uplift. They were discovered in 1927 and have yielded oil from the Ellis-Madison contact and the upper part of the Madison limestone.

Border-Red Coulee lies northwest of Kevin-Sunburst and extends into Alberta. It was first developed in 1929 and produces oil from a structural nose in sands of the Kootenai formation.

Cut Bank, found in the same area in 1929, produces oil from lithologic traps on a double anticlinal nose trending northwards. Kootenai sands are the reservoir beds here. This is the largest oilfield in Montana, with an output of about 4·1 MM brl/a.

In the north-central area of Montana there are several small oil- and gasfields associated with folds adjacent to the Sweetgrass Hills and Bearpaw Mountains. These include *Whitlash, Pritchard, Bear's Den, Flat Coulee, Bowes* and *Bowdoin*. Production has principally been of gas from Sunburst and Colorado sandstones.

The central fields in Montana are associated with the Big Snowy uplift. *Cat Creek* in the north consists of a number of elliptical asymmetric domes, intersected by *en échelon* normal faults trending northeast. The principal oil horizons here are sands among the Kootenai shales.

The *Devil's Basin* dome is at the top of the southern slope of the uplift. The *Gage* field, discovered in 1943, is on a minor flexure about halfway down the southern slope. The Jurassic rests here unconformably on the Mississippian and there is also an unconformity within the Mississippian. The Amsden (Upper Mississippian) limestone between these two unconformities is oil-bearing.

Several stratigraphic accumulations have also been found in central Montana, with Piper (Jurassic) and Tyler (Pennsylvanian) reservoirs. Oil and gas also occur in small fields in southern and eastern Montana in local folds with Cretaceous sandstone reservoirs. *Dry Creek* and *Cedar Creek* are the most important of these accumulations.

On the northeastern flank of the main Rocky mountain geosyncline, the Williston Basin covers eastern Montana and extends into Canada and parts of North and South Dakota.

In northeastern Montana, oil has been discovered along a 75-mile long trend on the west side of this basin. The reservoirs in this area are deep Lower Devonian Winnipegosis carbonates, and they probably represent the southeastern corner of a 1,400-mile long Devonian reef development, which stretches as far as the Rainbow Lake area of western Alberta and into northeastern British Columbia. Deeper reservoirs are found in the Ordovician Red River formation. The *Fairview* field is an important recent (1967) discovery.

In the same area, the *Tule Creek* group of fields, developed since 1960, produce oil from Upper Devonian Nisku dolomites. Trapping conditions in these fields may be[11] related to the development of "subsidence structures" as a result of the solution of Lower Devonian salt beds. (The *Bell Creek* field of southeastern Montana is described with the Powder River basin fields of Wyoming on page 242.)

4. The East Flank (North and South Dakota and Nebraska)

The Williston basin continues into the Dakotas. In **North Dakota,** two anticlinal accumulations, *Beaver Lodge* and *Tioga,* discovered in 1951-2, have a combined output which provided most of the state's production of 21·4 MM brl in 1971.

The Devonian Sanish sandstone reservoir of *Antelope* has a very high initial reservoir pressure, indicating that it is an isolated, completely oil-saturated lens deriving oil from the surrounding Bakken shales[12].

---

[11] R. Swenson, *Bull.AAPG*, 51(10), Oct. 1967, 1948,

    & J. Parker, *Bull.AAPG*, 51(10), Oct. 1967, 1929.

[12] G. Murray, *Bull.AAPG*, 52(1), Jan. 1968, 57.

In **South Dakota,** a few small oil accumulations produced only 255,000 brl of oil in 1972.

On the southeastern flank of the Rocky Mountain geosyncline, the Denver-Julesburg basin continues into **Nebraska,** where it is bounded by the Chadron-Cambridge Arch. Production comes from a number of small oilfields in this area, which provided most of the state's output of 9·2 MM brl in 1972.

5. The West Flank (Utah and Idaho)

In **Utah,** *Aneth* in the Uinta Basin is an oil accumulation which has resulted from trapping conditions provided by porosity changes on a structural nose. The output is nearly half the oil production of the whole state (25·9 MM brl in 1972). *Red Wash* is a similar combination trap, an east-west structural nose with stratigraphic trapping of both oil and gas in Cretaceous sandstones. Gasfields have been found at *Ashley, Cisco* and *Carbonera.* In the north of the state, the extreme southern edge of the Green River Basin contains the *Bridger Lake* oilfield with a Dakota sandstone reservoir. In northeast Utah, a productive trend discovered in 1970 comprises several Tertiary Green River sandstone fields, centred around the important accumulations of *Bluebell* and *Roosevelt.*

Idaho is one of the few United States with no gas or oil production.

6. The "Four Corners" Intermontane Area

This is the term used to describe the area between southwest Colorado, Utah, Arizona, Nevada and northwestern New Mexico, where there are a number of basins outside the main Rocky Mountains system.

In northwestern **New Mexico** the San Juan basin (which extends from Colorado), contains a number of large gas accumulations. The reservoirs are mainly Cretaceous Dakota sandstones; the largest gas accumulation so far found here is the *Blanco Mesaverde—Basin Dakota* field, a stratigraphical homoclinal trap covering more than 630,000 acres. Relatively small amounts of oil come from shallow Cretaceous Dakota sandstone fields such as *Rattlesnake, Bisti* and *Horseshoe* and deeper Pennsylvanian oil and gas have also been found. Thus, the anticlinal *Lust Strawn* accumulation[13] has a reservoir in a fractured biostrome of Pennsylvanian (Strawn Group) age. (The accumulations in southeast New Mexico fall within the West Texas province and are described on page 257).

In northwestern **Arizona** the *Dinah Bi Keyah* oilfield is of special geological interest[14]. Situated on the Lukachukai anticline on the southwestern rim of the San Juan Basin, it was discovered in 1967 as

---

[13] D. Thornton & H. Gaston, *Bull.AAPG,* 52(1), Jan. 1968, 66.
[14] J. McKenny & J. Masters, *World Oil,* Jan. 1970, p. 57.

an accumulation producing mainly from a fractured Pennsylvanian igneous sill.

In **Nevada,** there is only one producer of any importance, the *Eagle Spring* oilfield, discovered in 1954. The trap here has resulted from a combination of folding, faulting, truncation and overlap, and the principal reservoirs are Eocene Sheep Pass carbonates and porous zones in an Oligocene tuff[15]. Total Nevada output totalled only 93,000 brl in 1971.

[15] D. K. Murray & L. Bortz, *Bull.AAPG*, 51(10), Oct. 1967, 2133.

TABLE 70
"GIANT" OIL AND GAS FIELDS OF THE UNITED STATES (1)
(BY REGIONS)

Number*	Field	Original oil Reserves MM brl	Number*	Field	Original oil Reserves MM brl
	APPALACHIA		147	*Camrick*	*g*
7	*Allegany* (NY)	168	148	*Cement*	135
8	*Bradford* (Pa)	658	149	*Cushing*	455
	ARKANSAS		150	*Earlsboro*	140
9	*El Dorado* (Ark)	100	151	*Edmund, West*	130
10	*Magnolia*	200	152	*Elk City*	100
11	*Schuler* and *East*	107	153	*Eola-Robberson*	125
12	*Smackover*	525	154	*Fitts*	127
	ILLINOIS		155	*Glenn Pool*	320
55	*Clay City Consolidated*	300	156	*Golden Trend*	485
56	*Dale Consolidated*	100	157	*Healdton*	270
57	*Loudon*	330	158	*Hewitt*	205
58	*New Harmony Consolidated*	174	159	*Keyes*	*g*
59	*Salem Consolidated*	350	160	*Little River*	135
60	*Old Illinois*	675	161	*Mocane-Laverne*	*g*
	INDIANA-OHIO		162	*Oklahoma City*	770
61	*Lima* (Ohio)	925	163	*Putnam*	*g*
62	*Trenton* (Indiana) (all fields)	2,605	164	*Red Oak-Norris*	*g*
	KANSAS		165	*Seminole*	177
63	*Bemis Shutts*	350	166	*Sho-Vel-Tum*	900
64	*Chase-Silica*	260	167	*Sooner Trend*	100
65	*El Dorado* (Kan)	280	168	*St. Louis*	210
66	*Gorham*	100	169	*Tonkawa*	133
67	*Greenwood*	*g*			
68	*Hall-Gurney*	137		**ROCKY MOUNTAINS**	
69	*Hugoton*	*g*		**Colorado**	
70	*Kraft-Prusa*	120	170	*Rangely*	536
71	*Trapp*	220		**Utah**	
	KENTUCKY		171	*Aneth*	452
72	*Big Sandy*	*g*	172	*Red Wash*	135
73	*Big Sinking*	100		**Montana**	
	NEW MEXICO		173	*Cut Bank*	200
133	*Caprock and East*	140	173a†	*Bell Creek*	150
134	*Denton*	175		**North Dakota**	
135	*Empire Abo*	100	174	*Beaver Lodge*	103
136	*Eunice*	144	175	*Tioga*	155
137	*Hobbs*	196		**Wyoming**	
138	*Jalmat*	100	176	*Big Piney*	*g*
139	*Maljamar*	100	177	*Elk Basin* (also Montana)	400
140	*Monument*	250	178	*Garland*	110
141	*Vacuum*	285	179	*Grass Creek*	145
142	*Blanco Mesaverde*	*g*	180	*Hamilton Dome*	200
	OKLAHOMA		181	*Lance Creek*	120
143	*Allen*	120	182	*Lost Soldier*	150
144	*Avant*	107	183	*Oregon Basin*	187
145	*Bowlegs*	160	184	*Salt Creek*	510
146	*Burbank*	500	184a†	*Hilight*	100
			185	*Wertz*	100

* *Numbers refer to map on pp. 248-9.*
  g is a gas accumulation.
† *not shown on map—discovered since 1968.*

Source of data: AAPG "Geology of Giant Petroleum Fields", ed. M. Halbouty, 1970.

TABLE 71

"GIANT" OIL AND GAS FIELDS OF THE UNITED STATES (2)
(BY REGIONS)

Number*	Field	Original oil Reserves MM brl	Number*	Field	Original oil Reserves MM brl
	ALASKA			**NORTH LOUISIANA**	
1	Cook Inlet	g	74	Black Lake	107
2	Granite Point	175	75	Caddo-Pine Island	400
3	Kenai	g	76	Cotton Valley	162
4	McArthur River	160	77	Delhi	225
5	Middle Ground Shoal	200	78	Haynesville	175
6	Swanson River	250	79	Homer	100
6a†	Prudhoe Bay	10,100	80	Lake St. John	116
			81	Monroe	g
	CALIFORNIA		82	Rodessa	175
	(San Joaquin Valley)		83	Sligo	g
13	Belridge S.	200			
14	Buena Vista	615		**SOUTH LOUISIANA**	
15	Coalinga Eastside	395		**Onshore**	
16	Coalinga Nose (E.Ext.)	484	84	Bastian Bay	g
17	Coalinga Westside	237	85	Bay St. Elaine	153
18	Coles Levee N.	160	86	Bayou Sale	200
19	Cuyama S.	232	87	Bourg	g
20	Cymric	142	88	Caillou Island	500
21	Edison	120	89	Cote Blanch Bay W.	100
22	Elk Hills	1,300	90	Deep Lake	g
23	Fruitvale	105	91	Delta Farms	140
24	Greeley	116	92	Duck Lake	g
25	Kern River	615	93	Erath	150
26	Kern Front	147	94	Garden Island Bay	120
27	Kettleman North Dome	475	95	Golden Meadow	200
28	Lost Hills	127	96	Grand Bay	225
29	McKittrick Main Area	160	97	Hackberry East	90
30	Midway-Sunset	1,193	98	Hackberry West	134
31	Mount Poso	177	99	Iowa	128
32	Rio Bravo	135	100	Jennings	132
			101	Lafitte	220
	CALIFORNIA		102	Lake Arthur	g
	(Salinas Valley District)		103	Lake Barre	250
33	San Ardo	276	104	Lake Pelto	185
34	Cat Canyon W.	145	105	Lake Washington	300
35	Elwood	105	106	Leeville	150
36	Orcutt	145	107	Paradis	100
37	Rincon	115	108	Quarantine Bay	150
38	Santa Maria Valley	160	109	Timbalier Bay	300
39	South Mountain	148	100	Venice	215
40	Ventura	818	111	Vinton	132
			112	Weeks Island	237
	CALIFORNIA		113	West Bay	210
	(Los Angeles Basin)			**Offshore**	
41	Brea-Olinda	350	114	Bay Marchand, Blk. 2	600
42	Coyote East	100	115	Eugene Is., Blk. 32	g
43	Coyote West	230	116	Eugene Is., Blk. 126	125
44	Dominguez	255	117	Grand Isle, Blk. 16	175
45	Huntington Beach	934	118	Grand Isle, Blk. 47	100
46	Inglewood	300	119	Main Pass, Blk. 35	100
47	Long Beach	880	120	Main Pass, Blk. 69	300
48	Montebello	186	121	South Pass, Blk. 24	750
49	Richfield	160	122	South Pass, Blk. 27	310
50	Sante Fe Springs	615	123	S. Timbalier, Blk. 135	**
51	Seal Beach	195	124	Vermilion, Blk. 39	g
52	Torrance	190	125	West Delta, Blk. 30	400
53	Wilmington	2,600	126	West Delta, Blk. 73	**
53a†	Dos Quadros	100			
				GULF STATES	
	CALIFORNIA (North)			**Alabama**	
54	Rio Vista	g	127	Citronelle	120
				Mississippi	
			128	Baxterville	200
*Numbers refer to map on pp. 248-9.			129	Cranfield	g
			130	Gwinville	g
†Not shown on map—discovered since 1968.			131	Heidelberg	110
			132	Tinsley	220
**Reserves not yet defined.					

Fig. 41. "GIANT" OIL AND GAS

Numbers refer to the
Reproduced by permission of the Amer
of Giant Petroleu

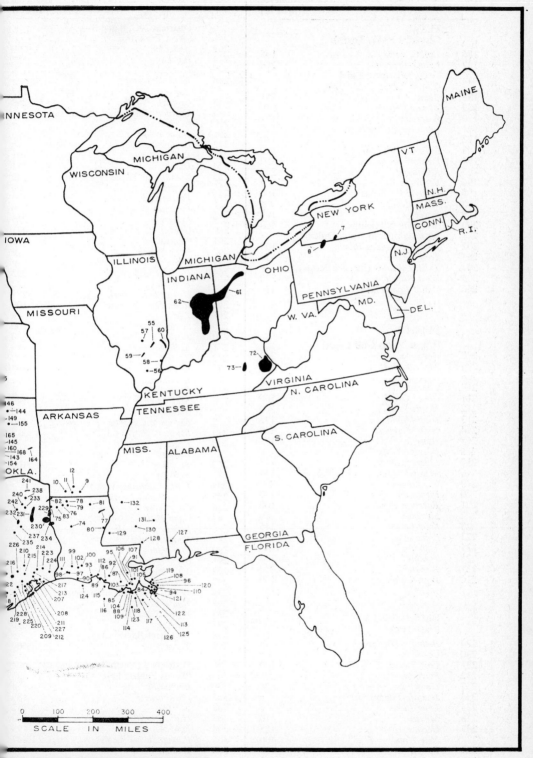

MULATIONS OF THE UNITED STATES

en on pages 247, 250 and 251.
sociation of Petroleum Geologists from "Geology
" 1970. Ed. M. Halbouty.

s

250

Number*	Field	Original oil Reserves MM brl
	TEXAS **District 1 (S.W. Texas)**	
186	*Darst Creek*	145
187	*Luling-Branyon*	164
	Texas Districts 2 and 4 **(Lower Texas Gulf Coast)**	
188	*Agua*	261
189	*Alazan, North*	100
190	*Borregas* (all fields)	151
191	*Government Wells N.*	100
192	*Greta*	130
193	*Kelsey* (all fields)	110
194	*LaGloria & South*	150
195	*Plymouth*	200
196	*Red Fish Bay*	g
197	*Refugio* (all fields)	150
198	*San Salvador*	g
199	*Saxet*	100
200	*Seeligson* (all fields)	460
201	*Stratton* (inc. in Agua Dulce)	
202	*Tijerina-Canales-Blucher*	100
203	*Tom O'Connor*	450
204	*Tulsita Wilcox* (inc. N. Pettus)	g
205	*West Ranch*	280
206	*White Point, East*	110
	District 3 **(Upper Texas Gulf Coast)**	
207	*Anahuac*	266
208	*Barbers Hill*	130
209	*Chocolate Bayou*	g
210	*Conroe*	605
211	*Goose Creek*	160
212	*Hastings*	500
213	*High Island*	175
214	*Hull-Merchant*	240
215	*Humble*	172
216	*Katy* and *North*	g
217	*Liberty South*	143
218	*Magnet-Withers* (all fields)	134
219	*Old Ocean*	200
220	*Pierce Junction*	140
221	*Raccoon Bend*	100
222	*Sheridan*	80
223	*Sour Lake*	145
224	*Spindletop*	180
225	*Thompson*	345
226	*Tomball*	104
227	*Webster*	450
228	*West Columbia*	190
	Districts 5 and 6 **(East Texas)**	
229	*Bethany-Waskom*	g
230	*Carthage*	g
231	*East Texas*	5,110
232	*Fairway*	201
233	*Hawkins*	526
234	*Joaquin-Logansport*	g
235	*Long Lake*	g
236	*Mexia*	110
237	*Neches*	185
238	*New Hope*	100
239	*Powell*	135
240	*Quitman*	112
241	*Talco*	250
242	*Van* (and shallow)	460

Number*	Field	Original oil Reserves MM brl
	Districts 7–C, 8 and 8–A **(West Texas)**	
243	*Andector*	140
244	*Big Lake*	132
245	*Block 31*	163
246	*Brown-Bassett*	g
247	*Cogdell*	157
248	*Cowden, North*	260
249	*Cowden, South* (inc. in Foster)	
250	*Coyanosa*	g
251	*Diamond-M*	495
252	*Dollarhide*	164
253	*Dora Roberts*	101
254	*Dune*	151
255	*Emma* and *Triple N*	120
256	*Emperor*	100
257	*Foster* and *Johnson*	320
258	*Furhman-Mascho*	100
259	*Fullerton*	275
260	*Goldsmith*	471
261	*Gomez*	g
262	*Grey Ranch*	g
263	*Hamon*	g
264	*Headlee* and *North*	202
265	*Hendrick*	272
266	*Howard Glasscock*	283
267	*Iatan-East Howard*	100
268	*J.M.*	g
269	*Jameson*	100
270	*Jordan*	100
271	*Kelly-Snyder*	1,188
272	*Kermit*	120
273	*Keystone*	303
274	*Levelland*	250
275	*Lockridge*	g
276	*McCamey*	140
277	*McElroy*	350
278	*Means* and *North*	120
279	*Midland Farms* (all)	212
280	*Pegasus*	137
281	*Penwell*	103
282	*Prentice*	100
283	*Puckett*	g
284	*Rojos Caballos* (West)	g
285	*Russell* and *North*	130
286	*Salt Creek*	100
287	*Sand Hills*	197
288	*Seminole* and *West*	200
289	*Shafter Lake*	100
290	*Slaughter*	440
291	*Spraberry Trend area*	280
292	*TXL* and *North*	285
293	*Waddell*	110
294	*Waha, West* (Ellenburger)	g
295	*Ward, South*	125
296	*Ward-Estes, North*	307
297	*Wasson*	650
298	*Worsham-Bayer*	g
299	*Yates*	1,500
	Districts 9 (North Texas) and 10 **(Texas Panhandle)**	
300	*Boonsville*	g
301	*Burkburnett*	189
302	*Electra*	182
303	*Hull-Silk-Sikes*	100
304	*K.M.A.*	195
305	*Panhandle* (all fields)	1,647
306	*Walnut Bend*	100

Source of data: AAPG "Geology of Giant Petroleum Fields", ed. M. Halbouty, 1971.

CHAPTER 9

The United States (2)

IV. TEXAS

THE state of Texas is one of the world's major oil areas, producing about 35% of the output of the entire United States. In 1971, in fact, the 1,223 MM brl that came from Texas made up a production greater than that of Libya, and nearly as great as that of Venezuela, in that year. Furthermore, the proved liquid oil reserves of Texas (16·1 B brl at end-1971) amounted to nearly 45% of the total for the whole of the United States outside Alaska, indicating its even greater potential productivity, which was held in check only by the "prorationing" laws. The "allowable" factor for production in Texas, fixed periodically by the Texas Railroad Commission (TRC), which for administrative and statistical purposes divides the state into a number of districts (Table 72) was for many years about 30% of the maximum economic rate. However, when fuel shortages are expected this figure is increased as necessary: thus, in August, 1970, it reached 70%, and more recently it has been 100% *(see footnote, p. 347)*.

The first commercial oil was found in Texas in 1889 in Nacogdoches County, and since then many hydrocarbon accumulations of every type have been discovered, including one of the largest oilfields in the world. There are no fewer than 99 known "giant" oilfields and 22 "giant" gasfields in Texas. About 172,000 oil-wells and 31,000 gas and distillate wells are currently functioning in the state, including the world's deepest oil and gas wells, while 7,000-8,000 new wells are drilled there every year.

TABLE 72

TEXAS RAILROAD COMMISSION DISTRICTS—OIL OUTPUT, 1972

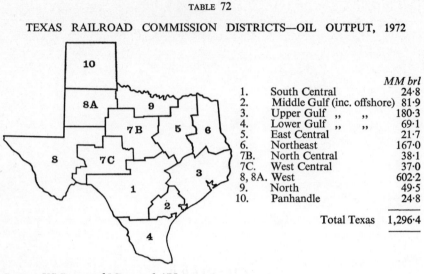

		MM brl
1.	South Central	24·8
2.	Middle Gulf (inc. offshore)	81·9
3.	Upper Gulf ,, ,,	180·3
4.	Lower Gulf ,, ,,	69·1
5.	East Central	21·7
6.	Northeast	167·0
7B.	North Central	38·1
7C.	West Central	37·0
8, 8A.	West	602·2
9.	North	49·5
10.	Panhandle	24·8
	Total Texas	1,296·4

Source: US Bureau of Mines and API.

The oilfields of Texas can best be considered in groups related to major tectonic features. (These do not coincide, of course, with the TRC—Texas Railroad Commission—Districts, except very generally).

1. NORTH TEXAS (TRC Districts 7B and 9)

A series of Palaeozoic fold fields lies along the crest and flanks of the Red River uplift on the southern flank of the Ouachita-Amarillo uplift in North Texas.

Petrolia was the first of these accumulations to be discovered, in 1907. The surface structure here is a dome, coinciding with a topographic high in the Permian Wichita (Red Beds). Below the Permian beds on the flanks of the structure is a great thickness (some 2,500ft) of Pennsylvanian strata; but towards the crest this thickness is much reduced as a result of an uplift of Arbuckle Ordovician limestone over an underlying ridge.

Several other anticlinal fields of this type have been found— *Burkburnett, Electra, KMA* and *Walnut Bend*. Production in these fields comes from Pennsylvanian and Permian sandstones. The discovery of an Ordovician reservoir in the *KMA* field in 1940 resulted in considerable deeper development in this area.

An exception to the generally anticlinal accumulations of North Texas is the large *Boonsville Bend* gasfield which was found in 1950 as a stratigraphic trap due to lensing in an area of homoclinal dip. *Hull-Silk-Sikes* (1939), an oil accumulation in a combination trap of sand lenses on an anticlinal "nose", produces more than 4 MM brl/a.

The most important structural trend in North-Central Texas is the Bend Arch, which runs northwards for about 150 miles from the Llano-Burnett uplift. This is a subsurface uplift which has been superimposed on the gently northward-dipping homocline formed by the upthrust of the Llano Mountains.

The Bend Arch has a northward pitch of about 35ft/mile and gentle dips of 30-40ft/mile on either flank. Cross-folds and minor culminations occur on the main arch, and in these structures oil accumulations occur along the axis of the main trend. Varying degrees of porosity also influence the location of oil pools. The best reservoir rock has been the Lower Pennsylvanian Marble Falls limestone of the Bend series, but oil has also been found in stratigraphically higher Cisco and Strawn lensing sands and porous limestones.

The Bend Arch oilfields include the *Breckinridge* and *Ivan* domes in Stephens County, the *Ranger* and *Eastland* structures in Eastland County, and the *Strawn* dome in Palo Pinto County.

Most of the fields in this area have shown a rapid decline in output rate once production has started. Owing to the difficulty of locating porous zones on the general uplift, more dry holes have been drilled in the Bend Arch than in any other part of Texas. The source of the oil is probably the Pennsylvanian shales.

2. TEXAS PANHANDLE (TRC District 10)

The continuation of the Amarillo-Ouachita trend into the north-west corner of Texas is associated with a belt of hydrocarbon accumulations more than 100 miles long and 5 to 40 miles wide, which is known from its shape as the Texas Panhandle. It is one of the most important natural gas producing areas in the world, with an annual output of about 600 Bcf.

The basic structure here is a long anticlinal arch which has resulted from the deposition and compaction of sediments over the granites and metamorphic rocks of the buried Amarillo Mountains. There are a number of local crests in which concentrations of oil and gas occur, and two major faults running northwest and southeast.

The reservoir beds are Permian dolomites and the Pennsylvanian Grey-Lime and arkosic Granite Wash. Gas has been produced since 1916, and oil was first found in 1925. There are now a large number of oil pools in a band some 90 miles wide along the northern flank of the *Amarillo* uplift, from which oil is produced from a number of separate culminations; these pools are usually considered as forming together the *Panhandle* oilfield. The largest of the subsidiary pools are those of *Borger-Pantex* and *West Pampa*. The *Panhandle* field (*sensu lato*) produces about 25 MM brl/a of oil, and there is also a very large output of gas from reservoirs in the higher Permian dolomites. These

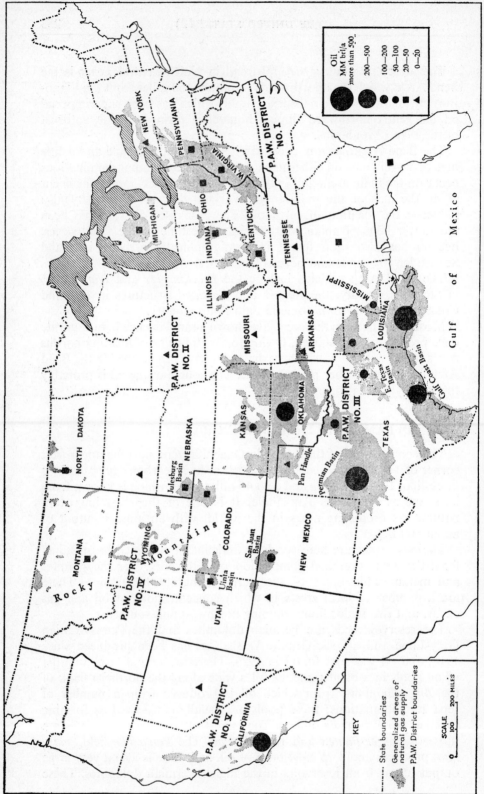

Fig. 42. MAJOR OIL AND GAS PRODUCING AREAS OF THE UNITED STATES

beds are arched across the Amarillo uplift to form an almost continuous gasfield connected with the *Hugoton* reservoir of Kansas (p. 233).

3. WEST TEXAS BASIN (including TRC Districts 7C, 8 and 8A)

A large Palaeozoic sedimentary basin lies on the western flank of the Bend Arch in West Texas and the southeast corner of New Mexico, from which about 670 MM brl/a of oil and large volumes of natural gas are currently produced. It contains no fewer than 57 "giant" fields in Texas, of which 47 have been classified as structural traps, 9 as stratigraphic or reef traps, and one is a combination trap—although several of the structural traps are also affected by local porosity and permeability variations. In southeastern New Mexico, there are a further nine major anticlinal oilfields.

Within the West Texas Basin, hydrocarbon accumulations are found in anticlines, fault structures, stratigraphic traps and reefs. Carbonate rocks are the predominant reservoirs, although oil and gas also occur in sandstones and siltstones. Production comes from Cretaceous to Cambrian beds at depths from 100ft down to 24,500ft, but the Permian carbonates form by far the most important reservoirs.

A broad, flat buried uplift runs from the Yates area northwest across the Basin; on its southwestern flank lie the smaller Delaware and Val Verde basins. This Central Basin platform was probably formed on a reef of the Permian Big Lime formation. After uplift, it was eroded and dolomitized by circulating ground waters and thus developed extraordinary porosity, before being sealed by the deposition of evaporites. Oil and gas accumulations in limestone or dolomite reservoirs lie on both flanks of the platform, and many individual wells have produced very large outputs of oil. Shallow Pennsylvanian beds and the underlying Ellenburger (Ordovician) are also locally productive reservoirs.

Typical of these accumulations is *Yates,* one of the largest oilfields in the United States. This is a bifurcating surface fold, with oil accumulations in porous zones of the Big Lime and also in shallower limestones and sands. A major fault cuts the pre-Triassic rocks along the southeast side of the field. Although it has produced more than 500 MM brl of oil since its discovery in 1926, *Yates* probably still holds nearly 1·0 B brl of recoverable oil.

Big Lake is another surface dome which reflects a more pronounced structure in the Permian beds. The most prolific of these are the "Texon" zones of the Permian Big Lime, but oil is also found in Pennsylvanian and Ordovician reservoirs. As with Yates, there is a strongly developed fault in the pre-Triassic rocks in this field, separating it from the central uplift. *Big Lake* has been producing oil since 1923 and is now approaching exhaustion.

Westbrook was the "discovery pool" of the province in 1920. It is an irregular anticline with a curving axis and some reversal on the east side. Production has come from dolomitic zones in the Clear Fork portion of the Big Lime.

Many other Palaeozoic oil accumulations lie along the flanks of the uplift; most of these are anticlinal traps.

The largest oilfield in the West Texas basin is *Kelly-Snyder*, with oil in a homoclinal stratigraphic accumulation and also reef-type reservoirs. Discovered in 1948, this field has an annual output of about 22 MM brl/a and its original producible reserves were estimated to have been about 1,188 MM brl, making *Kelly-Snyder* one of the largest US "giant" fields.

In the Val Verde and Delaware basins, several large anticlinal gasfields have been found, with Palaeozoic reservoirs usually at depths of between 15,000 and 22,000ft. The most important producing formations are Ellenburger (Ordovician) dolomites or Devonian carbonates. Probably the largest of these gasfields, discovered in 1952, is *Puckett*, with original gas reserves of some 6·5 Tcf. This field lies on the northern flank of the Val Verde Basin, on a northwest extension of a structural "nose". It has gas reservoirs in Permian Wolfcamp sandstone, and below an unconformity, in Devonian cherty limestone and Ordovician Ellenburger dolomite. The Ordovician reserves make this one of the largest gasfields in the United States.

Brown-Bassett in the east-central part of the Val Verde Basin is on a northwest segment of a faulted anticline. It has large gas reserves in a thick Ellenburger dolomite section, but the gas is 40-50% carbon dioxide, which greatly reduces its economic value.

Coyanosa in the easternmost part of the Delaware Basin has produced small amounts of Cretaceous oil from a shallow anticlinal structure but has very large gas reserves in deeper Pennsylvanian, Devonian and Ordovician beds.

Other major deep gasfields found in this area in recent years include *Clara Couch, Gomez, Sonora, Toyah, Atoka Penn, Indian Basin, Grey Ranch, Oates Northeast, Worsham-Bayer* and *Lockridge*. In addition to these deep Palaeozoic reservoirs, a number of relatively shallow Mesozoic stratigraphic traps occur along the north and west sides of the Delaware Basin[1].

The Central Basin Platform is separated by the Midland Basin from the Eastern Platform. The latter area is the site of oil and gas accumulations in reef-type reservoirs in Scurry County.

On the platform north of the Central Basin uplift, several large accumulations have been found, notably *Slaughter* and *Wasson*. The former is a stratigraphic trap and the latter is a dolomitized Permian reef accumulation which has also been affected by local folding.

---

[1] J. Hills. *AAPG*, "Natural Gases of N. America", 2, 1968, 1394.

In the Midland Basin, the large *Spraberry Trend* producing area runs roughly north-south. The reservoir here is the Dean-Wolfcamp (basal Permian) siltstone, and there are some 134,000 productive acres, with accumulation due to secondary fracturing which has produced a permeable zone on a homocline. Overall production from this area is about 20 MM brl/a.

The origin of the hydrocarbons in the West Texas Basin is a subject of controversy, since most of the Permian "redbeds" and evaporite deposits show that conditions were unfavourable to organic life. It is generally believed that the oil had a deeper source, coming up either along fault planes or by the overlapping of the Ordovician.

Exploration drilling is still being carried on vigorously in this area, in which some 2,050 wells were drilled during 1971, including a number to depths of more than 15,000ft. A computer-based system for retrieving data from the logs of the more than 135,000 wells that have been drilled to date in the West Texas Basin has recently been set up.

In southeastern **New Mexico,** there are several important oil accumulations on the northwestern extremity of the West Texas Central Basin uplift, or on the shelf areas beyond it. These include *Monument, Eunice, Jalmat, Vacuum* and *Hobbs,* all of which are anticlinal structures with Permian reservoirs. *Vacuum* is by far the largest oil producer, and *Jalmat* and its associated structures has important gas as well as oil production. However, the fields of northwestern New Mexico lie within a different petroliferous province and were described on p. 245.

4. BALCONES AND MEXIA-TALCO FAULT ZONES
(parts of TRC Districts 1, 2 and 4)

The sediments of eastern Texas dip southeastwards away from the Ouachita-Amarillo and Llano uplifts to form a large basin east of the Bend Arch. This basin has been subjected to regional uplifts and deformations and is cut by two roughly parallel major fault systems—the Balcones and the Mexia-Talco Fault Zones—made up of a series of *en échelon* faults, which have probably resulted from the subsidence of the Gulf Coast sediments to the south. The fault lines curve round the southeastern end of the Llano uplift of central Texas in a great arc.

Hydrocarbons are found in typical fault traps in several reservoirs, of which the most prolific are the Woodbine sand, the Edwards limestone of the Fredericksburg formation and the Cretaceous Austin Chalk. Sand lenses in the Navarro series also contain oil.

(a) Mexia-Talco Fault Zone Fields

In the Mexia and Talco fault zone, accumulations are found along a roughly north-south line of structural folds flanking the major strike faults. The westward flanks of the folds are the steeper and give evidence of "drag" into the downthrown fault-blocks.

The *Mexia* pool was remarkable in that, within a year of its discovery in 1921, output had reached the tremendous rate of 300,000 b/d; it is now, however, nearing exhaustion. The structure consists of an anticline about 7 miles long and 1½ miles wide, cut by a major fault on its west side. The crest of the anticline, which is asymmetric, lies to the east of the fault line, but is closer to it at depth than at the surface.

Some oil is found in the shallow Nacotoch sandstone, but principal production has been from the Cretaceous Woodbine sand. This bed is saturated with oil only on the upthrown or eastern side of the fault, where it is about 250ft higher than on the downthrown side, and sealed against the impermeable Austin Chalk and Eagle Ford shales. The source bed is probably the Eagle Ford shales or the shales in the Woodbine formation.

The *North* and *South Groesbeck* fields are two crestal accumulations on the continuation of the Mexia axis. Production in these fields also comes from Woodbine sands.

North Currie is an anticline about two miles long by a mile wide, with low structural relief, which is primarily a gas producer. Trapping has occurred on the eastern upthrown side of the fault, in Woodbine sands. The nearby *Currie* structure is more irregular.

The *Wortham* anticline crest lies east of the fault at the Woodbine horizon. The average dip of the fault plane here is 45°.

The *Richland* anticline is flat and irregular, probably much affected by "drag" into the fault zone.

Powell, which includes parts of the old *Corsicana* field, has oil in a fault-sealed "nosing" structure about eight miles long and a mile wide. The fault displacement is 650ft, and the fault steepens from 50° at the surface to 56° in the Austin Chalk. This large accumulation, whose overall output has exceeded 130 MM brl, is also nearly exhausted.

In a minor fault belt a few miles north of the Mexia zone (the Tehuacana fault zone) several other accumulations occur. *Nigger Creek* was the first pool to give oil from a fault structure up-dip from the Mexia fields. It lies in a "graben" between the Balcones and Mexia faults, and contains oil in the Woodbine formation sealed by a westward-dipping fault against the Austin Chalk. *Talco* in northeastern Texas was, when it was discovered in 1933, the largest of these fault fields. It is a structural "nose" sealed by a major fault, with main production from the lenticular sands of the Paluxy formation underlying the Woodbine. The field has produced nearly 350 MM brl of oil since its discovery.

(b) Balcones Fault Zone Fields

The *Luling* accumulation is a fault structure some 20 miles southeast of the main Balcones fault. Here, a homocline is intersected on the northwest, northeast and southwest sides by a system of faults whose average displacement is 450ft, and whose downthrow is towards the northwest.

An accumulation of oil was found here in 1922 in the Edwards limestone of Comanchean age, sealed by faulting against impermeable beds. In 1928, *New Luling* and *Roxana Jolly* were discovered in the same area, both further examples of fault accumulation. The Edwards limestone contains oil principally in its upper porous dolomitic zones, and is probably also a source rock.

Lytton Springs, discovered in 1925, is interesting in that production here has come from an igneous rock reservoir, although accumulation has also been affected by the general Balcones faulting. Oil has been found in the top 200ft of an igneous "serpentine" rock which is capped by impermeable Austin Chalk and Taylor marl. The igneous material is partly volcanic ash and lava, and partly decayed basalt. The sediments above the igneous mass are domed as a result of the extrusion.

Thrall and *Chapman* in northeast Texas have also produced oil from a decayed igneous rock, and the arching of the strata above the intrusion is probably the controlling structural feature.

5. THE SABINE UPLIFT AND EAST TEXAS BASIN
(parts of TRC Districts 5 and 6)

The Sabine Uplift is a broad dome-like subsurface structure about 80 miles in diameter, which strikes northwest—southeast across the northeastern districts of Texas and Louisiana and is the dominating structural influence for oil and gas accumulations in these areas. It probably had its origin in Comanchean times and uplift finished in the Mid-Tertiary. Several important fields associated with this uplift occur in eastern Texas, south Arkansas and north Louisiana.

One of the largest oil accumulations in the world, and the largest in the USA outside Alaska, is the *East Texas* field in the East Texas or Tyler basin, which lies on the western flank of the Sabine Uplift and is really part of the Gulf Embayment province. Discovered in 1930, the productive area covers some 140,000 acres. The structural conditions are very simple, the accumulation being due to the existence of a large-scale stratigraphic "pinch-out" trap—the Woodbine shoreline sands lying unconformably on truncated Lower Cretaceous strata.

Production comes from a thickness of some 1,000ft of Eagle Ford-Woodbine sandstones. The original recoverable reserves of this field were about 5·1 B brl, of which 1·3 B brl were unproduced at the beginning of 1969. Annual production is about 48 MM brl, depending on the current "allowable" factor.

Another important oil accumulation in the Tyler basin is *Van,* a faulted anticline with Woodbine sandstone reservoirs, from which nearly 350 MM brl of oil has been produced since discovery in 1929; the field still has an output of about 5·7 MM brl/a.

Fairway is a combination trap resulting from permeability variations in a Lower Cretaceous limestone (Pearsall Formation) on a structural nose. When this field was found in 1960, it had initial recoverable oil reserves estimated at 200 MM brl. *Joaquin-Logansport,* discovered in 1936, is a large gasfield extending into North Louisiana which is classified as a combination trap, since both anticlinal and stratigraphic trapping conditions are present. *Neches* is an anticlinal accumulation which was discovered in 1953.

6. THE GULF COAST EMBAYMENT IN TEXAS
(a) Southeast Texas and Upper Gulf Coast (TRC District 3)

The crescentic area of thick Cenozoic sedimentation which covers southeast Texas and South Louisiana forms one of the world's most important oil regions, with a production in 1971 of nearly 1,220 MM brl* —more than one-third of the total US output in that year.

The reasons for the oil richness of the Gulf Coast are not hard to find, since in this area all the conditions have existed which are necessary if petroleum accumulations are to be formed and preserved. Thus, throughout Cenozoic times, a huge volume of sediments eroded from the Rocky Mountains and High Plains was mixed with much organic matter and deposited in a geosynclinal depression† along the edge of the Gulf of Mexico. This basin sank at about the same rate as it was filled and thus typical "paralic" environment conditions were maintained for a lengthy period, resulting in the genesis of large volumes of hydrocarbons and their migration in the course of sedimentary compaction into many excellent sandstone reservoirs. This is also a halokinetic area, and the most important tectonic feature has been the relative upward movement of the large number of salt masses which are now found close to the present coast-line of the Gulf of Mexico and beneath the shallow waters of the Continental Shelf. (A smaller group of interior salt domes occurs in eastern Texas and northwest Louisiana—p. 266). Many "timely" traps were formed by the salt movement—anticlinal arches in overlying sediments, homoclines sealed against the salt masses, or fault traps in a variety of complex structures. Finally, the presence of evaporites and thick clay beds has served to seal the reservoirs and preserve their contents.

On the Gulf Coast, salt domes are now generally classified according to their top surface depth: where this is at 0-2,000ft the domes are "shallow", at 2,000-10,000ft they are "intermediate", and below 10,000ft they are "deep". Even the deep-seated domes are often characterised by slight but distinctive circular or ovoid surface expression and radial

---

* *From southern Louisiana plus TRC Districts 1, 2, 3, and 4.*
† *The Gulf Coast geosyncline falls outside the normal definition of a geosyncline, since the basin has not been subjected to orogeny.*

faulting—often with a central "graben"—as well as showing typical geophysical gravity minima.

There is no evidence of lateral pressure in this area in Cretaceous times, so the movement of the salt was probably caused by the vertical pressure of accumulating overburden and the differences of density between the salt and sediments; alternatively, it has been suggested that the salt masses may have grown downwards due to the more rapid sinking of the surrounding strata.

The age of the salt has been considered to be Comanchean, Triassic or Permian; of these alternatives, the last is the most likely, since great thicknesses of Permian salt are known in nearby West Texas. The distribution of salt domes may be related to the presence of underlying faults, but seismic surveys indicate a large-scale east-west pre-Miocene sub-surface salt uplift, off which many of the domes have probably developed,

The discovery and development of salt dome oil in the Gulf Coast area can be considered as falling into several distinct phases:

(i) *Cap rock production* began in 1901 at *Spindletop;* this was soon followed by similar discoveries at *Sour Lake, Batson* and *Humble.*

(ii) *Shallow super-cap sands* were then developed—of Pleistocene age in these fields, and of Miocene age in the later discoveries of *Jennings, Saratoga, Goose Creek* and *Orange.*

(iii) *Miocene flank sand* production began in 1914, on the north side of *Sour Lake* and the east side of *Humble.* Many other salt domes, such as *West Columbia* and *Damon Mound,* where no oil had been discovered in the cap and super-cap beds, were subsequently found to have flank sand reservoirs.

(iv) *Oligocene and Jackson (Eocene) flank sand* reservoirs were then discovered by post-1914 deeper drilling around many of the known salt domes.

(v) *Yegua (Middle Eocene)* reservoir production began in 1929, when still deeper wells were drilled into the flank sands of the *Humble* and *Pettus* domes.

(vi) *Offshore operations* after 1945 have led to the discovery of many important accumulations of oil and gas associated with piercement salt dome uplifts on the Continental Shelf of the Gulf of Mexico. Where these domes reach the surface, they may have 300ft or more of subsea topographic relief and rise sufficiently close to sea level to become sites for subsequent reef growths. The reservoir beds in these offshore fields are Miocene, Pliocene and even Pleistocene sandstones.

In the Gulf Coast area, geophysical surveys have been outstandingly successful in delineating buried salt domes and their associated structures. Since most of the sediments consist of relatively soft sandstones and clays, rotary drilling techniques were extensively used for early exploration in this area, and it was here also that micropalaeontology was first used on a large scale to differentiate similar strata.

In the southeastern part of Texas, the sediments dip gently southwards and the section thickens into the Gulf Coast geosyncline. Hydrocarbon accumulations are found in anticlines, fault structures, unconformity traps, stratigraphic pinch-outs along east-west trends, and most frequently over or around exposed or buried salt domes. There are more than 4,000 flowing and 9,000 pumping oil-wells in this area, from which 180·3 MM brl of oil were produced in 1972, principally from Miocene (Frio, Yegua, Wilcox) sandstones. There are also about 4,300 gas-wells.

(1) Coastal Salt Dome Fields

Only a few of the most interesting examples of the large number of hydrocarbon-bearing structures which have been found in the Texas and Louisiana sectors of the Gulf Coast are described below:

(i) *Spindletop* was the first major salt dome oilfield to be discovered, in 1901. The oil output from this field was spectacular—within only three years of discovery 30 MM brl/a was being produced from an area of only about 260 acres. This field was also remarkable as the site of the first commercial rotary drilling operation in the United States.

Spindletop can be considered as a typical Gulf Coast salt dome. There is a low surface hill of Lissie and Beaumont sands, and the salt mass is found at a depth of about 1,200ft. It is a mile or so in diameter, steep-sided and cylindrical; and the principal oil reservoir is the cavernous limestone cap rock above the gypsum layer overlying the salt. There is much native sulphur in the cap rock, which does not completely cover the salt.

Oil has also been obtained from super-cap and flank sands which are Pliocene, Miocene, Oligocene and sometimes re-worked Cretaceous in age.

(ii) *West Columbia* is an example of a piercement salt dome which produces only from its flank beds. The surface expression of this field is a low rim rising only a few feet above the surrounding prairie. The salt core takes the form of a truncated cone, circular in section with a flat top, which has a cap rock of anhydrite and gypsum. The salt has penetrated sediments from the Eocene upwards, and its motion over a long period of time has probably been related to local faulting.

(iii) *Barbers Hill*. This is a surface mound with typical drainage features, and is underlain by a circular-section piercement salt plug capped with gypsum, anhydrite and limestone. The surrounding sediments have steep dips, often as much as 70°. Oil comes from cap rock and flank sand reservoirs.

(iv) *Goose Creek*. There is little topographical evidence of this subsurface dome, and the discovery was made on the evidence of local gas shows. With development, subsidence has occurred over the dome, presumably because the removal of fluids from the super-cap reservoir sandstones has allowed the clays to compact.

(v) *Damon Mound* is a salt dome field in which production comes from flank reservoirs and is largely affected by lensing of the sands. There is an oval mound at the surface which rises 80ft above the surrounding plain. The shallow salt plug has a cap of gypsum, anhydrite and limestone; flank sandstones and limestone give production down to the Eocene Jackson.

(vi) *Hockley* is one of the largest salt domes on the Gulf Coast, but has produced relatively little oil. There is no surface mound, but instead a number of small hillocks. There are numerous sulphurous gas seepages, and the local vegetation is stunted.

At a depth of about 1,000ft a typical circular salt plug has been found, with a cap rock of anhydrite and limestone; at 4,000ft the diameter of the salt mass has increased to more than 6 miles. Oil has been found in the cap rock and in the flank sands down to the Eocene.

(vii) *Katy* is a simple anticline probably related to a deep-seated salt uplift. Its original recoverable gas reserves were more than 6 Tcf in Yegua (Middle Eocene) sandstone reservoirs, together with a large volume of condensate. The lithology of these reservoir beds is largely deltaic, so it has been suggested that the gas here was formed from terrigenous vegetable matter. However, since at *Conroe,* only 40 miles away the same beds contained 605 MM brl of recoverable oil, this theory does not seem tenable.

(viii) *Conroe,* discovered in 1931, is a broad gentle structure also probably related to deep-seated salt, which is crossed by several normal faults which form a central "graben" area. Production has come from the Eocene Cockfield sandstone.

(ix) *Hastings* is an anticlinal oil accumulation in the form of a much-faulted dome overlying a buried salt mass. The fault pattern is typical of Gulf Coast salt dome structures, with a central graben and many normal faults downthrown towards the graben. The major productive formation is the Frio sandstone, and this field is one of the largest oil producers in the area, with an annual output of some 10 MM brl.

(x) *High Island* has a surface mound and extensive gas seepages. In the subsurface, there is an overhanging cap rock which first produced oil from a calcite zone in 1922. Subsequently, prolific Lower Miocene and Middle Oligocene flank sandstone reservoirs were discovered.

(2) The Interior Domes

In Northeastern Texas, there is a group of interior salt domes in the synclinal area between the Mexia fault zone and the Sabine Uplift. These are usually cylindrical plugs over 5,000ft high with a diameter of $\frac{1}{2}$-$1\frac{1}{2}$ miles within 3,000ft of the ground. The top surface of the salt is usually flat and the flanks dip steeply. Many fragments of Cretaceous rocks are found associated with the salt.

The first interior dome to produce oil (1929) was that of *Boggy Creek* in Cherokee County. Oil was found in the Woodbine sand of basal Cretaceous age, and intensive search has since located a number of other salt dome oil accumulations in this area.

(b) Southwest Texas and Lower Gulf Coast of Texas
(TRC Districts 2 and part of 4)

The Gulf Coast producing trends continue in an arcuate pattern paralleling the coastline of Southwest Texas. Hydrocarbon reservoirs are mainly Frio-Vicksburg (Oligo-Miocene) sandstones, Yegua and Wilcox (Lower Eocene) sandstones, and Edwards Limestone (Cretaceous), the age of the reservoir formation increasing with the distance of the fields from the coast. The principal traps are anticlines and fault closures related to normal faults paralleling the coastline. There are also associated stratigraphic traps.

One of the largest gasfields to be found in recent years is the *Northeast Thompsonville* field, near the Mexican frontier, covering an area of $6\frac{1}{2}$ by 2 miles, and producing from the Wilcox sandstone. The structure here is a simple closure on the high side of a north-south fault. There are a number of other similar "deep" Wilcox sandstone gasfields along a 200-mile trend from the Rio Grande to Houston, all producing gas from reservoir formations of similar ages at depths below 8,500ft.

The most important oilfields (*Tom O'Connor-Great, Seeligson, Agua Dulce-Stratton*) lie on a Frio (Lower Oligocene) producing trend paralleling the shoreline. They are large anticlines which produce important volumes of both oil and gas.

Seeligson is by far the largest accumulation in this group of fields, with an annual output of about 25 MM brl of oil and 120 Bcf of gas. *Tom O'Connor* (discovered in 1934) produces more than 10 MM brl of oil and 40 Bcf of gas yearly.

V. LOUISIANA AND THE SOUTHEAST STATES

LOUISIANA is second only to Texas as an oil-producing state, with an output which is growing rapidly, mainly due to offshore discoveries, and which totalled 947·8 MM brl in 1971. As many as 3,017 oil and gas wells were drilled in Louisiana in that year, of which more than 800 were on the Continental Shelf of the Gulf of Mexico. The state has undisputed ownership of Zone 1—the offshore area nearest to the coastline (still undefined); Zones 2 and 3 are in dispute between Louisiana and the US Federal Government, which has established undisputed ownership of the outermost Zone 4 area.

Nearly 95% of Louisiana production comes from the south, where there are 36 "giant" oilfields and 7 "giant" gasfields, i.e. 43 out of the state's total of 54 "giants". Of these southern "giants" 30 are onshore and 13 are offshore accumulations.

Many onshore and offshore fields are definitely associated with salt uplifts; a number of the other major accumulations are probably structurally associated with salt at depth. Most of the fields have been discovered by seismic surveys, many in the last few years.

(a) North Louisiana

There are 11 "giant" fields in North Louisiana, mainly on the flank of the Sabine Uplift, which is the dominant tectonic feature in this area. These accumulations are related to deep movements, and have Eocene, Cretaceous and occasional Jurassic reservoirs.

The *Monroe* gasfield is the largest gas accumulation in Louisiana, and one of the largest in the United States, covering an area of about 425 sq miles. Gas is produced from two reservoirs, the higher being a chalk of Navarro (Upper Cretaceous) age, and the lower being part Woodbine sand and part Comanchean (Lower Cretaceous). Accumulation here is predominantly due to stratigraphic overlap and lensing on a regional uplift. *Sligo* is also an important anticlinal gas accumulation, with Mesozoic reservoirs.

The largest oilfield in the area is *Caddo-Pine Island,* a large, low anticline discovered as long ago as 1905 on the Texas state boundary by random drilling, and still producing nearly 6 MM brl/a. This is a "bald-headed" structure near the top of the Sabine Uplift, in which oil has accumulated in Comanchean (Lower Cretaceous) beds on the flanks of a truncated anticline, while gas with smaller volumes of oil comes from Upper Cretaceous strata arched over the anticline.

The *Rodessa* field is an elongate anticline running northeast-southwest into both Arkansas and Texas. Trapping here has been controlled by a major fault, so that accumulation is in the uplifted block to the north-west of the fault. A prolific field in the 1930's, with oil coming from several Comanchean sandstones, the output from this field has now declined to about 0·7 MM brl/a.

Black Lake, discovered in 1964, is a combination trap formed by biohermal growth in a Lower Cretaceous limestone over a structural nose. Reserves have been estimated as 105 MM brl.

Other important fields in North Louisiana are *Cotton Valley,* an anticline with additional lenticular accumulation and Jurassic gas and oil production; *Delhi,* a stratigraphic oil accumulation in Upper and Lower Cretaceous sandstones due to varying permeability brought about by truncation on a southeast-dipping homocline; and *Haynesville,* an anticline discovered in 1921 by surface geology.

The first production of oil from Tertiary beds in north Louisiana was at *Urania* in 1924. This is a homoclinal flexure with irregular Wilcox sandstone reservoirs. The *Homer* field, discovered in 1919, is a broad dome traversed by an east-west fault with a down-throw to the south.

T

Production comes from Nacatoch and Oakes (Austin) sandstones. *Bull Bayou* and *Zwolle* are similar accumulations.

In northwestern Louisiana, there is an area of interior salt domes, lying about 150 miles from the Gulf Coast. These domes occur along a northwest-southeast trend and are similar to the coastal salt masses, although generally not productive of oil. Their characteristic surface expression is a central basin encircled by low hills.

The *Vacherie* dome in this area is remarkable in that a thickness of about 130ft of Washita conglomerate lies above the cap rock. This conglomerate is of Cretaceous age and appears to have been carried up to the surface by the rising salt mass.

(b) South Louisiana

The Gulf Coast embayment in South Louisiana covers an area of 49,000 sq miles, two-fifths of which lies on the present Continental Shelf. It contains a thickness of up to 60,000ft of Mesozoic and Cenozoic sediments.

The geological history of the area is one of regression, the coastline of the Gulf of Mexico having retreated southwards to leave great thicknesses of progressively younger beds in successive bands which roughly parallel the present coast and dip southwards towards the basin. Oil and gas have been found in reservoirs which become younger from north to south. Thus, in the south-central part of the state, Wilcox Eocene beds form the reservoirs, and are generally found at considerable depths. Nearer to the coast, where most of the major oil and gas accumulations occur, the reservoirs are mainly sandstones of Lower, Middle and Upper Miocene age. In the offshore fields, Middle and Upper Miocene, Pliocene and Pleistocene reservoirs are successively found, becoming younger in proportion to the distance from the present coastline.

These Cenozoic reservoirs are all sandstones, some marine and others deltaic. Attempts have been made to find a relationship between the nature of the sandstone reservoir rock and the type of hydrocarbon it contains, but no general rule is evident, other than that the gas accumulations tend to occur rather more in marine-facies reservoirs than otherwise, and are more frequent on the western or more unstable sector of the Continental Shelf; whereas oil accumulations tend to favour deltaic sandstones and the stable Shelf sector.

Many of the onshore accumulations are associated with salt dome structures, which as in Texas give rise to characteristic surface topography. Thus, the shallow *Belle Island* salt dome is one of a line of five called the Five Islands, which appear at the surface to be symmetrical hills rising some 100ft from a flat, marshy plain. The Belle Island dome is the only one of this group where oil has been found, and consists of a subcircular mass of salt about 8,000ft in diameter, over which there is a moderately thick cap rock. Surface solution has produced central-basin lakes. It was one of the first salt dome structures drilled on the Gulf Coast (in 1896), due to the local gas seepages; although originally exploration here was unsuccessful, it probably inspired the discovery of *Spindletop* a few years later.

The first commercial oilfield to be established was the *Jennings* dome, drilled in 1901 and followed rapidly by other similar discoveries at *Anse la Butte, Bayou Bouillon, Welsh, Vinton* and *Hackberry*. Some of these fields were not fully developed until deeper drilling found larger reservoirs many years later. Of the accumulations mentioned above, *Jennings* still has a small output, while the *Hackberry East* and *West* domes each produce more than 4 MM brl/a.

At least 150 piercement salt domes have been discovered in South Louisiana—using the term to signify any salt uplift which has actually been reached by drilling. "Growth faults" are also important structural features in this area. The Miocene provides about half the known productive reservoirs, followed by the Oligocene, Pliocene, Eocene and Pleistocene in descending order of frequency.

Only a few of the many South Louisiana accumulations can be described here. Among the onshore fields, *Washington* may be taken as a typical Eocene producer. It is a fault-closed anticline, with oil in Eocene (Cockfield) sandstone reservoirs.

Rayne has a Frio (Oligocene) reservoir, in an anticline cut by a major growth fault, with estimated ultimate recoverable reserves of 50 MM brl of oil and condensate and 1·5 Tcf of gas.

West Tepetate is of interest as an example of accumulation in the downblock of a fault. This is exceptional in the Gulf Coast area where most fault accumulations are in the upper block. It seems that an uplift due to underlying salt movement produced here an elongated anticline, and oil moving updip from the deeper part of the Gulf has been trapped in this anticline against the fault plane which parallels the Gulf shoreline.[2]

Johnson Bayou has a 3,200ft Lower Miocene sandstone producing section in a simple anticlinal trap. It produces nearly 1·4 MM brl/a of oil and condensate and about 8 Bcf/a of gas.

Bastian Bay was discovered in 1941, and covers some 1,200 producing acres. The structure is a double-faulted anticline, overlying a salt uplift. The reservoir is a series of Upper Miocene sandstone beds and the field produces mainly gas, although surrounded by several major oilfields with much greater structural relief. This is one of the largest gasfields in South Louisiana, with an annual output of more than 100 Bcf, and in addition about 700,000 brl of crude oil and 3·3 MM brl of condensate. Other important onshore accumulations in this area are *Lake Barre, West Bay* and *Bayou Sale*.

---

[2] F. Bates & J. Wharton, *Bull.AAPG*, 32, Sept. 1948, 1712.

A number of large offshore hydrocarbon accumulations have been found in South Louisiana since 1949, mostly associated with salt domes or anticlines overlying buried salt masses.

Thus, the *Bay Marchand—Timbalier Bay—Caillou Island* salt dome complex is more than 28 miles long and up to 12 miles wide, with three domes derived from a common Triassic-Jurassic salt mass penetrating to within 2000–3000 ft of the surface and forming a variety of associated traps in sandstones of Pliocene to Miocene age. Production from these fields totals some 100 MM brl/a. *Johnson Bayou* is a simple anticlinal trap with a 3,200ft thick Miocene reservoir producing crude oil, gas and condensate.

Eugene Island, Grand Isle, Main Pass, South Pass, South Timbalier, Vermilion and *West Delta* are other major offshore oil or gas accumulations with Tertiary sandstone reservoirs lying in one or more "blocks"— the concession sectors into which the shallow waters have been divided. Of these, the largest producers are *South Pass Blks 24* and *27* and *West Delta Blk 30,* each of which has an output of oil exceeding 20 MM brl/a. *South Pass Blk 24* is the largest of these fields, and the largest accumulation so far found in Louisiana.

(c) The Southeast States

An output of 61·4 MM brl in 1972 shows that **Mississippi** is now an important oil-producing state.

Most production comes from the interior salt dome basin in the south of the state. Wilcox sandstone reservoirs are found here along a continuation of the same trend that gives Eocene (Wilcox) oil in Texas and Louisiana, while deeper Jurassic (Smackover) calcareous sandstones are the main producers in a 150-mile long arc which parallels the Gulf Coast. The Jurassic Norphlet formation is also an important reservoir.

The largest oilfield is *Pachuta Creek,* discovered in 1968, with Smackover reservoirs. *Tinsley, Baxterville* and *Heidelberg* are other oilfields—complexly faulted anticlines over salt uplifts, with Eocene and Upper Cretaceous sandstone or chalk reservoirs. Several smaller fault-trap oilfields have also been found with accumulation in the upthrown block on the Gulf side of the fault.

Two important gasfields are *Cranfield,* a slightly faulted near-circular dome overlying a deep-seated salt plug with much gas in Tertiary and Cretaceous sandstone reservoirs, and *Gwinville,* a similar structure. Other gas accumulations, discovered in 1969, are *Thomasville* and *Tchula Lake,* both with Smackover reservoirs.

The eastern sector of the Gulf Coast geosyncline in **Alabama** contains one "giant" accumulation at *Citronelle,* an anticline with a Jurassic reservoir. Relatively small amounts of oil and gas are also produced in **Florida** and **Georgia.** The Eocene (Wilcox) and Jurassic (Cotton Valley, Smackover) are the most important reservoir formations in these states.

VI. CALIFORNIA

CALIFORNIA is the third largest oil-producing state of the USA, with an output in 1972 of 348 MM brl of crude oil from nearly 42,000 wells.

Oil has been found in formations ranging in age from Pliocene to Jurassic, with occasional "basement" production from fractured schists. However, by far the largest proportion comes from thick Tertiary sandstone reservoirs.

The fields are usually large anticlinal folds which parallel the mountain ranges and are often cut by faults. As many as 43 "giant" oilfields have been discovered in California, so that it is second only to Texas in this respect. There are a few non-associated gasfields, but most of the gas produced comes from the oilfields. Many of the structural trends run out to sea, and offshore exploration is therefore being carried out intensively, particularly in the Santa Barbara channel, where a major Pliocene oilfield (*Dos Quadros*) was discovered in 1969.

The onshore oilfields lie in four intermontane basins in the area between the Sierra Nevada Mountains, the Coast Ranges and the sea. Two periods of orogeny combined to produce the ranges and their associated folds and faults: these were in the late Jurassic and Middle Pleistocene. North-south pressure applied to the underlying "basement" blocks has produced some of the longest shear faults in the world (e.g. the San Andreas rift, 600 miles long).

Extensive oil seepages are common, due to the many folds and the absence of massive cap rocks. The preservation of the Californian accumulations is due to the rapid deposition of sedimentary cover above them, while their high productivity per unit area results from the thickness and multiplicity of the reservoir sandstones.

The source rock is usually considered to be the highly organic Maricopa or Monterey shales, which are nearly always in close contact with the reservoir sandstones.

(a) San Joaquin Valley

The most northerly of the Californian basins was the first to be developed as a result of its numerous surface seepages. It is about 250 miles long and 50-60 miles wide, and contains a number of important anticlinal fields which are the result of the dominant NW-SE folding, as well as fields which are examples of unconformity and fault trapping. In general, the fields in the northern part of this basin are gas accumulations, while the southern part contains a number of "giant" oil accumulations.

i. Coalinga-Kettleman North

These accumulations occur on an anticline developed on an eastward-dipping monocline. At *Coalinga,* the trap has been formed by the

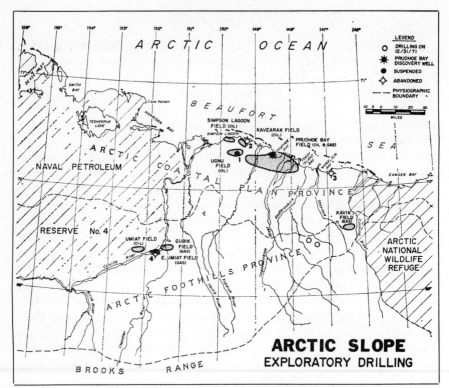

Fig. 43. THE NORTH SLOPE OILFIELDS OF ALASKA

Source: Bull. AAPG, Sept. 1972.

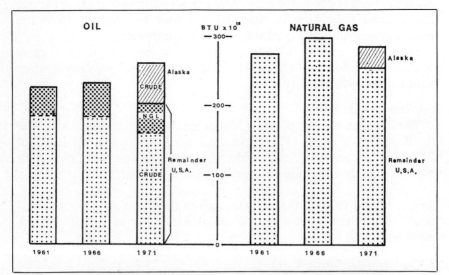

Fig. 44. USA—OIL AND GAS RESERVES (AT END OF YEAR)

Source: BP Statistical Review.

overlap of Etchegoin shales on Miocene sands and of Temblor sands on Kreyenhagen shales.

Oil is found in several reservoirs, the most important of which has been the Vaqueros (Lower Miocene) sandstone. There is also some deeper Cretaceous production. The *Coalinga Eastside* and *Westside* fields were found in 1890 and 1910 respectively, and the *Coalinga Nose* in 1938. The latter contains oil in sandstone lenses on the "nose" of the main structure. The three *Coalinga* fields together produce more than 17 MM brl/a. *Kettleman North Dome* also has important gas production.

ii. Lost Hills-Belridge

These fields lie on a southeastward extension of the same line of folding as the Coalinga-Kettleman structures. The producing horizons are Pliocene and Miocene (Monterey or Upper Temblor) sandstones. There is also production from deeper pre-Miocene sandstones.

iii. Kern River

This accumulation was found in 1899, on the east of the San Joaquin valley. There is some uncertainty as to the true subsurface structure as early drilling records were faulty. There is a broad, low anticline present, but accumulation has probably been due to faulting. Production comes from Etchegoin and Tulare sands. The field has produced nearly half a billion barrels of oil in its lifetime, and still produces nearly 20 MM brl/a.

iv. McKittrick

This is the oldest Californian field, discovered in 1887, and is an example of fault accumulation. There is a low-angle overthrust fault with many minor thrusts and a series of steep strike faults. Oil comes from steeply dipping Santa Margarita sandstone inliers surrounded by Maricopa shales.

v. Elk Hills

This elongate anticline, 17 miles long by 7 miles wide, with three culminations, was discovered in 1911; regional thrusting from the southwest probably produced the fold. The prolific reservoir horizons are sandstones in the San Joaquin (Etchegoin) formation. The field produces about 2·1 MM brl/a, but still has reserves of at least 1·0 B brl, which are kept as a strategic naval fuel reserve.

vi. Midway-Sunset

The controlling features in this very large accumulation, second in size only to *Wilmington* in California, are the unconformities between the Pliocene beds and the underlying Miocene shales, and also those between the individual Pliocene beds. The process of accumulation has been aided by local folding. *Midway-Sunset* has been producing oil since 1894 and is one of the larger "giants", with a cumulative production of more than 1·0 B brl of oil, and an output still close to 26 MM brl/a.

Tar-filled sands occur above the oil reservoirs at shallow depths and it is thought that these have resulted from the oxidation of upward-moving hydrocarbons by surface waters so that now they act as seals.

vii. Other notable oilfields in this area are *Buena Vista,* which is an anticlinal accumulation (discovered as long ago as 1909) with some lenticular sandstone trapping; and *Cuyama, Greeley, Cymric* and *Kern Front,* which are large faulted anticlines.

(b) Central California (Salinas Valley)

The most important oilfield in this area is the anticlinal fold of *San Ardo,* discovered in 1947, which produces about 17 MM brl/a of oil.

(c) Santa Maria District

Oil was discovered here as long ago as 1902 as a result of drilling near the extensive surface seepages. Within the basin between the San Rafael and Santa Inez Mountains, there are two east-west anticlinal trends which form the lines of the Solomon and Casmalia Hills on the north and the Purisma Hills on the south.

In the north, the *Casmalia* field lies on an easterly plunging anticline, while the *Santa Maria* field is on a cross-faulted anticline at the structurally highest point of the Solomon Hills. Production comes from fractured shales in the Monterey series. The largest producer in the area is *Cat Canyon,* which lies on the same trend line on a fold plunging to the north-west, and gives oil from sands slightly younger than the shale horizons of the other fields, at the rate of about 4·3 MM brl/a.

In the south, the *Lompoc* field near the western extremity of the Purisma Hills, produces oil from fractures in the Monterey shales, which were probably the source rock for the oil in this area.

(d) Ventura District

The first successful oil-wells in California were drilled in 1875 in the southern part of Ventura County, where there are extensive seepages and deposits of asphalt.

The basin between the Santa Inez and Santa Monica Mountains extends on the west some way out to sea. Considerable shallow-water drilling has been carried out successfully off the coasts of Santa Barbara and Ventura Counties, and as mentioned on p. 269, a major offshore field has now been discovered (*Dos Quadros*).

Within this basin, the axes of the folds are east-west, in contrast to the generally prevailing northwest-southeast trends. The oil accumulations occur in groups which coincide with fold zones.

Thus, one line of oilfields has been found along the narrow strip of land between the Santa Inez Mountains and the coast, some of them extending offshore. These fields are anticlines with east-west axes, which are generally steeply-folded and faulted.

Ventura Avenue is the most important of these fields. Discovered in 1916, it is a slightly asymmetrical, east-west trending faulted anticline, 16 miles long, which plunges in both directions. Production comes from a large number of Lower Pliocene sandstone beds, mainly in the thick Pico formation. This field, which has produced more than three-quarters of a billion barrels of oil, still produces about 13 MM brl/a of oil.

Rincon, west of the Ventura Avenue fold, lies on the same structural trend. The eastern part of this fold is on land, the western part lies offshore. Oil comes from the same Lower Pliocene formations.

Summerland is one of the oldest of these coastal fields, with Pico and Vaqueros sandstone production; *Capitan* gives oil from Vaqueros and Sespe sands.

A number of small pools have been found on both sides of the Clara River, along an east-west line of folding. The accumulations are anticlinal and often associated with complex faulting. The oil reservoir sandstones here are Lower Miocene and Oligocene in age. The most important of these fields are, from west to east, *Sulphur Mountains, Ojai, Santa Paula* and *Sespe.*

On the south side of the Clara River there is an eastward extension of the anticlinal trend associated with the coastal fields. A number of relatively small anticlinal oilfields have been found here, among them *South Mountain, Bardsdale, Shiels Canyon, Torrey Canyon, Tapo Canyon, Wiley Canyon* and *Elsmere.*

Of these, the *South Mountain* pool is the most important. Discovered in 1915, it is a structure *en échelon* with the neighbouring folds and is cut by several intersecting faults. Oil comes from thick reservoir sands in the Sespe (Oligocene) formation at the rate of about 3·2 million brl/a.

(e) Los Angeles Basin

Many very productive oilfields have been found within the Los Angeles basin, which is bounded on the north by the Santa Monica Mountains, on the south by the San Joaquin Hills and Santa Ana Mountains, on the east by the Repetto and Puente Hills, and on the west by the sea. A thick series of Miocene and Pliocene sediments accumulated in this basin as a result of the rapid sinking of the Tertiary sea-floor, and were subsequently folded and faulted. Oilfields are densely scattered over the whole basin, which has been one of the most productive areas of the United States, with oil recoveries of generally more than 100,000 brl/acre. Four principal lines of accumulation occur, each of which is associated with a major structural fold trend running parallel to the general NW-SE trend of the Tertiary fold belts of California.

i. Inglewood Fault Fields

A line of fault and fold accumulations extend from Beverly in the northwest to Newport in the southeast. The reservoirs are formed from a great thickness of Pliocene and Miocene sands. From north to south these fields include: *Beverly Hills, Inglewood, Potrero, Athens-Rosecrans, Dominguez, Long Beach* and *Wilmington, Alamitos Heights, Seal Beach, Huntingdon Beach* and *Newport.*

The *Wilmington* field is currently the largest oil producer both in California and in the United States, with an output of about 78 MM brl/a. About two-thirds of this oil comes from the seven main producing zones in the shallow offshore fault-block extensions of the old *Long Beach* field, which are operated from several man-made islands in the City of Long Beach harbour area.

Long Beach itself was first discovered in 1921, and had produced nearly 862 MM brl of oil by mid-1969. It still produces about 4·2 MM brl/a of oil from a large, asymmetrical, northwest-southeast trending anticline whose southeast extremity is truncated by faults. Oil and gas accumulations occur in a great thickness (more than 3,000ft) of sandstone reservoir beds, interbedded with Pliocene Repetto clays. Although *Wilmington* is essentially an oilfield with initial reserves of some 2,600 MM brl, it also had initial gas reserves of more than 1·0 Tcf.[3]

Inglewood is an anticline which has been faulted after the accumulation of oil had taken place. Some oil is produced from sandstones in the fault "gouge" zone, as well as from the anticlinal Pliocene (Pico and Repetto) reservoir beds. The field, first found in 1924, still produces 6·6 MM brl/a.

Huntington Beach is a fold on the edge of the sea in which a double parallel N.W.-S.E. fault has produced additional fault-block traps. Oil has been produced since 1920 from the Repetto (Pliocene) and Puente (Miocene) formations, and output is still more than 22 MM brl/a.

ii. Whittier Fault Fields

A smaller group of oilfields is closely associated with the Whittier fault zone, which parallels the eastern mountain boundary of the Los Angeles basin. Accumulations usually occur on the downthrown side of faults in monoclinal structures. Reservoirs occur in Lower and Middle Pliocene sandstones, but the principal accumulations are in the Puente (Miocene).

From north to south these fields include *Salt Lake, Los Angeles, Montebello, Whittier, Puente* and *Brea-Olinda.*

Whittier is a simple fault accumulation, in which a steep Pliocene monocline is sealed by faulting against Puente shales.

At *Puente* an unconformity plane between the Pliocene and Puente

---

[3] S. V. Hamilton, *AAPG*, "Natural Gases of N. America", 1968, 1, 164.

has been folded into an anticlinal fold and overthrust to the south. Puente shales and sandstones give oil in the core of the inlier.

iii. Santa Fé Pools

Parallel to the Whittier trend but lying further to the west is a minor structural trend on which occur several faulted anticlinal fields. These are, from north to south, *Sante Fé Springs, West and East Coyote,* and *Richfield,* all discovered before 1920 and all still producing important volumes of oil yearly.

The *Sante Fé* pool is the most important of these, and is an almost circular domal structure. It has been one of the most prolific fields for its size in the world, and has produced at least 600 MM brl of oil from a great thickness of Pliocene and Miocene sands; the surface area of the field is only 1,515 acres.

iv. Torrance Pools

The most westerly of the structural lines, lying west of the Inglewood Fault zone between Venice and Torrance, is associated with the *Venice, Lawndale* and *Torrance* oilfields.

The *Venice-Playa del Rey* field is on a broad anticline whose flanks dip more gently than most Californian structures. Oil comes from the Repetto series, and also from a deeper zone of angular schist fragments embedded in a sandstone.

The *Lawndale* pool is small, but geologically of special interest in that production has been obtained from sand lenses quite independently of any structure.

At *Torrance* there is a gentle, southeast-plunging faulted anticline, although accumulation may also have been effected along the Miocene-Pliocene unconformity. Oil comes from Lower Pliocene and Upper Miocene beds.

Gas Production

Apart from the production of associated gas from its oilfields, California also produces some non-associated gas from Upper Cretaceous and Tertiary (Eocene and Pliocene) reservoir beds in the Sacramento and San Joaquin valleys, as well as importing considerable volumes by pipeline from Canada. By far the largest gas-producing field is *Rio Vista.*

The structure of this field, which was discovered in 1936, is a broad, faulted dome, with production from six sandstone zones of Eocene to Paleocene age[4].

Due to its severe "smog" problem, the state of California requires the burning of natural gas wherever possible in place of fuel oil, and compulsory catalytic "clean-up" of automobile exhaust emissions will be introduced in 1975.

---

[4] E. Burroughs *et al., AAPG,* "Nat. Gases of N. America", 1968, 1, 93.

VII. ALASKA

THE search for oil in Alaska—a state with roughly twice the area of Texas—began as long ago as 1898, but prior to 1957 the only commercial success was at *Katalla,* 200 miles southwest of Anchorage, where a small, shallow Tertiary accumulation was found which produced a total of only 154,000 brl of oil in the period 1902-1933. Since 1957, however, it has been established that there are as many as ten sedimentary basins in Alaska which are potentially oil-bearing, and several important discoveries of oil and gas have been made both in the Cook Inlet Basin and, more recently, in the Arctic Slope province.

The Cook Inlet Basin is a topographical depression approximately 250 miles long by 65 miles wide, which is bounded on the northwest and southwest by large thrust faults downthrown to the basin[5]. It contains a thickness of some 25,000ft of Tertiary (Eocene-Pliocene) sediments which are underlaid by igneous, metamorphic and sedimentary Mesozoic and Palaeozoic rocks. Most of the Tertiary beds seem to have had a fluviatile origin, having been deposited from the large rivers flowing into the basin from the continent lying to the west and northwest. Hydrocarbons have been found in a number of sands and conglomerates in the lower part of the Tertiary section, termed the Kenai beds. The source of the oil is believed[6] to be the underlying Jurassic shales.

The accumulations so far discovered are closed anticlines with steep flank dips, complicated by thrust and transverse faults.

In **South Alaska,** the first oilfield to be found (in 1957) was *Swanson River,* in the Kenai Peninsula. This is an elongated anticline cut by several transverse normal faults with downthrows to the north. The area of accumulation is only the crestal portion of a large fold, so it has been suggested[7] that *Swanson River* is a typical "starved anticline". The oil reservoir is in the Hemlock Zone sandstone, and there are also several gas reservoirs in this sandstone series as well as in the Upper Kenai formation, which is an important gas reservoir bed in all the South Alaskan fields. *Swanson River* oil is greatly undersaturated with gas, so a gas repressurization programme was initiated in 1962. The original oil reserves in this field were estimated to be 250 MM brl, and output is about 11 MM brl/a.

Three similar oilfields were subsequently discovered in the period 1963-65 at *Middle Ground Shoal* in Cook Inlet (reserves 200 MM brl), *Granite Point* (reserves 175 MM brl) and *McArthur River* (reserves 160 MM brl).

Gasfields have also been found in South Alaska at *Kenai* (a broad,

---

[5] F. Osment *et al.,* 7 WPC Mexico, 1967, PN 2–6.
[6] M. Hill, 6 WPC Frankfurt, 1963, 1, 39, PD-3.
[7] T. Kelly, *AAPG,* "Natural Gases of N. America", 1, 9, 1968.

elliptical, faulted anticline), *Sterling, West Fork* and *Beluga,* all with Kenai formation sandstone and siltstone gas reservoirs. The gas contains about 99·5% methane and besides being used locally for electricity generation, is being liquefied and exported by tankers to Japan in the form of LNG.

The proved reserves of the Cook Inlet Basin are not less than 500 MM brl of oil and 3 Tcf of gas. Production is increasing rapidly; thus, in 1971 it was 78·7 MM brl of oil and 229·4 Bcf of gas (compared with only 11·1 MM brl and 12·5 Bcf in 1965). In 1972, oil output was 79·5 MM brl.

In North Alaska, the Arctic Slope province covering about 125,000 sq miles is underlain by the Palaeozoic and early Mesozoic beds which make up the east-west Broken Range mountain chain (the "Colville geosynclorium"). The folds decrease in complexity northward across Cretaceous foothill belts towards the Arctic Coastal Plain, where gently deformed marine and non-marine Tertiary beds outcrop. About two-thirds of this province is potentially oil-bearing territory, and an area of 37,000 sq miles was set aside in 1923 as US Naval Petroleum Reserve No. 4. Exploration resulted in the discovery in 1958 near the southern bend of the Colville Rim of the medium-sized *Umiat* oilfield (reserve 70 MM brl), and the *Gubik* gasfield (reserves 300 Bcf). These are anticlinal accumulations with Cretaceous sandstone reservoirs[8].

The North Slope

During 1968, a very important new petroleum province was discovered on the Alaskan North Slope in the Beaufort Sea coastal area, to the east of the previously known Cretaceous productive zone. Here, exploration wells drilled near *Prudhoe Bay* (between the mouths of the Kuparuk and Sag rivers) found the largest oil accumulation ever discovered in North America, which is estimated to contain reserves of at least 10·1 B brl of recoverable oil and 26 Tcf of gas. It consists of a succession of stratigraphic traps formed on a structural "nose" as a result of the truncation of south and southwest dipping reservoir formations by a Lower Cretaceous unconformity. The accumulations are bounded to the north by east-west faulting and steeper dips, and to the east and northeast the oil-bearing Triassic Sadlerochit sandstones and conglomerates (which form the major reservoir) and the Carboniferous Lisburne carbonates are overstepped by the Lower Cretaceous shales and clays. The Jurassic Kuparuk River sandstones form a further reservoir below the unconformity[9, 10].

Since the Prudhoe Bay discovery, several other large anticlinal accumulations have been found. Cretaceous oil discoveries have been

[8] G. Oates *et al., AAPG,* "Nat. Gases of N. America", 1968, 1, 3.
[9] A. Thomas, Geol. Soc. London, Meeting, Dec. 9, 1970.
[10] D. Morgridge and W. Smith, *AAPG,* "Stratigraphic Oil and Gas Fields", 1972, 489.

made at *Ugnu,* 36 miles east of the Prudhoe discovery well, *Kavearak, Simpson Lagoon* on the coast, and *Arco West Sak,* 25 miles west of Prudhoe Bay, and at *Kavik* (gas), just west of the Canning River. Exploration operations in the whole area between the Colville and Canning Rivers are likely to increase in the future, once agreement has been reached on an acceptable outlet for Alaskan oil. Up to 1973, this issue was still being debated, since environmentalists objected to each of the pipeline routes suggested. Three alternative routes for moving oil to distant markets seem possible: the first is the construction of a pipeline to the port of Valdez in southern Alaska and thence transport by tanker to the US West Coast, and the second is the much more ambitious project of building a transcontinental pipeline to supply markets in the east of Canada and the United States. The third possible route is by tankers specially equipped to defeat the ice (as was shown to be possible by the exploratory voyage of the *Manhattan* in 1969), or even by sub-ice tanker submarines, to markets in Japan and Western Europe as well as on the US East Coast.

Among the many problems involved in these exercises is the extreme shallowness of the waters off the North Slope coast—only 60ft for many miles from the coast—whereas a loaded modern tanker needs a draught of at least 75ft; while on the other hand the construction of pipelines along which oil can flow freely through areas of permafrost and extreme low temperatures is a difficult and expensive operation.

TABLE 73

USA—THE LARGEST OILFIELDS

Field	Original Reserves MM brl	Cumulative Production to 1/1/69 MM brl	Remaining Reserves as at 1/1/69 MM brl
Prudhoe Bay (Alaska)	10,100	—	10,100
East Texas (Texas)	5,100	3,809	1,291
Wilmington (California)	2,600	1,235	1,365
Panhandle (Texas)	1,647	1,198	449
Yates (Texas)	1,500	520	980
Elk Hills (California)	1,300	279	1,021
Midway Sunset (California)	1,193	1,024	169
Kelly-Snyder (Texas)	1,188	437	751

CHAPTER 10

Canada and Mexico

CANADA

OIL and gas have been produced in Eastern Canada since 1861, and in Western Canada since 1905. It was not, however, until after 1947 that a series of new discoveries in Alberta made Canada a hydrocarbon producer of world importance. With a rapidly growing population (which roughly doubled between 1941 and 1971), Canadian consumption of oil and gas has been increasing in recent years at the rate of nearly 7% p.a.; in fact, since 1966 Canada has been the world's largest *per capita* consumer of petroleum products.

In 1971, crude oil production totalled 482·8 MM brl, and marketable gas production exceeded 2·6 Tcf. Oil now provides about 57% and natural gas 18% of Canadian primary energy consumption.

The proportion of primary energy accounted for by liquid oil has doubled since 1950, and the proportion due to natural gas has multiplied by a factor of more than six over the same period. The successful exploration and development of the petroleum resources of Western Canada, and the building of several transcontinental oil and gas pipelines to link the producing areas of the West with the consuming areas of the East and of the USA, accounted for this rapid progress.

However, while Canada contains about 22% of the total North American proved oil reserves (excluding Alaska), it still produces only about 12% of the continent's liquid oil output. A considerable proportion of potential Canadian oil production (probably more than half) remains "shut in" for economic reasons, since it is more expensive to produce than Venezuelan or Middle East crudes. Nevertheless, a large and increasing proportion of the Canadian annual oil output is currently exported to the USA, where production costs in some areas are even higher.

For similar reasons, only about 60% of the domestic oil demand is normally supplied from Canadian indigenous sources, while the balance is made up of imported crudes.

At the end of 1971, the liquid oil (crude plus condensate) reserves in Canada amounted to 10·1 B brl, and marketable natural gas reserves were about 55·5 Tcf (Table 74). Most of the recoverable oil reserves are contained in Palaeozoic carbonate reservoirs of Mississippian and Upper Devonian ages. The Upper Cretaceous is also an important reservoir horizon.

TABLE 74

PETROLEUM PRODUCTION IN CANADA, 1971

	Crude oil* production MM brl	Marketable gas production Bcf	Sulphur 1000 long tons
Alberta	360·98	2,226·5	4,600
Saskatchewan	88·51	85·8	3
Manitoba	5·67	—	—
British Columbia	25·55	343·1	64
North West Territories	1·06	1·1	—
Eastern Canada	1·00	20·1	—
Total	482·77	2,676·5	4,667

* not including 89·3 MM brl of natural gas liquids.

TABLE 75

CANADA—PROVED OIL AND GAS RESERVES AT YEAR-END

	Crude Oil* (B brl)	Nat. Gas liquids (B brl)	Total liquid Petroleum (B brl)	Natural Gas (Tcf)
1966	7·8	1·3	9·1	43·5
1968	8·4	1·6	10·0	47·6
1971	8·3	1·8	10·1	55·5

* It is estimated that a further 6·3 B brl of "non-conventional" crude oil could be recovered from within "an economic radius" of the Athabaska oil sands plant at Fort McMurray.

EASTERN CANADA

Exploration and development activity has always been much less intense in Eastern Canada than in the west. It has largely been concentrated in southwestern **Ontario,** where oil and gas have been produced on a small scale since as long ago as 1861 in the area between the Great Lakes. The structures here are generally small anticlinal closures on a southwards dipping homocline. Production is obtained from minor old-established fields such as *Petrolia, Oil Springs* and *Becher,* where the reservoir beds are Palaeozoic (Devonian to Cambrian) sandstones and limestones. Most of the oil has come from shallow Devonian reservoirs, and most of the gas from Silurian beds.

A small amount of oil is also produced in **New Brunswick** from the *Stony Creek* field, which is an anticlinal oil and gas accumulation in deltaic Mississippian sandstones. On the Gaspe Peninsula of **Quebec,** some oil has been found in Devonian reservoirs in small anticlines near the York and St. John Rivers.

The small oil content in Eastern Canada of the Ordovician limestones which, as the Madison formation, are so prolific in the Ohio/Indiana area of the USA, is due to the general absence of permeability.

Considerable interest is currently being shown in offshore exploration in the Ontario section of Lake Erie, in the Hudson Bay region, and on a wide sector of the Atlantic Ocean Continental Shelf of the maritime provinces. The first offshore discoveries were two gasfields at the western end of Lake Erie—*Tilbury* and *d'Clute*—which are both Guelph (Middle Silurian) biohermal reservoirs. On the east side of the lake, the Grimsby and Whirlpool (Lower Silurian) channel sands also produce commercial gas. The first exploration success which has been reported from the Atlantic Continental Shelf was a Cretaceous gas and oil discovery on the western tip of Sable Island, some 175 miles east of Halifax, Nova Scotia.

Output in 1971 was about 1·0 MM brl of oil and 10 Bcf of gas. The remaining proved oil reserves of Eastern Canada are only about 9 MM brl of oil.

WESTERN CANADA

1. ALBERTA
The hydrocarbon accumulations of Alberta may be considered in two groups—those occurring in the structurally complex foothills zone of the Rocky Mountains and those which have been found beneath the gently dipping homocline of the Interior Plains.

(a) The Foothills Zone
A number of large gas accumulations, some with associated oil, have been found along the narrow and sinuous foothill belt which separates the Rocky Mountains from the Alberta Plains. The fields here are essentially complex fault structures, overthrust from the west, with strikes parallel to the mountains, and closures formed by fault planes and the plunges of the fault-blocks. Most of the gas accumulations occur in Mississippian dolomitic limestones, although there are also some Triassic and Devonian producers[1].

The *Turner Valley* field, discovered in 1905, is one such fault-block accumulation, which lies between two major thrust faults. The beds on the eastern edge of the block have been folded and faulted, resulting in a shallow anticlinal structure being thrust over the fault-sealed

---

[1] G. C. Wells, *AAPG*, "Nat. Gases of N. America", 2, 1229, 1968.

U

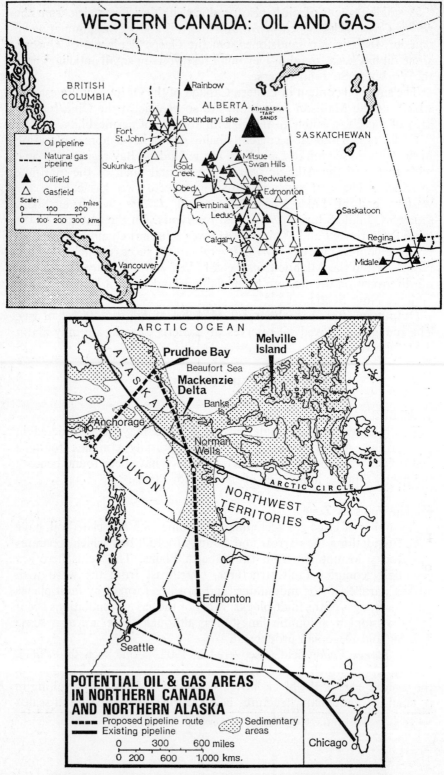

Figs. 45, 46. OILFIELDS OF WESTERN AND NORTHERN CANADA

Source: Petroleum Press Service.

homocline of Mississippian limestone, in the porous zones of which gas and oil have accumulated.

The *Pincher Creek, Westerton* and *Jumping Pound* "wet" gasfields are other examples of this type of accumulation in porous Mississippian dolomites.

Altogether, there are at least 20 similar Mississippian gasfields with strikes paralleling the mountains lying within a narrow belt between *Mountain Park* in the north and *Lookout Butte* in the south. *Panther River* has a Devonian reservoir, and *Mountain Park* produces gas from Triassic rocks. The initial marketable gas reserves of these fields are estimated to have been at least 6 Tcf.

(b) The Alberta Plains

The part of Alberta lying east of the 6th Meridian and northeast of the Foothills Zone consists of an area of some 190,000 sq miles of mainly Mesozoic and Palaeozoic strata, forming a gently dipping homocline. There are two principal types of hydrocarbon accumulation —Palaeozoic reefal limestone oilfields and Mesozoic sandstone accumulations, mainly of gas.

(i) Palaeozoic Limestone Fields

Leduc, 15 miles southwest of Edmonton, was the first to be found (in 1947) of a series of prolific reefal limestone oilfields, the discovery of which has been the principal reason for the rapid development of the Canadian oil industry since the war. It is a typical biohermal accumulation, with oil in two porous and permeable Upper Devonian reef dolomites which are surrounded and sealed by impervious shales. The upper reservoir is the Nisku D2 reef of the Winterburn Group, the lower the Leduc D3 reef of the Woodbend Group. The dominating trapping agency is the local high permeability, although there is also a structural factor in the D2, due to the "draping" effect of the beds overlying the bioherm. *Leduc,* with its southern *Woodbend* extension, also has a higher Cretaceous (Blairmore) sandstone reservoir, and had produced some 300 MM brl of oil up to 1969.

There are in fact several productive reefs in this part of Alberta, many of which have been found to be associated with important Palaeozoic oil accumulations. To the south of Leduc is a zone of Upper Devonian "foredeep" atoll reefs (the Leduc-Rimbay chain) with important oil accumulations at *Wizard Lake, Bonnie Glen, Westerose* and *Homeglen-Rimbay;* this line of atolls is separated from the Bashaw barrier reef to the east by what must have been an open-water area. Further to the east, but with a roughly parallel trend, is the Fenn-Big Valley reef.

To the north of Leduc, there are oil-bearing reefs in the Middle and Upper Slave Point formations, and in the Swan Hills member of the Middle Devonian Beaverhill Lake group. Important oilfields in this area are *Golden Lake, Acheson, Big Lake, Kaybob* and *Redwater,* among others. The Windfall-Simonette and Sturgeon Lake reefs of the Upper Devonian Woodbend Group contain large gas accumulations at *Windfall, Pine Creek* and *Berland River.*

Another group of reefal fields—the *Rainbow-Zama* "pinnacle" reef accumulations in northern Alberta, discovered since 1965, now account for as much as 10% of the province's crude oil reserves. *Rainbow* is an accumulation in Middle Devonian Keg River reef limestone; within the Rainbow Basin reserves of some 630 MM brl of recoverable oil have been proved in a number of pools, the largest of which contains at least 150 MM brl.

A further group of Devonian reef accumulations lies in the *Zama-Virgo* Basin, further to the north. A total of 175 MM brl of recoverable oil has been proved here in some 100 pools, of which the largest holds roughly 9 MM brl of oil.

Both Rainbow-Zama and Zama-Virgo are evaporite basins in which "pinnacle" reefs and atolls are surrounded by carbonate banks and bordered to the west by a broad barrier reef. The "pinnacle" reefs grew up to 800ft thick at Rainbow near the basin axis, and up to 500ft at Zama, and form mounds of dolomite and limestone with flank dips of 30°- 40°, and cross-sectional areas of up to a square mile. "Drape" traps also occur in overlying Upper Devonian carbonate-sand beds which owe their structure to the solution and collapse of the Muskeg salt beds immediately above the reefs. Multiple-stack seismic reflection surveys have been very successful in locating these salt-collapse structures and thus the underlying reefs[2].

(ii) Mesozoic sandstone reservoirs

These typically contain gas accumulations (in all more than 20 Tcf of gas has been found) in various types of lithologic or stratigraphic trap, formed by lenticularity or the "shale-out" of sandstone reservoirs on a general homocline. The most important field is *Pembina,* in which gas was first discovered in 1905. Subsequent exploration in 1953 showed that this is also in fact a major oilfield, with some 1·7 B brl of oil in a stratigraphic trap resulting from the updip change of lithology of the Upper Cretaceous Belly River and *Cardium* Sandstones. This is the only field in Canada in the world "giant" class, and has a producing area of some 755,000 acres.

Hydrocarbon accumulations in Cretaceous sandstones are found in a number of reservoirs overlying deeper Palaeozoic reef accumulations, and also in Permo-Pennsylvanian, Triassic and Jurassic formations (major examples are *Acheson* and *Leduc-Woodbend*).

Table 76 lists the most important of the approximately 300 oilfields

---

[2] J. Robertson, *World Oil,* Sept. 1969, 87.

of Alberta and gives details of producing formations and reserves. Most of the other fields have recoverable reserves of less than 25 MM brl apiece. There are also more than 160 gasfields with about 700 pools, but about 500 of these have recoverable reserves of less than 25 Bcf each.

Eight "giant" gasfields are known: *Medicine Hat-Hatton,* which lies across the Alberta-Saskatchewan boundary, was actually the first hydrocarbon accumulation to be discovered in Western Canada, gas being found in a test well for coal as long ago as 1890. Later discoveries here led to the development of a major gasfield after 1951, with initial recoverable reserves of 2,030 Bcf from a productive area of 775,000 acres —certainly one of the largest gasfields in Canada. The gas comes from the shallow (c. 1,500ft) Medicine Hat Sandstone reservoir of Upper Cretaceous (Colorado) age, having been trapped in permeable zones of the sandstone body by what was originally an updip pinchout. Other large gasfields are *Crossfield* (2,652 Bcf in Mississippian limestones), *Edson* (2,150 Bcf), *Waterton* (1,412), *Cessford* (1,447), *Westerose South* (1,256), *Harmattan-Elkton* (1,118) and *Pembina* (1,092). (The relatively new discoveries in British Columbia of *Clarke Lake* and *Beaver River* may also each contain more than 1·0 Tcf of recoverable gas).

TABLE 76

ALBERTA—PRINCIPAL OILFIELDS

Field	Estimated original recoverable reserves MM brl	Main reservoirs
Acheson	106	Cret. sandstone; D_3 reef
Bonnie Glen	424	Cret. sandstone; D_3 reef
Carson Creek North	113	Beaverhill Lake reefs
Fenn-Big Valley	249	D_2 and D_3 reefs
Golden Spike	285	D_2 and D_3 reefs
Judy Creek	487	Beaverhill Lake reefs
Kaybob	120	Beaverhill Lake reefs
Leduc-Woodbend	329	Cret. sandstone; D_2 and D_3 reefs
Mitsue	168	Devonian sandstone
Nipisi	182	Devonian sandstone
Pembina	1,742	Cret. sandstones
Redwater	780	D_3 reef
Sturgeon Lake South	150	Triassic sandstone; D_3 reef
Swan Hills	931	Beaverhill Lake reefs
Swan Hills South	392	Beaverhill Lake reefs
Turner Valley	121	Cretaceous sandstone; Mississippian dolomite
Virginia Hills	174	Beaverhill Lake reefs
Westerose	111	D_3 reef
Willesden Green	100	Cretaceous sandstone
Wizard Lake	243	D_3 reef
Zama-Rainbow	801	Muskeg and Keg River reefs

(iii) Athabaska Tar Sands

A huge belt of viscous oil-soaked Lower Cretaceous sandstone lies between the Peace River area of northern Alberta and Lloydminster in Saskatchewan. The so-called Athabaska "tar sands" extend over an area of some 13,000 sq miles and contain at least 625 B brl of heavy, high-sulphur crude oil, from which nearly 300 B brl of so-called "synthetic" or "non-conventional" crude oil might theoretically be extracted. In this context, it is of interest to note that the Canadian Petroleum Association estimate that 6·3 B brl of "non-conventional" crude oil could be recovered from within "an economic radius" of the Fort McMurray plant.

The technical problems of extracting oil from the sands have up to the present made "synthetic" crude uneconomic, but improvements in techniques combined with the decline in US reserves may make the present pilot production operations attractive for larger-scale development in the foreseeable future (p. 355).

2. Saskatchewan

The first discovery of hydrocarbons in Saskatchewan was gas in the Lloydminster area in 1944. In 1952, a light crude oil accumulation was found at *Ratcliffe,* and thereafter a number of oilfields were discovered, mainly in the southwestern and southeastern parts of the province. Most of the accumulations are in stratigraphic and erosional traps, with Jurassic sandstone or Mississippian limestone reservoirs.

There are now about sixty producing oilfields in Saskatchewan, many with multiple pools. Of these, the most important are shown in Table 77. There are three giant fields, of which *Weyburn* is the largest.

TABLE 77

SASKATCHEWAN—MAJOR OILFIELDS

	Discovery date	Original recoverable reserves MM brl	Principal reservoir bed
Battrum	1955	55	Jurassic sandstone
Dolland	1953	77	Jurassic sandstone
Fosterton	1952	54	Jurassic sandstone
Inston	1954	51	Jurassic sandstone
Midale	1953	109	Mississippian limestone
Steelman	1954	232	Mississippian limestone
Weyburn	1955	332	Mississippian limestone

Mississippian oil accumulations have also been found in southern Saskatchewan on the northern limit of the Williston Basin at *Elswick, Neptune, Freda Lake* and *North Hoosier.*

The *Antelope* field, which lies across the border of Saskatchewan and Alberta, produces about 1 Bcf/a of non-associated gas from a basal Mississippian sandstone.

Devonian reef oil (in Birdbear carbonates) was found for the first time in 1966 at *Hummingbird,* 100 miles south of Regina. The accumulation here is a domal structure which is thought to have resulted from multiple-stage salt solution and collapse. Other productive reservoirs are two shallower Mississippian carbonate beds.

Kisbey was discovered in 1968, about 100 miles northeast of *Hummingbird.* The reservoir here is of Devonian age. Other Devonian accumulations have been found at *Ceylon* and *Smiley,* and deeper Ordovician oil has been discovered at *Beaubier.* The Winnipegosis (Middle Devonian, Elk Point formation), which contains oil in Alberta, Montana and North Dakota, underlies 30 million acres of central and southern Saskatchewan and is a target for deep exploration.

In **Manitoba,** exploration which began in 1951 has located three important oilfields in the southwest corner of the province: *Daly* (1951) with original recoverable reserves of 20 MM brl; *North Virden-Scallion* (1953) with 57 MM brl; and *Virden-Roselea* (1953) with 43 MM brl. The principal reservoir formation in each case is the Mississippian Lodgepole limestone.

3. British Columbia—Peace River Area

The area lying roughly between the Peace River, the Fort Nelson River on the north, and the Smoky River on the south, is now a major gas-producing region. Exploration and development only began in the 1950's, but the proved gas reserves are now at least $9\frac{1}{2}$ Tcf.

The part of northwestern Alberta lying west of longitude 118°, together with the northeastern corner of British Columbia east of the Rocky Mountains, comprises an area of some 80,000 sq miles. It is a continuation of the great southwest-dipping homocline that forms the Alberta Plains; but only in this area has a wide belt of folding and minor thrust faulting been developed on the east side of the Rocky Mountains foothills. Deep subsidence has produced a thick sedimentary section with many potential reservoirs, and the presence of a submerging granite arch along a northeast-southwest trend has affected Palaeozoic sedimentation and helped to create Mesozoic "drape" structures.

Gas accumulations occur in structural traps—fault fields and anticlinal folds with axes paralleling the foothill belt—and also in various types of stratigraphic and lithologic traps, including reefal accumulations, updip sandstone "pinchouts", etc.

The age of the reservoir formations ranges from Middle Devonian dolomites to Late Cretaceous sandstones. At least 26 major fields, each with gas reserves of more than 100 Bcf have been found[3].

Devonian reefal dolomites account for about 45% of the proved gas reserves; Triassic dolomites and sandstones hold a further 35%, and

---

[3] L. M. Clark *et al.,* *AAPG,* "Nat. Gases of N. America", 2,683, 1968.

the remaining 20% is contained in Cretaceous and Jurassic sandstones, Mississippian and Permian carbonates, and Permian and Mississippian sandstones.

Stratigraphic traps resulting from the updip truncation of porous beds below erosional unconformities have been responsible for several important gas accumulations such as *Buick Creek, Rigel,* etc.

The *Fort St. John* gasfield was the first important accumulation to be discovered in British Columbia, and has been in production since 1951. It consists of a gently folded anticline with a principal axis at depth trending west-northwest, or nearly at right angles to the surface trend. There are five productive reservoirs ranging in age from Early Cretaceous sandstones to Permian sandstone and dolomite. Original gas reserves in this field were 278 Bcf.

The important gasfields of *Laprise, Bubbles, Jedney* and *Beg* are combination traps with Triassic (Baldonnel) dolomite reservoirs.

Dolomitized Late and Middle Devonian reefal carbonates form the reservoirs in the northern fields, such as *Clarke Lake,* in which individual wells have very high productivities. The *Beaver River* gasfield is a major structural accumulation.

Some of the gasfields also contain important oil accumulations in at least one reservoir—e.g. *Aitken Creek, Blueberry, Boundary Lake, Fort St. John* and *Rigel.* Most of the oil-containing reservoirs are Triassic sandstones or dolomites; but *Aitken Creek* and *Rigel* have Lower Cretaceous sandstone oil reservoirs and *Blueberry* has a Mississippian dolomite reservoir. *Boundary Lake* (discovered in 1955) had produced 44·2 MM brl of oil by mid-1968, much more than any other field in this area. *Peejay* and *Milligan Creek* had produced 18·0 and 15·6 MM brl respectively by the same date, from Triassic sandstones.

Several exploratory offshore wells have been drilled without commercial success on the Continental Shelf of British Columbia in the Hecate (Queen Charlotte) and Vancouver Island (Tofino) basins (Tertiary and Upper Mesozoic sediments).

4. Northern Canada

Oil was first discovered in the Northwest Territories in 1919 at *Norman Wells,* where surface seepages are derived from exposed and fractured Upper Devonian shales, but the reservoir is an underlying lenticular reef limestone. This field was developed as a wartime measure, but shut down as uneconomic in 1946.

Sporadic exploration has been carried out for a number of years in the Far North and Arctic Island area, and an exploration well was drilled in 1961/2 on Melville Island. Only a little gas was found, and other unsuccessful wells were subsequently drilled on Cornwallis and Bathurst Islands. However, the discovery of oil in 1968 on the Alaskan North Slope greatly stimulated interest both in the Arctic Islands and in the Mackenzie River Delta area, with its Continental Shelf extension in the Beaufort Sea. A successful "wildcat" was completed at *Atkinson Point* (110 miles northwest of Inuvik) early in 1970, and was reported to have made substantial discovery of good-quality oil in a Lower Cretaceous sandstone reservoir, similar to one of the reservoirs at Prudhoe Bay, while several important gas accumulations have been found in the Mackenzie River Delta.

Two major sedimentary basins underlie the Arctic Islands. The Melville Basin covers 226,000 sq miles and contains a thick Palaeozoic section ranging from Ordo-Silurian carbonates to Upper Devonian clastics. The earlier unsuccessful wells drilled on Melville, Cornwallis and Bathurst Islands were sited in this basin.

The Sverdrup Basin, on which attention is now being concentrated, is a major halokinetic basin covering 173,000 sq miles and containing mainly Mesozoic strata (from Middle Pennsylvanian to Early Tertiary in age). The age of the evaporites (mainly gypsum) is probably Pennsylvanian. This basin is separated in the southeast from the Canadian "Shield" by the gently dipping Palaeozoic beds of the Melville Basin (which form the Arctic Lowlands) and in the north from the Arctic Ocean by a narrow uplift, beyond which lies a largely offshore Tertiary basin. There are numerous outcrops of solid and tarry bitumens in this area; major deposits of Lower Triassic "tar sands" are known on northwest Melville Island. Large anticlines have been discovered, and there are many potential reservoirs in the maximum thickness of about 30,000ft of sediments present.

Large volumes of gas have been found in Triassic sandstone reservoirs at *Drake Point* on the Sabine Peninsula, in the northeast of Melville Island, and at Kristoffer Bay on *King Christian Island*. Exploration drilling is being carried out to test some of the large structures known on other islands in the Sverdrup Basin, notably on the Hoodoo uplift on *Ellef Ringnes,* less than 100 miles from the North Pole, where a gas discovery was made in 1971. The first oil discovery, in 1972, was the *Romulus* well drilled on the Fosheim Peninsula of Ellesmere Island[4].

South of the Arctic Plains lies a large area of the Yukon and Northwest Territories as yet unexplored for petroleum. There are three main structural features: the Anderson Plain, a westward-dipping homocline with low-amplitude anticlinal folds; the Horton Plain, a flat, featureless area; and Coppermine Arch, a northwest-trending complex anticlinal feature.[5,6] Exploration is being initially concentrated around Fort Simpson and the Horn River Basin. In any case, very large capital expenditures will be needed to transport oil or gas from these remote areas to centres of consumption, and the technical problems involved

[4] J. Chilton, E. Bietz and C. Bulmer, *World Oil,* April 1972, 53; and May 1972, 118.
[5] J. Kornfeld, *World Oil*, March 1970, 95.
[6] M. Stanton, *World Oil*, May 1969, 101.

are also daunting. The most probable pipeline route southwards would be along the Mackenzie River valley, and a combination outlet for the Canadian and Alaskan Prudhoe Bay discoveries is obviously a possibility.

MEXICO

THE first oil-well was drilled in Mexico in 1869 near the Furbero seepages in Vera Cruz, but commercial production was not established until 1904, by the La Paz exploration well in San Luis Potosi. Subsequent development was very rapid, however, so that in the next few years many large accumulations had been discovered in the southern part of the country, along a trend which was called the "Golden Lane" (*Faja de Oro*). By 1921, the output of oil from these fields had reached its maximum level of 193 MM brl/a, and Mexico then ranked second only to the United States as an oil producer, with an output of more than 25% of current world production. A sharp decline followed, due to the rapid exhaustion of some of the highly permeable limestone reservoirs. In 1938, when the industry was nationalised, the output had fallen to only 39 MM brl, but has since increased again to totals comparable to those of the "boom" years. Thus, oil and condensate production in 1971 amounted to 177·2 MM brl, but since this total scarcely exceeded that of the previous year it is evident that further growth will be slow.

Petroleum accounts for 90% of Mexico's total energy requirements, about 73% coming from oil and 17% from the rapidly increasing output of natural gas (665 Bcf in 1970); most of this was associated gas, although a number of important non-associated gasfields have also been discovered in recent years.

There are four main areas of hydrocarbon accumulation, which are, from north to south: the *Northeastern Area or Burgos basin,* with relatively little oil but important gas production from Tertiary sandstones; the *Tampico-Nautla embayment,* which has accounted for nearly 90% of the past oil production and from which about 62% of the current output still comes, from Cretaceous (and some Jurassic) limestones, both onshore and offshore; the *Vera Cruz basin,* where the oil reservoirs are also Cretaceous limestones, but there is additional Oligocene sandstone gas production; and the *Isthmus of Tehuantepec* area, where Miocene reservoirs and structures related to salt uplifts now account for about 37% of the total oil output, and nearly 50% of the gas.

(i) Northeastern Area fields

The Burgos basin accounts for less than 1% of Mexican crude oil production, but as much as one-third of the natural gas. The area forms part of the Rio Grande embayment of the Gulf Coast geosyncline of Southern Texas and Louisiana, and there is an eastward downdip variation of facies from brackish water to shallow water and then marine type sediments, paralleling the present coastline. Structural traps are mainly gentle folds with northwest-southeast axes. The main gas-producing formations are Eocene and Oligocene sands, and a number of important gas accumulations have been found, of which the most notable is *Reynosa* on the Texas border. The *Brasil* gasfield, southeast of Reynosa, is nearly as large, while *Treviño,* a few miles to the east, is the largest oilfield—a longitudinally faulted anticlinal trap. Several new gasfields have been discovered in this area in recent years, including *Campo Llanura, René* and *Pinta;* they are all anticlinal accumulations with Oligocene or Eocene sandstone reservoirs.

(ii) Tampico-Nautla Embayment

The Tampico-Nautla Embayment is in part an extension of the main Gulf of Mexico Tertiary geosyncline, and in part (in the west) the "foredeep" of a separate Paleocene basin paralleling the Sierra Madre Mountains. The oilfields in this area account for more than 60% of the current national oil production, and about 17% of the total gas output. They may be considered in three groups, from north to south:

(a) Northern Fields

These occur on the southern-plunging prolongation of the Sierra Tamaulipas Mountains anticline. There are two directions of folding— the main Sierra Tamaulipas axial trend bifurcates into northeast and northwest trends.

The principal oil reservoirs are the Cretaceous Tamaulipas limestone, the Agua Nueva limestone and the base of the San Felipe limestones; but Upper Jurassic oolite production has also increased in importance in recent years.

The Tamaulipas limestone was formed at the same time as the El Abra limestone, the main reservoir bed in the southern Mexican fields, but here it represents a deeper-water facies. It is a very dense limestone, and oil production depends largely upon the existence of secondary porosity and permeability, due to fissuring, solution or jointing. Because of this factor, wells drilled into this reservoir vary enormously in their productivity, even within short distances of each other. The Agua Nueva limestone, which is dense and flaggy, similarly depends for its productivity on secondary reservoir features.

The oil source rock has been variously considered to be the Kimmeridge section of the Jurassic, the black shale bands in the Tamaulipas limestone, or the basal limestones and black shales of the Agua Nueva formation. A Jurassic source rock would require subsequent vertical migration across 1,200-2,000ft of compacted beds, and is therefore unlikely. More probably, both the Tamaulipas and the Agua Nueva shales acted as source rocks.

The first oilfield to be found in this area, where oil seepages are common, was at *Ebano* in 1904, and since then a number of accumulations have been discovered lying in an irregular arc to the west of Tampico; these are collectively termed the *Ebano-Panuco* fields. The traps are culminations on local flexures of the plunging prolongation of the Sierra Tamaulipas Mountains uplift, and are affected by faulting. The most important accumulations are: *Panuco, Topila, Ebano, Cacalilao, Corcorado, Chijol, Quebracho* and *Altamira*. The oil is heavy and has a high sulphur content.

Further to the north and near the coast are the anticlinal accumulations of *Tamaulipas-Constitucion* (discovered in 1956) and the smaller *Barcodon* anticline. An important offshore discovery made in 1968 in this area was the *Arenque Sur* field, 30 km east of Tampico. This is a structural culmination on a 50 km long north-south trend; the San Andres limestone reservoir is of Late Jurassic age. This field is therefore both structurally and stratigraphically important.

(b) Tuxpan Fields

Asphaltic oil seepages are common in this area, often being associated with igneous intrusions. Andesitic and basaltic plugs and dykes of post-Oligocene age often locally increase the porosity and permeability of the reservoir rocks by the fissuring they have produced, and add a high proportion of carbon dioxide to the natural gases.

Most of the oilfields lie northwest and west of Tuxpan along a narrow, arcuate, anticlinal ridge which stretches for 51 miles from Dos Bocas in the northeast to San Isidro in the south. The principal accumulations along this "Golden Lane" are: *Dos Bocas, San Geronimo, Juan Casiano, Amatlan, Zacamixtle, Toteco, Cerro Azul, Potrero del Llano, Cerro Viejo, Tierra Blanca, Chapapote Nunez, Alamo, Jardin, Paso Real* and *San Isidro*. These fields are structural culminations along one huge undulating fold, separated from each other by faults or local saddles.

The limestone ridge, which forms the principal oil reservoir, is about a mile wide, and is somewhat asymmetrical in cross-section, with a dip steeper on the west than on the east. It is composed of some 8,000ft of massively bedded El Abra (Lower Cretaceous) limestone, which probably developed as a barrier reef growing on an underlying structural uplift. This bed outcrops at the surface in the Sierra Boca del Abra range, and consequently the oilfields are affected by a powerful natural water drive. Some oil is also found in the relatively thin overlying San Felipe limestone.

Porosity in the El Abra limestone is principally due to its content of hollow fossil casts of molluscs, rudistids and corals; much of the limestone is in fact actually made up of shell breccia. Fracturing, faulting and solution cavities have greatly added to the natural porosity and permeability of the rock, and explain the enormous volumes of oil that have been obtained from wells in "flush" production. The limestone has most probably been the source rock for the oil which it contains.

Development of the "Golden Lane" oilfields began in 1910, and up to 1933 they were the most important factor in Mexican oil production, since practically every well drilled into the reservoir was a large oil producer. Some wells produced tremendous flows of oil—Cerro Azul No. 4 became world-famous for its initial output of 260,000 b/d, and the first Dos Bocas well flowed out of control for two months at an estimated rate of 200,000 b/d. Since the water drive is very efficient, however, the wells also tend to produce salt water suddenly when the oil-water level reaches them.

The output of the "Golden Lane" oilfields has declined in recent years, but other similar accumulations have been discovered on roughly parallel limestone ridges both to the west and, more recently, to the east in the Gulf of Mexico. Thus, the *Tres Hermanos* field, northwest of Tuxpan, is now one of the largest oil producers in Mexico; *Morallilo* and *Solis* are other important onshore accumulations. Offshore, there have been several discoveries since 1963 along what is now termed the "Marine Golden Lane"—*Arrecife Medio, Isla de Lobos, Tiburon, Esturion, Bagre* and *Atun*. Of these, *Atun* appears to be the largest accumulation, producing about 20,000 b/d in 1969 from an El Abra-type reservoir lying at a depth of about 9,700ft.

(c) Poza Rica Fields

The "Golden Lane" ridge continues in an arc running southeastwards between Tuxpan and Nautla. The *Poza Rica* field about 30 miles south of Tuxpan, was discovered in 1928. The reservoir here is a porous patch of "Tamabra" limestone—a mixed Lower Cretaceous facies of Tamaulipas and El Abra—situated on a local anticlinal "nose." This field is now the second largest producer in Mexico (with an output of some 51,000 b/d), and other similar accumulations on the same structural trend have been found at *Soledad, Miquetla, Jiliapa, Remolino* and *San Andres-Hallazgo*.

A parallel ridge, nearer to the coast, has produced a number of less important oil accumulations, while a continuation of the "Marine Golden Lane" trend south of *Atun* has been found to contain oil in structural culminations offshore at *Morsa* and *Escualo*. The *Furbero* field, southwest of Poza Rica, is an interesting example of production from crystalline rocks. Here, a dyke of gabbro has been intruded into Eocene shales, baking and fracturing them so that they have formed a reservoir for later Cretaceous or Jurassic oil migrating along the faults.

(iii) Vera Cruz (or Papaloapan) Embayment

This basin lies to the southeast of the Tampico-Nautla embayment, but differs from it stratigraphically through facies changes and thickness variations in the Cretaceous beds, and also because of the presence of Tertiary transgressions. The oil reservoirs here are again Cretaceous limestones, but there is also an important gas output from Oligocene sandstones.

The first oilfield to be discovered in this area in 1953 was *Angostura;* other accumulations are *Cocuite, San Pablo, Rincon Pacheco* and *Mirador.* The combined output of these fields accounts for only about 1 % of the national total.

(iv) Isthmus of Tehuantepec-Tabasco Area

The most southerly oil-producing region is the northern part of the Isthmus of Tehuantepec. In the western part of this area, the Isthmus Saline Basin contains Tertiary sediments which have been affected by northwest- and northeast-striking fold systems; salt domes occur at the intersections of these trends. The age of the salt is probably Jurassic and oil is found in Miocene sandstones above and on the flanks of the domes, many of which have sulphur deposits in their cap rocks. The most important fields are *Cinco Presidentes* (now one of the largest oil producers in Mexico), *Magallanes, La Venta* and *Ogarrio. El Plan* was the first accumulation to be discovered in this area, in 1923, and is still a minor producer; *Santa Ana* was the first Mexican offshore field to be found.

Further east, in the Tabasco area and the Macuspana Basin, the influence of salt intrusions on oil accumulations becomes progressively less, so that Miocene sandstones in local anticlines form the reservoirs. *Mecoacan* and *Tupilco* are important oilfields in the Tabasco region, and *José Colomo-Chilapilla* and *Ciudad Pemex* are the most notable accumulations in the Macuspana Basin. The *José Colomo* field, found in 1951, is a large east-west striking anticline which originally produced light oil and gas from a Miocene sandstone reservoir; this was found to be only one of a series which contain overall at least 2·5 Tcf of gas.

The total proved reserves of hydrocarbons in Mexico at the end of 1968 were 2,744 MM brl of crude oil, 423 MM brl of condensate, and 11·8 Tcf of natural gas. By the end of 1971, oil reserves amounted to 2,837 MM brl, a reserves/production ratio of only 18:1. In spite of the relatively high level of oil and gas production achieved in recent years, there is therefore anxiety about meeting the future demand, which will probably be the equivalent of 3,600 MM brl of oil for the period 1971-80.

Hence, apart from an increased effort in the major productive areas, both onshore and offshore, exploration is also being actively carried out in other parts of the country, such as the Yucatan Peninsula, the

Fig. 47. OILFIELDS OF MEXICO

Source: Petroleum Press Service.

State of Chiapas contiguous to Guatemala, and north-central Mexico, where there is an area of about 400,000 sq km of marine sediments which may possibly have not been so greatly affected by local igneous and tectonic activity as to be devoid of petroleum accumulations.

Two important new oilfields (*Cactus* and *Sitio*) were discovered in Chiapas in 1972, in which year the total crude oil and condensate production in Mexico totalled 185 MM brl—still 11 MM brl short of annual consumption, which can only be met by increased imports.

CHAPTER 11

South America and the Caribbean

SINCE Venezuela accounts for about 80% of the oil and 70% of the natural gas produced from the whole of South America, any discussion of the history of petroleum in this continent is inevitably related to its history in Venezuela. Table 78 illustrates this preponderance and also shows that, while between 1960 and 1970 world crude oil output increased by some 117%, the overall increase in South America in those years was only 36%. This was chiefly due to economic factors which slowed down the rate of growth of Venezuelan production relative to the extremely rapid growth of Middle East and North African outputs during this period.

In the pre-Columbus era, the Indians of Venezuela used asphalt from the "menes" or seepages of Lagunillas and Guanoco for caulking their canoes, and the Incas of Peru also used seepage oil. The Peruvian oil industry was commercially established on a small scale even before that of Venezuela, having an output of 275,000 brl in 1900. Commercial oil production was also established in Trinidad in the early 1900's. Although drilling commenced in Venezuela in 1883, it was only in 1914 that the first large oil accumulation was found. However, the development of the Venezuelan oil industry was thereafter extremely rapid. While less than 0·5 MM brl was produced in 1920 compared with 2·82 MM brl in Peru, by 1930 the output of Venezuela was 136·7 MM brl, compared with only 52·8 MM brl from all the other oil-producing countries of South America, which by that year included Argentina, Bolivia, Colombia and Ecuador, in addition to Peru and Trinidad.

Colombia was for many years the second most important South American oil producer, but by 1960 it had been outstripped by Argentina. In the Caribbean, the output from Trinidad, where complex geological conditions are common, remained nearly static between 1940 and 1956, when the first offshore accumulation was found, but has since increased again largely due to Continental Shelf discoveries. Progress in Bolivia and Brazil has been relatively rapid in recent years, and the new fields found in the interior of Ecuador will greatly add to the production from that country once adequate outlets are available. However, until Venezuelan exploration once more enters an active phase, South American oil production will inevitably continue to decline as a proportion of that of the world.

From the geological point of view, the sedimentary and tectonic history of South America has been controlled by the distribution and interplay of a number of major geotectonic units[1]:

(1) The stable core of the continent is made up of three Pre-Cambrian "cratons": the Guyana "Shield", the Central Brazilian "Shield" and the coastal Brazilian "Shield".

(2) Interposed between these and superimposed on former Pre-Cambrian geosynclines are four "intracratonic basins": the Amazonas, Parnaiba, Sao Francisco and Parana basins.

(3) The Deseado and Patagonian massifs and the Argentinian Pampean mountains are formed by smaller "island Shields".

(4) A series of "pericratonic basins" (Llanos-Iquitos-Acre-Beni-Chaco-Pampas) lie along the western borders of the main sedimentary basins and form plains areas.

(5) Mobile orogenic belts garland the western borders of the "pericratons" and "island Shields".

Most of the large oil accumulations of South America so far discovered have been found in Tertiary or Mesozoic reservoirs, and in structures which are generally associated with the Tertiary folding movements which have formed the Andean mountain ranges; however, in the Argentine, epeirogeny and block-faulting has been the dominant tectonic factor, while in Bolivia Palaeozoic accumulations also occur.

TABLE 78

SOUTH AMERICA—OIL PRODUCTION
(thousands of brl)

Year	Argentina	Bolivia	Brazil	Chile	Colombia	Ecuador	Peru	Trinidad	Venezuela	Total South America
1900	—	—	—	—	—	—	275	—	—	275
1910	20	—	—	—	—	—	1,258	143	—	1,421
1920	1,651	—	—	—	—	60	2,817	2,083	457	7,068
1930	9,002	56	—	—	20,346	1,553	12,449	9,419	136,669	189,494
1940	20,609	288	2	—	25,593	2,349	12,126	22,227	185,570	268,764
1950	23,353	616	339	629	34,060	2,632	15,012	20,632	546,783	644,056
1960	64,232	3,574	29,613	7,231	55,770	2,730	19,255	42,357	1,041,708	1,266,470
1970	143,410	8,820	59,980	12,410	73,000	1,500	26,269	51,043	1,353,418	1,729,850

Note: Statistics from Cuba have not been obtainable in recent years, so are not included. The 1970 output was estimated to have been 1·4 MM brl.

[1] H. J. Harrington, *Bull.AAPG*, 46,(10), Oct. 1962, 1773.

x

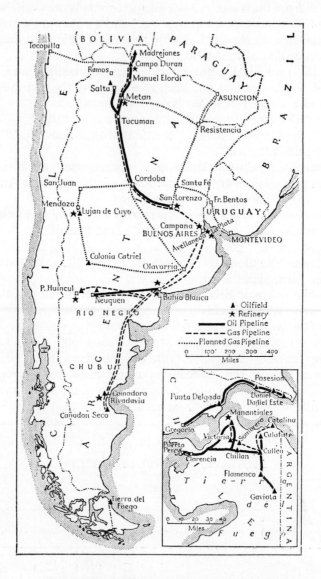

Fig. 48. OILFIELDS OF ARGENTINA AND CHILE

Source: Petroleum Information Bureau.

TABLE 79

SOUTH AMERICA AND W. INDIES—OIL PRODUCTION, 1971

	MM brl
Argentina	154·5
Bolivia	13·2
Brazil	62·2
Chile	12·9
Colombia	78·2
Cuba	1·4
Ecuador	1·6
Peru	22·6
Trinidad	47·1
Venezuela	1,295·4
Total	1,689·1

TABLE 80

SOUTH AMERICA AND W. INDIES—ESTIMATED PROVED OIL RESERVES, end-1971

	MM brl
Argentina	2,000
Bolivia	262
Brazil	855
Chile	120
Colombia	1,626
Ecuador	6,070
Peru	510
Trinidad	1,053*
Venezuela	13,740
	26,236

Source: World Oil, Aug. 15, 1972. * onshore and offshore

ARGENTINA

Oil was first produced in Argentina in 1910, but development was slow until the 1950's, when the Colombian oil production rate was finally outstripped and Argentina became established as second only to Venezuela as a South American oil producing country.

The rate of increase in output has, however, been slower in recent years, due to disputes between the state-owned YPF (Yacimientos Petroliferos Fiscales) and the former contracting companies. Nevertheless, oil output reached 158·5 MM brl in 1972, a rate of production which was nearly sufficient to meet the country's internal requirements; the stimulus to exploration which has resulted from the implementation of the new Hydrocarbons Law is also likely to result in a marked expansion in future output. This will be necessary in view of the forecast requirements in 1980 of 214 MM brl of crude oil.

Several sedimentary basins, totalling onshore about 1,213,000 sq km in area, contain fault-block type structures, generally with Mesozoic reservoirs. Production comes mainly from the Upper Cretaceous Chubutiano formation, but there are also several older reservoir beds.

The principal basins, their areas, and the groups of fields so far discovered in them, with the ages of their reservoir beds, are respectively as follows:

TABLE 81

ARGENTINA—BASINS AND OILFIELDS

North-west Basin (502,760 sq km)	—Salta fields (Permo-Carboniferous)
Guyo Basin (128,480 sq km)	—Mendoza fields (Triassic)
Neuquén Basin (139,800 sq km)	—Neuquén and Rio Negro fields (Neocomian, Dogger)
San Jorge Basin (north flank) ⎱ (total area 91,200	—Commodoro Rivadavia and Chubut fields (Upper Cretaceous); also less important Paleocene and Jurassic reservoirs
San Jorge Basin (south flank) ⎰ sq km	—Santa Cruz field (Upper Cretaceous)
Austral or Magellanian Basin (118,760 sq km)	—Santa Cruz and Tierra del Fuego fields (Upper and Lower Cretaceous)
Salado Basin (160,880 sq km)	
Colorado Basin (70,880 sq km)	

The proportionate oil output from the different geographical provinces is approximately: Santa Cruz 30%, Mendoza 20%, Chubut 22%, Rio Negro 14%, with the other oil-producing areas of Salta, Neuquén and Tierra del Fuego together providing the balance of 14%.

Of the basins, San Jorge has accounted for about two-thirds of the Argentine cumulative production and still has about two-thirds also of the balance of proved onshore reserves of about 2·0 B brl of primary and secondary oil. The most important oilfield in this basin is *Commodoro Rivadavia,* which has been the major Argentine producer since its discovery in 1907.

In this part of Patagonia, there is a very thick series of continental deposits which have been interrupted by three wedge-shaped intercalations of marine sediments. The oldest wedge, whose edge faces east, is of Jurassic and Lower Cretaceous age; the second, which pinches out towards the west, is Upper Cretaceous; and the last, which also pinches out towards the west, is of Middle Tertiary age.

This predominance of continental conditions complicates any interpretation of the local geology, owing to the irregularity of the deposits and the paucity of fossils. It is thought that the oil was formed in coastal lagoons along the fringe of an Early Cretaceous or Jurassic continent, passing later up fault planes to the present reservoir beds. These are mainly sands and shales of the Chubutiano and Salamanquean formations of Upper Cretaceous age.

There is a sharp structural contrast in this area between the strongly folded western zone (Sierra San Bernardo, etc.) and the gently dipping coastal area. There are many faults, and each fault-block acts as an independent unit for petroleum production. It is possible that they represent the foundered remains of an ancient structural arch.

Some shale beds are locally productive due to fracturing caused by fault movements and mingling with disintegrated sandstones. Many of the fault faces in this field have been sealed by plastic shales dragged down by the faulting movements.

In the Mendoza basin, the discovery of the *Tupungato* field in 1934, and its subsequent development, have been a major factor in the growth of the Argentine oil industry. This field is about 30 miles south-southwest of the city of Mendoza, in the eastern foothill zone of the Andes. The surface structure is a Pliocene dome, with a faulted west flank. The structure differs at depth from its surface expression owing to the presence of numerous low-angle thrust faults. Oil is found in fractured volcanic tuffs lying in Upper Triassic sediments. There is also some shallower Pliocene oil.

Also in Mendoza, the *Cacheuta* field has produced relatively small quantities of oil since 1896. The structure is a faulted monocline, and the producing horizon is an Upper Triassic sandstone. Some oil has also come from the *Sosneado* and *Lunlunta* structures, and from *Barranca, Carrizal, Ventana* and *Pinta de las Bardas,* small fields to the northeast of Mendoza city.

In Neuquén, the *Plaza Huincul* fields produce oil from Cretaceous and Jurassic sandstones in complex buried structures. These are small pools on the crest or flanks of the Otena-Challaco uplift which crosses southeast Neuquén. They include *Centra* and *Oesta Octogono, Laguna Colorado, Bajo los Baguales* and *Challaco.*

In Salta province, production has been from Trias-Permian reservoir beds in faulted anticlines at *Tartagal-Lomitos-Tranquitas, Ramos, San Pedro* and *Aguas Blancas. Campo Durán* is an important condensate field, discovered in 1951.

In the Jujuy area near the Bolivian frontier (at Barro Negro des Tres Cruces and elsewhere) there are many seepages from dolomitic limestones on the edge of the Brazilian "Shield." A test well in this area, drilled in 1969 at *Caimancito,* is reported to have been successful in finding a Cretaceous reservoir. Further north, very close to the frontier, Devonian oil has been found at *Macueta.*

In the Argentine section of Tierra del Fuego there are 11 small fields, some of which produce only gas. The most important of these are *Rio Chico, San Sebastian* and *La Sara.*

There are also promising oil possibilities in the 400,000 sq miles of the Argentine offshore areas, in the extensions of several of the onshore sedimentary basins[2]. In particular, more than half the main San Jorge petroliferous basin lies offshore in the Gulf of San Jorge.

The Austral or Magellanian basin, which has onshore oil in Southern Santa Cruz province and on Tierra del Fuego, is also promising, but weather conditions make exploration here difficult.

---

[2] W. Ludwig *et al., Bull.AAPG,* 52, Dec. 1968, 2337.

The first offshore exploration concessions were granted in Argentina at the end of 1967 in two main areas—Samborombon Bay, and off Bahia Blanca, and there has since been considerable marine exploration activity.

BRAZIL

Most of the land area of Brazil consists of an old pre-Cambrian "Shield" which has been covered by a varying thickness of gently dipping Palaeozoic and Mesozoic sediments. Oil prospects are concentrated in a number of Cretaceous basins on the periphery of the "Shield", in which there is an adequate sedimentary thickness for oil to have been generated and to have accumulated.

The total sedimentary area of the country is probably not less than $3\frac{1}{2}$ million sq km, much of which is still inaccessible. The Parana basin on the west of the Shield contains Permo-Triassic and Devonian sediments. In the Lower Amazon basin there are Devonian shales and limestones, and in the Upper Amazon basin Tertiary sediments occur which may be folded into closed structures resulting from the Andean orogenies. However, it is in the sedimentary basins or embayments contiguous to the Atlantic coast that exploration has so far been concentrated, resulting in the discovery of several important oilfields.

Until recently, most of the oil produced by the state company, Petroleo Brasileiro, came from the Reconcavo-Tucano basin (the Bahia fields), mainly from *Agua Grande,* with lesser production from *Candeias, Taquipe, Dom João* and *Buracica.* These are generally asymmetrical anticlinal structures with their principal oil accumulations in Lower Cretaceous sandstones. Of these, the 'A' sandstone is the most important, consisting of a number of bar-shaped lenses deposited on ancient shelves and shoal ridges. Petroleum migration in this area was probably controlled by the topography of the original depositional basin[3]. A programme of secondary recovery by water injection was initiated in these fields in 1965 to offset their natural decline.

In the Sergipe basin, oil accumulation has been shown[4] to be due to favourable structural evolution during the Late Cretaceous combined with local lateral permeability barriers or reservoir "pinch-outs". There are suitable sealing evaporites and organic source rocks near the reservoir beds. The *Carmópolis* field, discovered in 1963, is an east-west elongated dome covering a proved area of some 45 sq km. Relatively shallow oil is obtained from eight Lower Cretaceous sandstones. The original reserves (1963) were 1·2 MM brl; the field is now declining and produces 21,400 b/d.

---

[3] E. Bauer, *Bull.AAPG*, 51, 28, 1967.
[4] E. Meister and N. Aurich, *Bull.AAPG*, June 1972, 1036.

Siririzinho was drilled in 1967 on a north-south "high" in the same area, and now produces 7,000 b/d from similar reservoirs, while *Piachuilo* produces less than 2,000 b/d.

The *Miranga* field, which was found in Bahia late in 1965, now produces about 20% of the national output and is by far the largest Brazilian oilfield.

In 1971, Brazilian oil production reached a peak of 62·3 MM brl— still small by world standards, but very much larger than the 0·3 MM brl produced in 1950. However, even this rate of production can now satisfy less than a third of the country's rapidly expanding requirements. Proved recoverable oil reserves are about 855 MM brl.

Exploration is currently being concentrated far to the north and north-west of the Bahia fields in the states of Rio Grande do Norte and Maranhao, where several promising discoveries have been reported, notably in the Barreirinhas and Sao Luis basins. Exploration has also commenced on the Continental Shelf in the southerly Espirito Santo basin; two offshore oilfields (*Guaricema* and *Caioba*) are expected to go into production in 1973.

An experimental oil-shale distillation plant has been constructed at São Mateus do Sul in Parana for processing samples of the extensive Lower Permian Irati oil-shales, which contain up to 12% of distillable oil. It has been estimated that the Brazilian oil-shales ("xistos") which extend over much of the southern part of the country from northern Uruguay through the states of Rio Grande do Sul, Santa Catarina and Parana to Sao Paulo are second only to those of the United States in extent. Besides the Permian oil shales, there are others of Tertiary, Cretaceous and Devonian age[56].

TABLE 82

BRAZIL—OIL PRODUCTION, 1971
(MM brl)

Alagoas fields:	0·29	*Bahia fields:*	
Sergipe fields:		*Agua Grande*	8·78
		Aracas	8·33
Carmopolis	7·53	*Buracica*	7·04
Riachuele	0·92	*Candeias*	2·08
Siririzinho	2·16	*Dom Joao*	4·50
Others	0·27	*Boa Esperanca*	0·69
	———	*Miranga*	12·58
	10·88	*Taquipe*	3·23
		Others	3·85
			———
			51·08

Total production 62·20 MM brl

5 V. Padula, *Bull.AAPG*, 53(3), March 1969, 591.
6 J. Rasmuss, *World Oil,* Oct. 1967, 118.

BOLIVIA

A small volume of oil has been produced in Bolivia since 1925 from a group of steep, overthrust folds within the narrow, southern "Sub-Andean Trough", where the strike of the Eastern Cordillera changes from north-south to northwest-southeast, and the mountain range makes its closest approach to the Brazilian "Shield"[7]. The largest of these accumulations is *Camiri,* which has a Devonian (Iquiri) sandstone reservoir, originally discovered in 1927 and extended with the discovery of the *Guairuy* pool in 1947. Bolivian oil production for many years did not exceed about 3·0 MM brl/a, but in 1960 an important new Devonian discovery was made at *Caranda* near Santa Cruz. This was followed by the development of a number of other structures with mainly Devonian sandstone accumulations in the same area—*Tatarenda, Rio Grande, Naranjillos,* and *Monteagudo.* In the last-named field there are also Cretaceous and Triassic reservoirs. These Santa Cruz fields now produce about two-thirds of the total Bolivian crude oil output.

Further south, close to the Argentine frontier, several Devonian sandstone accumulations have been found—*San Alberto, Madrejones* and *Bermejo-Toro.*

The completion in 1966 of the Trans-Andean pipeline, which rises to nearly 15,000 ft above sea-level in the course of its passage from Santa Cruz to Arica on the Pacific Coast of Chile, has resulted in a sharp increase in oil production from Bolivia, and in 1971 this amounted to about 13·2 MM brl. Bolivian oil is very low in sulphur content, and therefore much in demand on the West Coast of the USA. Reserves are estimated to be about 262 MM brl.

The Bolivian government nationalised the oil industry in 1969, and all exploration and production operations have since been carried out by YPFB (Yacimientos Petroliferos Fiscales Bolivianos). It is also interesting to note that in 1972, with the completion of a pipeline from the Santa Cruz area to the Argentine frontier, Bolivia became the first gas-exporting country in South America.

CHILE

Oil was discovered in Chile for the first time in 1946 in Lower Cretaceous sandstones in the *Cerro Manantiales* field on Tierra del Fuego. Several other small Lower Cretaceous accumulations have since been found in the extreme south of Chile, both in the northern part of Tierra del Fuego and on the mainland, in the Springhill district of Magallanes Province near the eastern entrance of the Magellan Straits. The trend of the Andes range in this region is northwest-southeast, and there is a geosynclinal area on the northeast of this axis. The Late

---

[7] W. Lamb & P. Truitt, 6 WPC, 1, 40 PD 3, 1963.

Cretaceous basin was bisected by a northeast-southwest trending uplift[8], so that the sedimentation histories of the beds deposited on either side of this ridge are different.

Oil production reached 13·7 MM brl in 1964, and thereafter the natural decline of the fields has been offset by the discovery of a few new accumulations. The 12·9 MM brl of oil produced in 1971 is still sufficient, however, to meet about half the national requirements.

There is no outstanding field, several accumulations being roughly of the same size—eg. *Calafate, Cullen, Canadon, Daniel, Daniel Este* and *Posesion*. All of these have individual production rates of between about 1·3 and 2 MM brl/a.

About half the total Chilean crude oil output now comes from the mainland, and half from Tierra del Fuego. All the reservoirs are sandstones of the Cretaceous Springhill formation.

Due to high gas/oil ratios, the Chilean output of associated gas is substantial, amounting to about 255 Bcf/a, of which about 70% is reinjected.

COLOMBIA

Petroleum is produced in Colombia from three sedimentary basins in which structural deformations have resulted from Tertiary orogenesis. These are the Magdalena River basin, the Zulia extension of the Maracaibo basin, and, most recently discovered, the Putumayo basin.

Most of the oil produced in the past has come from the Magdalena River valley area, which is an elongated, roughly north-south present-day basin, lying between the central and eastern Colombian Andes. This includes only the small central part of a much more extensive Cretaceous basin, and also the western, shallow margin of an Eocene-Oligocene basin.

Structurally, the Magdalena valley has been described as a "graben", or perhaps more correctly as a "half-graben". Asphaltic and naphthenic oils are obtained from a group of fields with Tertiary (mainly Eocene-Oligocene) sandstone reservoirs in the central part of the area, while lighter, paraffinic oils are produced from smaller Cretaceous limestone reservoir fields. It has been suggested[9] that the heavier oils in this basin formed first, and then migrated to the margins, or shelf areas, while the lighter oils remained confined to traps in the central part of the basin.

The *Infantas* field, the first commercial accumulation to be discovered (in 1918), is a sharply folded faulted anticline, running roughly north-south, with a productive area about seven miles long by one mile wide, cut by a series of east-west cross-faults. There is an overthrust on the

---

[8] R. Charrier & A. Lahsen, *Bull.AAPG*, 53(3) March 1969, 568.

[9] L. G. Morales, *AAPG*, "Habitat of Oil", 1958, 641.

eastern edge of the field. *La Cira,* found in 1926, is a broad anticline to the north-west of *Infantas,* about five miles long by four miles wide, and overthrust on the eastern flank.

Production in these two fields comes from Tertiary (Eocene-Oligocene) sandstones which are divided up by faults into a series of separate productive zones; the main reservoirs are believed to be interconnected at depth.

Cantagallo was first discovered in 1943 and produces from Eocene sandstones. It comprises an eastward-dipping monocline bounded on the northwest by a major fault.

Casabe was discovered as a result of a 1941 seismic survey in a very swampy area; it comprises a large subsurface fold bounded by a major fault. Production comes from Colorado and Mugrosa (Oligocene) sandstones, and the upper part of the Toro-La Paz (Eocene) formation.

Oil was found at *Velasquez* in 1946 in similar Eocene sandstone reservoir beds. The structure here is a relatively simple homocline dipping to the southeast, which is cut by normal faults. In the same area, the *Provincia* field, discovered in 1962, is now the largest producer in this basin.

In the lower Magdalena valley (sometimes also called the Caribbean basin), the *El Difícil* anticline was discovered in 1943, with gas, condensate and oil in an Oligocene-Eocene limestone reservoir. Another oil accumulation in this area is at *Cicuco,* while *Jobo-Tablan* is a gasfield which supplies natural gas by pipeline to Cartagena. (However, the total consumption of natural gas in Colombia is still only about 20 Bcf/a out of a total production of some 88 Bcf/a: the balance is flared.)

The second oil-producing basin is the Barco region, close to the Venezuelan border, where oil was first found in 1933 and commercial production began in 1938. This forms the western or Zulia extension of the Maracaibo Basin[10]. Accumulations in this area are in Tertiary and Cretaceous reservoirs, generally in faulted fold structures.

Petrolea North Dome is an asymmetrical, intricately faulted anticline with production from several Cretaceous sandstones, mainly the La Luna and Cogollo formations[11]. *Carbonera* is a small structure further east, producing heavy oil from Eocene sands; *Rio de Oro* is a narrow, compressed anticline with an Upper Cretaceous sandstone reservoir.

Tibu, 25 miles north of Petrolea, which was discovered in 1941, is a large, well-marked anticline. Production was first obtained from Eocene sand lenses, but a more productive Lower Cretaceous reservoir has since been found.

The most recently discovered field in this area is *Rio Zulia,* which was for some years the largest producing field in the country (with a peak output from Tertiary reservoirs of 12·7 MM brl in 1966).

[10] H. G. Richards, *Bull.AAPG,* 52, Dec. 1968, 2324.
[11] F. Notestein *et al., Bull.Geol.Soc.Amer.,* 1944, 1166.

During the last few years, a new petroleum province has been opened up in the remote Putumayo basin of southern Colombia, near the frontier with Ecuador, where all supplies—including drilling rigs—had originally to be flown in.

The *Orito* field (discovered in 1963) and several other similar accumulations in the same area, have oil mainly in Cretaceous and fractured "basement" reservoirs. Production began in March, 1969, upon completion of the pipeline across the Andes to Tumaco on the Pacific coast, and during the rest of that year *Orito* produced 17·3 MM brl of oil; production also started from *Caribe, Lovo* and *Puerto Colon*.

The total output of oil from Colombia was 64·8 MM brl in 1968 (a decrease of 8% since 1967 and 11% since 1966), coming from 2,186 wells (of which 2,025 were on artificial lift). By 1971, output had risen to 78 MM brl, and it is possible that after several years of semi-stagnation caused by legislative and labour difficulties, production may grow rapidly in the near future, as the Putumayo fields are developed.

Recent successful exploration drilling at *Carare* and *Tenetrifa* may mean that further oil accumulations will be found in the Llanos area, the great plains east of the Andes.

The total proved oil reserves of Colombia have been estimated at 1,626 MM brl, with 2·8 Tcf of natural gas, but these figures are almost certainly too low in view of these results and the discoveries in the Putumayo basin.

TABLE 83

COLOMBIA—OIL PRODUCTION, 1971

	MM brl
Orito	20·14
Rio Zulia	9·17
Payoa	2·97
Infantas	1·58
La Cira	6·07
Tibú	4·29
Provincia	7·48
Casabe	2·69
Yarigui	3·26
Velasquez	4·25
Palagua	2·84
Other fields	13·41
Total	78·15

ECUADOR

Oil has been produced for many years in Ecuador from several relatively small fields on the west coast of the Santa Elena Peninsula, an area of very complex geology, with local dips which are often greater than 70°, and intrusions of dolerite and great masses of chert.

Figs. 49, 50. OILFIELDS OF COLOMBIA AND ECUADOR

Source: Petroleum Press Service.

Most of the oil output has come from *Ancon* in the southern part of the peninsula, where the dips are somewhat gentler at the surface, although here, too, the rocks are much disturbed by tear-faults. Oil has been produced in this area since 1921 from very shallow (less than 1,000 ft deep) lenses of sandstone which occur in the Socorro (Upper Eocene) series of clays, sands and silts. Deeper oil also comes from fissures in a very thick brecciated sandstone of Middle or Upper Eocene age. The structure is essentially a series of complex fault blocks. Similar accumulations in Eocene reservoirs have been found in the *Cautivo, San Raimundo, Petropolis* and *Carmela* fields in the same area. The oil output of these fields is about 1½ MM brl/a and diminishing yearly.

The proved oil reserves of the coastal fields are now only equivalent to a few years' current production, although offshore exploration has been commenced in an effort to find possible extensions. Consequently, the recent oil discoveries in southern Colombia aroused interest in the possibility that equivalent accumulations might exist in northern Ecuador. Considerable exploration activity has been concentrated in the Oriente region of the Putumayo Basin. This is a north-south asymmetrical downwarp whose main axis is about 60 miles east of the Andes Mountains, which contains a number of anticlinal structures with downthrown faults on the east flank. Important Mesozoic oil discoveries have been made in this area at *Lago Agrio* and at *Bermejo, Parahuacu, Shushufindi* and *Atacapi*.

Of these, the *Lago Agrio* structure is about 10 miles long and 2 miles wide, and the reservoirs are thick Caballos and Villeta (Lower and Upper Cretaceous) sandstones; its reserves have been estimated at 300 MM brl. The development of these fields has required the construction of a third Trans-Andean pipeline to the coast between Esmeraldas and San Lorenzo, but as a result of their discovery the oil reserves of Ecuador have been greatly increased, to perhaps as much as 6·0 B brl. The gas reserves have also been increased as a result of an offshore discovery *(Amistad)* in the Gulf of Guayaqil.

PERU

Seepage oil and asphalt were used in Peru in pre-Conquest times, and there are records of shallow wells being drilled as early as 1864. By 1900, the output of oil amounted to 275,000 brl/a from a number of small accumulations discovered on the coastal plain.

These oilfields occur in a belt of much faulted Tertiary rocks, which extends southwards from the Ecuador border. A thick (up to 25,000ft) series of sediments, ranging in age from the Eocene to the Pliocene, rests either on Cretaceous rocks or metamorphosed Pennsylvanian beds.

Production generally comes from the Eocene (Parinas and Saman sands and conglomerates), the Oligocene Heath formation, or the Miocene Zorritos sands and conglomerates.

The main tectonic features are the Western Cordillera fold and fault zone on the east[12], and a sub-Pacific fault which parallels the coast to the west, following the edge of the Continental Shelf. The Tertiary beds between have been folded and extensively block-faulted, so that oil accumulations occur in faulted anticlines of varying complexity.

La Brea-Parinas, first discovered in 1869, has produced most of the Peruvian oil. The sedimentary section here consists of intermingled shales and sands which probably have provided both source and reservoir rocks, and there are a number of separate pools, the oil recovery rates from which vary with the local sand/shale ratios[13]. The best recovery rate is found in the *Bonanza* pool, which has a sand/shale ratio of 0·62; other pools are *Puerto Rico, Bronco, Honda, Chivo, Siches* and *Malacas*[14].

Further to the northeast, along the Pacific coast, two other Tertiary oilfield areas are at Cabo-Blanco—Lobitos (the *Interlob-Lima* fields) and Zorritos (*Los Organos-Hualtacal*). In 1970, the Lima coastal fields produced 10·2 MM brl of oil as compared with only 4·4 MM brl from La Brea-Parinas, where no less than 7,700 shallow wells have been drilled. In addition, about 0·9 MM brl of oil was produced from *Los Organos.*

An increasing amount of Tertiary oil is now being obtained from offshore wells drilled on the Continental Shelf west of Talara. The most important of these marine fields are *Litoral, Humboldt* and *Providencia,* which together produced some 9·7 MM brl of oil in 1970. Cretaceous production totalling about 1·3 MM brl/a is also being obtained from the *Aguas Calientes* and *Maquia* anticlines in the eastern interior region. The total output of oil from Peru amounted to 22·6 MM brl in 1971; there was also an output of about 66 Bcf of associated gas, of which as much as 60% was flared for lack of outlets.

TRINIDAD

The first oil-well was drilled on the island of Trinidad in 1866, but commercial production was not established until 1902. Output thereafter increased steadily, until in the late 1930's Trinidad became the largest oil producer in the British Commonwealth, with a yearly production equivalent to about 1% of the world total. However, due to the complex nature of the local stratigraphy and the small size of the individual accumulations, the onshore oil output has been declining in recent years, and future increases must certainly come from the offshore areas around the coasts which are now being developed. Although the production from Trinidad in 1968 was the highest-ever (69·9 MM brl) it had fallen to 47·1 MM brl by 1971. The average

---

12 S. Szekely, *Bull.AAPG*, March 1969, 553.
13 W. Youngquist, *AAPG*, "Habitat of Oil", 1958, 708.
14 J. Rasmuss, *World Oil*, May 1967, 123.

production of oil from each of the approximately 3,500 onshore wells is less than 40 brl/d, and the proved recoverable onshore reserves (550 MM brl) are equivalent to little more than ten years of current production.

The average gas/oil ratio is more than 2,300 cf/brl and some 121 Bcf of associated gas was produced in 1970. Much attention has been given to secondary recovery operations to increase the low (13-14%) primary oil recovery factor, and about 14% of the gas output is currently reinjected, while in addition, a number of water or steam injection operations are under way.

The geology of Trinidad is closely related to that of East Venezuela, since the southern part of the island is a continuation of the Maturin Basin, while the Northern Range of Mountains is a continuation of the Venezuelan Andes. The "basement" here is probably composed of the northward-slanting rim of the South American Guayana "Shield", with its intensely folded gneisses, schists and pyrogenetic rocks.

The east-west trend of the Northern Range (probably due to rift faulting) is deceptive, since the general structural trend is really north-east-southwest. This direction agrees with the average strike of the pre-Oligocene beds both in Barbados and along the north of Venezuela as far as Puerto Cabello.

The island is characterised by an extremely intricate system of folding and faulting. Thrusting seems to have come from both north and south, so that the least deformed portion is a central strip with closely folded and broken beds on either side. The differential folding intensity may also be related to the age of the beds, since, with almost continuous folding and faulting movements in action, the most affected beds would be those which have been longest subjected to these forces.

The stratigraphy of Trinidad is very complex, and has been further complicated by the varied nomenclature applied to equivalent beds as a result of rapidly changing facies, coupled with the unsystematic early development of the industry.

Indications of gas and oil are widespread, and take the form of nearly every known type of "show". Deposits of asphalt and wax are common, mud volcanoes are frequent, and a pyrobitumen similar to impsonite occurs in Eocene and Cretaceous beds. Oil-stained sands, liquid oil seepages, and bituminous marls are all in evidence.

"Mud volcanoes" with associated gas "shows" are also plentiful, consisting essentially of a welling-up of gas and mud, with a certain amount of frothing. Occasional more violent movements occur, even producing islands which may later disappear, as in 1911, 1926, and, most recently, in 1964. The "Devil's Woodyard" is an example of a periodic "mud volcano", which in quiescent periods is used as a cricket pitch.

In the extreme south of Trinidad, "mud volcanoes" occur along straight lines which probably represent faults. Diapiric mud dykes have been found to traverse as much as 5,000ft of sediments, and may have enabled trapped oil and gas to escape from depths as great as this.

Oil accumulations have been found in the southwestern part of the Southern Basin, which is a broad, low, present-day valley, on the western side of which lies the extraordinary Pitch Lake. This is a mass of semi-solid bitumen, presumably representing the last stages of inspissation and destruction of an enormous oilfield, which covers an area of 137 acres in the south of Trinidad and from which material has been systematically removed for many years, the lake being apparently replenished from a subterranean source. Oil production in this area comes mainly from Miocene sandstone reservoirs which are complicated by the local interfingering sands and clays. There is also some deeper Cretaceous production.

Most of the accumulations are anticlinal, probably structurally related to diapiric mud movements, but production is also often found in the neighbouring synclines.

The reason for the concentration of oil in the southwestern region is probably the stratigraphic trapping resulting from the interdigitation of the Forest-Cruse sandstones with the clayey Naparima beds. There are also two intersecting structural trends in this area which are probably related to underlying faults, and which have influenced local oil accumulations. One runs roughly east-west and is intersected by the Los Bajos-Palo Seco trend which runs west-northwest to east-southeast. (The intervening Siparia synclinal area is also productive of oil from sandstone lenses.)

Among the more important oilfields, *Fyzabad* (discovered in 1920) consists of a series of interconnected domes along a roughly east-west axis broken by faulting. The *Forest Reserve* dome (1913) is structurally a mud plug uplift. However, the mud was injected when the axial uplift was already in existence, and overflowed in a southerly direction. *Bernstein* is a similar smaller domal accumulation to the east. The *Cruse* field lies to the west of Forest, and the *Apex* field is slightly offset to the south and east. Most of the oil in these fields has come from sandy, deltaic Miocene lens-type reservoirs. In the Penal region east of *Apex,* the small *Barrackpore* field is a faulted fold; here, the Cruse sandstone reservoir bed has passed into a clay, and production comes from younger sands.

Palo Seco (1929) lies about two miles from the crest of the Morne Diablo uplift. This field is of interest in that better production has been obtained by drilling towards the syncline than towards the anticlinal axis. The probable reason is local thrusting, which shattered the crestal reservoir rock and allowed its oil to escape. The main oil-bearing reservoir here is the Top Cruse (Miocene) sandstone.

Point Fortin is a strong fold, overthrust to the south with production from the Forest and Cruse beds.

Los Bajos, southeast of Point Fortin, lies on the Los Bajos-Palo Seco trend, and produces from the Forest and Cruse series on the upthrown side of the fault.

Brighton is a gentle dome (first discovered in 1908) on the crest of which the Pitch Lake is situated. The top La Brea and Moruga beds are followed by a clayey Forest-Cruse complex, and nearly all these rocks are saturated with a heavy, sulphurous oil. The field has been extended offshore in recent years.

Vessigny, to the southeast of Brighton, is another anticline which produces oil from Forest sands.

To the south, the *Parrylands* field gave initial Forest production, but a much more important Cruse development has been found at *Guapo.* La Brea "tar sands" outcrop in this area.

In the *Oropouche Lagoon* area, light oil and gas have been found in the Ste. Croix formation, but in small volumes only.

In the southeast corner of Trinidad, the Siparia syncline abuts on the Lizard Springs uplift, in an area of thrusting, slumping, and sedimentary vulcanicity. Asphaltic oil has impregnated the basal Moruga bed, and has formed accumulations on the flanks of subsidiary uplifts related to strike faults and unconformities. From the most important of these reservoirs, at *Guayaguayare* (originally discovered in 1902), lighter, paraffinic oil is obtained, principally from the Miocene Gros Morne sands.

Apart from the extension of the Brighton field, the only offshore accumulation actually in production is *Soldado* (23·9 MM brl in 1970), which lies offshore of the southwest coast; gas is now being injected into the Upper Cruse sands in the crestal area of this field. Considerable offshore exploration has been carried out in recent years around the coasts of Trinidad, where thick sedimentary sequences exist. In late 1968, a large accumulation of condensate and gas was discovered off Mayaro on the east coast (*Point Radix*); three oilfields have since been found in this area (*Teak A, Teak B* and *Samaan*) from which it is expected that a production of 150,000 b/d may be achieved by end-1973. A large gasfield has also been found off the southeast corner of the island (*South East Galeota*) and the establishment of a LNG exporting facility is planned, with exports at the rate of 200 rising to 400 MMcf/d to the southern USA planned for 1976.

It was at one time thought that the oil in Trinidad was related in origin to the lignites which occur in the Upper Tertiary formations. However, nearly all these lignites have since been proved to lie stratigraphically above the oil reservoir beds and to be separated from them by unconformities.

The most probable petroleum source beds are the Upper Miocene marls and clays which are often foraminiferal and inderdigitate with the reservoir sands. There is also some evidence of vertical migration, so that other, deeper source rocks may have existed.

Y

TABLE 84

TRINIDAD—OIL PRODUCTION, 1971

	MM brl
Soldado	21·6
Palo Seco	6·4
Forest Reserve	3·9
Guayaguayare	3·3
Fyzabad	2·1
Brighton Marine	1·9
Coora/Quarry	1·4
Penal	1·1
Other fields	4·4
	47·1

VENEZUELA

Oil was known to the Indians of Venezuela long before the Spanish Conquest. As in the Middle East, asphalt was used for the caulking of boats and the lining of handwoven baskets, while liquid oil was collected from surface seepages for medicinal and illuminating purposes. The first barrel of Venezuelan oil to be exported was sent to Spain in 1539 by one Don Francisco de Castellanos "to alleviate the gout of the Emperor Charles V". In the 17th century, buccaneering vessels used to enter Lake Maracaibo to execute repairs with the "pitche" available from the seepages along the edge of the lake, and Humboldt gave an account of the asphalt deposits of this area (as well as those of Trinidad) in the report of his travels, published in 1807.

The first modern oil concession was granted in Venezuela in 1866, and in 1878 seepage-oil production was established at La Alquitrana, near Cúcuta. The first shallow cable-tool well was drilled in 1883 and produced a few gallons of oil daily for many years. Four shallow wells were drilled near seepages on *Capure Island,* near Pedernales, in 1890, and heavy oil was also found by drilling at *Guanoco* in 1913. In 1914, the first commercial accumulation was discovered at *Mene Grande,* east of Lake Maracaibo, a field which is still producing important quantities of oil. Thereafter, the *Bolivar Coastal Fields* were rapidly developed, to be followed by the discovery of many other prolific oilfields in Western, and later Eastern, Venezuela. Originally, the oil found was nearly all of Tertiary age, in Miocene and Eocene reservoirs; but with the discovery of a Cretaceous limestone reservoir at *La Paz* in 1944, a new and continuing phase of deeper exploration commenced.

For many years, about 600 wells were drilled every year in Venezuela, many of them in the shallow waters of Lake Maracaibo, which was one of the first offshore areas of the world to be developed. However, in the early 1960's, the intensity of exploration drilling declined, due to unfavourable political and fiscal conditions. As a consequence of this curtailment of exploration activity, Venezuelan oil production has grown only very slowly in recent times. Thus, in 1968, production was 1,319 MM brl, (of which 82% came from Western Venezuela and 18% from Eastern Venezuela), while in 1971, output was actually smaller, amounting to 1,295 MM brl.

Costs of production are high in Venezuela, since the average production per well is only about 300 b/d, compared with perhaps 4,500 b/d in the Middle East. The reluctance of the Venezuelan Government to grant new concessions over a period of more than a decade, coupled with the high sulphur contents of Venezuelan crudes, which render them less acceptable in pollution-conscious areas, are other factors which have restrained development.

A new system of awarding "service contracts" by the State oil corporation (CVP) to private oil companies has been initiated, and time will show whether the terms offered are sufficiently attractive to stimulate exploration activity.

Venezuela was for many years second only to the United States as a world oil producer. The Soviet Union now holds that position, and both Iran and Saudi Arabia have outstripped Venezuelan output. Until recently, however, Venezuela was still the world's largest oil exporter, and in 1968 exported about 900 million brl of oil to 36 countries, but mainly to the United States East Coast area and to Canada. However, because of the US quota system, Venezuela's share of US oil imports has been steadily dropping in recent years from the peak of 52% reached in 1959, while in 1971 Saudi Arabia, Iran, Kuwait and Libya all exported more oil than the 816 MM brl of Venezuela.

The overall primary oil recovery factor is only about 20%, but efforts are being made to improve this figure (and thus the producible reserves) by the introduction of gas injection and steam flooding schemes. Large volumes of associated gas (about 45% of the total output) are currently being injected into producing reservoirs and a further 20% of the total gas output is being usefully consumed in one way or another. Even so, more than one-third of the total natural gas production of about 1,650 Bcf/a is still being flared for lack of outlets.

Because of the curtailment of exploration activity in recent years, the proved oil reserves of Venezuela have been declining slowly from a peak of about 18 B brl in 1960 to approximately 13·7 B brl at the end of 1971, or only some ten years' current production. However, from Table 90 it will be seen that the *ultimately producible* oil resources of Venezuela are estimated to amount to more than 69·4 B brl, of which only some 30 B brl had been produced by the end of 1971. When large-scale exploration is resumed it is likely that the reserves/production ratio will be fairly rapidly increased.

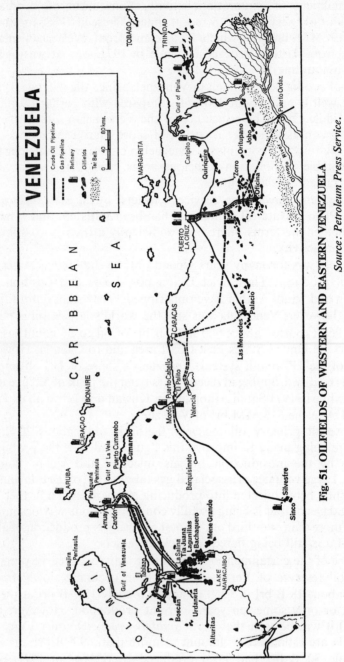

Fig. 51. OILFIELDS OF WESTERN AND EASTERN VENEZUELA

Source: Petroleum Press Service.

Several hundred oil accumulations have been found in Venezuela, at least 40 of which are US-type "giants" containing at least 100 MM brl apiece (Table 85). About 20 of these fields have produced more than 100 MM brl since their initial development.

TABLE 85

THE "GIANT" OILFIELDS* OF VENEZUELA

Maracaibo Basin	Maturin Basin	
Bolivar Coastal	Aguasay	Oscurote
Boscan	Boca	Oveja
Centro	Chimire	Pedernales
CL	Dacion	Quiriquire
La Concepcion	East Guara	San Joaquin
La Paz	West Guara	Santa Ana
Lamar	Jusepin	Santa Rosa
Las Cruces	Las Mercedes	Soto
Los Claros	Leona	Temblador
Mara	Limon	Trico
Mene Grande	Mata	Yopales
North Lamar	Merey	Zapatos
Urdaneta	Morichal	
	Nardo	
Barinas Basin	Nipa	
Silvestre	Oficina	
Sinco	Oritupano	*US classification

These oilfields lie in four major sedimentary basins—those of Maracaibo, Falcon, Barinas and Maturin. Two further basins—those of the Gulf of Venezuela and Cariaco—have as yet been little explored. The total surface area of the basins is about 347,800 sq km, forming 43% of the total area of Venezuela (912,050 sq km). (Table 88 shows a computation of the volumes of sediments in these basins.)

The principal tectonic features of Venezuela are the Lake Maracaibo geosyncline in the west, the Guayana highlands in the southeast (part of the Brazilian "Shield"), and the strongly folded and metamorphosed Andean mountain chains in the north, which run roughly east-west and continue into Trinidad.

In West Venezuela, a north-south geosynclinal trough was formed between the uplifted belts of the Sierra de Perija on the west, the Serrandia de Trujillo on the east, and the main Venezuelan Andes range on the south. The present-day Lake Maracaibo forms part of a continuously sinking basin in which an enormous volume of sediments has accumulated in Mesozoic and Tertiary times. Faulting has affected the sediments, with down throws towards the lake, and lines of folding paralleling the mountain ranges, the steeper flanks of the structures being also lakewards.

In Central and East Venezuela, between the "Shield" and the mountains, a huge structural and topographic basin is filled with great thicknesses of Tertiary rocks in the east, while in the west, Palaeozoic strata occur at relatively shallow depths. This major basin is subdivided into the Maturin and the smaller Barinas (Apure) basins by the El Baul uplift. Tectonic pressure has been generally southwards from the Caribbean area, so that the lines of folding of the basin sediments roughly parallel the edge of the "Shield".

TABLE 86

VENEZUELA—ESTIMATED RESOURCES OF PRINCIPAL OILFIELDS

	MM brl
Bolivar Coastal Fields	16,500
Lamar	1,000
La Paz	905
Quiriquire	810
Boscan	675
Mene Grande	615
Oficina	610
Mara	425
Urdaneta	350
Nipa	340
Chimire	315
West Guara	260
Santa Rosa	240
Mata	230
Los Claros	220
Sinco	215
Jusepin	210
Centro	200
Dacion	200
East Guara	195

Source: A. Martinez "Chronology of Venezuelan Oil", 1968, p. 187.

THE MARACAIBO BASIN

The Lake Maracaibo oilfields currently account for more than 60% of the national output, and the Zulia fields for a further $16\frac{1}{2}$%, so that this is by far the most important oil-producing area in Venezuela. Sediments accumulated here in an oxygen-deficient sinking basin or embayment frequently cut off from the sea[15]; and, as in the Gulf of Mexico and the Caspian Sea, conditions were nearly ideal for petroleum generation. As compaction proceeded in the centre of the basin, migration of hydrocarbons appears to have been outwards towards the edges of the basin where many structural traps occur. The main accumulations have been trapped, however, along unconformities where Oligo-Miocene beds transgressively overlie Eocene strata.

Most of the early production from the Lake Maracaibo area came from Miocene sandstone reservoirs, with smaller amounts from Eocene beds. The discovery in 1945 of Cretaceous oil in the La Paz-Mara area, northwest of the Lake in the State of Zulia, opened up a new phase of development of deeper Cretaceous limestone reservoirs beneath Lake Maracaibo and elsewhere. This phase of exploration is still continuing, and many wells have been drilled beneath the waters of the Lake.

---

[15] A. C. Redfield, *AAPG* "Habitat of Oil", 1958, 968.

The oilfields can be grouped as follows:

(a) Eastern Maracaibo Fields

Extending for more than 100 km along the eastern shore of Lake Maracaibo is the group of oilfields which first made Venezuela an oil producing and exporting nation of world importance. The discovery well was drilled at La Rosa in 1916, and since then an almost continuous belt of highly productive fields have been found stretching along the shore of the Lake and extending far out under its waters. These together form the *Bolivar Coastal Field ("BCF")*, which is now recognised as essentially one huge oil accumulation, although divided up for statistical and administrative purposes into a number of separate units. From north to south these are: *Ambrosio, La Rosa-Cabinas, Punta Benitez, Tia Juana, Lagunillas, Pueblo Viejo* and *Bachaquero*—with several recent extensions such as *South-west Bachaquero, Ceuta* and *Lama*. Of these, *Lagunillas*, discovered in 1916, has been one of the very few fields in the world to have produced more than three billion barrels of oil, much of this from wells drilled some distance out in the Lake. "BCF" includes at least two dozen (US) "giant" oilfields; its overall ultimate resources have been estimated at not less than 16½ billion brl.

Production here was originally derived from Oligocene-Miocene sandstones, which transgressively overlap Eocene shales. The reservoir beds are locally called the La Rosa and Lagunillas series, corresponding to the Agua Clara and Cerro Pelado formations elsewhere in Venezuela. The most prolific reservoir is the Lower Lagunillas sandstone, and the traps have generally been formed by simple up-dip "pinch-out" processes or the sealing effects of asphaltic oils in south-west dipping homoclines.

Smaller but important Eocene production has also been obtained (since 1940) from the Pauji, Misoa and Trujillo sandstones, and in the Lana area from the deeper Paleocene Guasare beds. The Eocene accumulations are found in truncation and overlap traps similar to those occurring in the overlying Miocene beds.

From 1957 onwards, exploration was carried westwards from "BCF" into the waters of the Lake, and further accumulations were consequently found (e.g. *Centro*), which are essentially extensions of the eastern coastal fields. Deeper Cretaceous limestone reservoirs have subsequently also been discovered in the same area.

A few miles south of "BCF" is *Mene Grande*, which was the first oilfield to be discovered in Venezuela, and which is still a major producer. The structure here is an anticline—the south extension of the southward-plunging Misoa "nose". Oil has been found in beds ranging from Eocene to Pleistocene in age, but the main reservoir is the Miocene Isnotu formation, a lateral equivalent of the Lagunillas sandstone. Very large seepages of heavy oils and asphalt are associated with this field, probably the result of migration up faults. The Miocene entrapment is due to up-dip sealing by inspissated heavy oils.

Other fields near Mene Grande include the relatively small *Misoa, El Menito* and *Barua* structures, and the *Motatan* field, discovered in 1952, which has oil in a fractured sandstone reservoir.

(b) The Tarra Fields

About 250 km southwest of the Lake, a group of fields is associated with a major anticlinal uplift, some 46 miles long at the surface. Near the Colombian border, this fold is asymmetric, with a steep to overturned west flank and a gentle east flank complicated by a thrust fault, along the strike of which many seepages occur. The fields here include *Las Cruces* and *Los Manueles,* producing from Eocene sandstones; and the *West Tarra* field situated on a broad anticline, which has oil in the Cretaceous Tibú limestone.

(c) Rio de Oro Fields

To the north and northwest of the Tarra area, a rather remote group of fields awaits full development after the provision of further pipeline outlets. Accumulations here include the *Rio de Oro* anticline, *Rosario,* which has oil in fractured Cretaceous limestones; and the elongated *Alturitas* anticline which is cut by many faults and has Paleocene-Eocene and deeper Cretaceous reservoir beds. *San Jose* is also a Cretaceous producer.

(d) The West Maracaibo Fields

These comprise a series of faulted anticlines lying roughly parallel with the Perijá tectonic axis. It was at *La Paz,* an early Eocene producer, that the first Cretaceous discovery was made (in 1944). (Cretaceous reservoirs have since produced huge quantities of oil in other parts of Venezuela—principally from limestones of the Cogollo or Capacho (Maraco) formation, the La Luna bed, and the basal Colón limestone.)

La Paz is an interesting example of an oilfield where exploration in depth has resulted in the discovery of three quite separate reservoirs of widely differing ages. Accumulation here is governed by the existence of a well-defined and faulted asymmetrical anticline, with a northwest-southeast axis and a steeper northwest flank. Tertiary (Paleocene) oil was discovered in 1922 as a result of drilling in a surface seepage area; Cretaceous production was subsequently found in 1944, and still deeper "basement" production in 1953 (see p. 321). It is estimated that the ultimate resources of this field are not less than 905 MM brl.

The high productivity of the Cretaceous wells at La Paz, a consequence of the thick reservoir section and high permeability due to extensive fracturing, caused a search for Cretaceous oil to be initiated in neighbouring areas, some of which had previously produced only small amounts of Tertiary oil. Discoveries include *Los Tetones* and *Kilometro 24,* which initially had small Eocene production but subsequently became part of the important *Mara* field, an asymmetric, faulted anticline expressed at the surface in Eocene outcrops and reflected at depth in the Cretaceous reservoir limestones. The steep flank is to the northwest, as with La Paz.

Another example is the *La Concepcion* anticline which was initially an Eocene producer from the Ramillete and Punta Gorda sandstones and the Paleocene Guasare formation—but later became a much more important producer from at least three deeper Cretaceous reservoir beds. Other important Cretaceous limestone oilfields in this group are at *Sibucara* and *El Mojan.*

A third, still deeper, oil reservoir formation in these West Maracaibo fields comprises the top 1,000ft or so of the crystalline "basement" rocks, which have considerable secondary permeability as a result of tectonic fracturing. The potentialities of this reservoir were first discovered while deepening a well at *La Paz* in 1953.

(e) The Heavy-Oil Fields

A series of uplifted structural blocks on the western shore of Lake Maracaibo dip generally southwestwards, and are separated by transverse faults; accumulations of heavy oil are trapped above and below the Oligocene-Eocene unconformity—in the Oligocene Icotea sands and the Eocene Misoa sandstones. The resultant oilfields include *Boscan, Los Claros* and *Urdaneta.* The latter forms one of the largest accumulations in Venezuela.

The main source of the Tertiary oil in West Venezuela was probably the dark Eocene shale, notably the Colón shale. The Oligo-Miocene accumulations are always related to an Eocene unconformity, so it is possible that the oil has migrated into the upper beds from the underlying Eocene. The Cretaceous oil is thought to be indigenous to the formations in which it is now found, with possible lateral migration from source to reservoir beds. In particular, the dark-coloured La Luna and Cogollo limestones and shales are considered to be likely source rocks.

The Gulf of Venezuela Basin

This is a roughly rectangular area about 200 km by 125 km in its maximum dimensions, which is separated from the main Maracaibo Basin by the Oca fault. The dominant tectonic feature is block-faulting rather than folding, and there appears to be a platform area in the west and a syncline on the east. Most of the tectonic units present are thought[16] to have been formed in Mio-Pliocene times, but little exploration drilling has yet taken place in this area.

---

[16] G. Coronel, 7 WPC, Mexico, 1967, PD-9 (3).

The Falcon Basin

A narrow trough began to form during late Eocene time in the area northeast of the Lake Maracaibo depression, in which a great thickness of Eocene, Oligocene and Miocene strata accumulated. Miocene uplifting and folding then followed, resulting in the formation of a number of complex structures. The only oilfield of note yet found in this relatively minor basin is *Cumarebo,* a faulted asymmetrical anticline discovered in 1931, with oil in Miocene sandstones of the Mosquito (Socorro) and Cuajarao (Urumaco) formations. Other relatively small accumulations have been discovered in the east of the basin at *El Mene de Acosta* and *Abundancia,* where the reservoirs are Oligocene sands and the structures are faulted, elongated anticlines. In the west, the axis of the gently folded *Mene de Mauroa* field continues to form several other small oil-bearing anticlines (e.g. *Media,* and the abandoned *Hombre Pintado* and *Las Palmas* fields) and there is stratigraphic trapping along the unconformity between the Agua Clara (Upper Oligocene) and La Puerta (Miocene) beds.

The Barinas Basin

A large area of southwest Venezuela is blanketed for the most part by recent river-borne sediments. The first oilfields found in this basin were *Silvestre* (1948) and *Sinco* (1953), a few miles south of Barinas town. Oil occurs in basal Eocene sandstones in faulted structural noses where there is adequate cover. Other minor accumulations are *Palmita, Estero* and *Hato Viejo.*

THE MATURIN (EAST VENEZUELAN) BASIN

Although seepage-type production had been known in the Orinoco Delta area of Venezuela for many years, this region was not seriously explored for oil until the late 1930's. The first commercial discoveries were made in 1937, and this area now produces about 18% of the national output.

The sedimentary basin which covers a large part of East Venezuela extends also over the southern part of the island of Trinidad and the intervening Gulf of Paria. It is bounded by the east-west trending Cordillera de la Costa and Serrania del Interior mountain ranges on the north, the Guayana "Shield" on the south, and the El Baul "swell", separating it from the Barinas basin, on the west. The eastward boundary of the Maturin basin probably now lies beneath the Atlantic Ocean; it has been connected with the open sea during most of its history, since the sediments become generally less marine to the west and southwest. There have been many transgressions and changes of facies, with structural deformations related to tectonism of the mountain chains. Oil is found in several Tertiary sandstone reservoirs, mainly of Mio-Oligocene age, but also extending upwards to the Plio-Pleistocene.

Traps are sometimes stratigraphic and sometimes structural, often being related to the numerous normal, low-angle "growth faults" which have been formed by tectonism within the basin[17]. These faults, which are not exposed at the surface, are generally arcuate in plan, the concave side lying towards the southern "Shield".

The oil accumulations may be grouped as follows:

(1) Anaco (or San Joaquin) Group

These fields lie on a structural "high" running across the northern flank of the basin. They comprise a series of anticlinal or domal closures associated with a northeast-southwest trending thrust fault zone. The principal fields of this group—first discovered in 1937—are *Santa Ana, San Joaquin, Guario, Santa Rosa* and *El Toco;* in addition, there is an accumulation at *San Roque* which is essentially a stratigraphic trap, and at *El Roble,* which is situated on a northwest-plunging structural nose. In all these fields, light oil comes from a series of Oligocene and Lower Miocene sandstones (Merecure and Oficina formations). Few test wells have been drilled into the underlying Cretaceous, so that no deeper oil has yet been found.

The *San Joaquin* field has been cited[18] as an almost perfect example of oil trapping on a dome, with gas cap formation.

(2) Oficina Group

Many oilfields have been discovered towards the centre of the Maturin Basin, lying in an area roughly elliptical in shape, with maximum dimensions of about 100 miles east-west and 50 miles north-south. The major structural features are two fault lines of late Miocene age, one striking east-west and the other northwest-southeast, with the downthrows in each case basinwards. Oil is found in a number of sandstone members of the Oligocene Oficina formation, which rests unconformably on the Cretaceous.

The most important oilfields in the Oficina Group are listed below:

TABLE 87

VENEZUELA—OFICINA GROUP OILFIELDS

Aguasay and East Aguasay	Limon	Oscurote
Boca	Lobo	Oveja
Chimire	Mangos	Silla
Dacion	Mata	Soto
Ganso	Miga	Trico
East and West Guara	Nardo	Yopales
Guere	Nigua	Zapatos
Guico	Nipo	Zorro
Guario	Orion	Zumos
Leona	Oritupano	

[17] S. Cebull, *Bull.AAPG*, May 1970, 54(5), 730.
[18] H. Funkhouser *et al., Bull.AAPG*, 32, 1851, 1948.

There seems some evidence of a systematic variation of the specific gravity of the oil in this area with the position of the accumulation relative to the centre of the basin—the heavier oils being found towards the southern edge, and the lighter oils northwards towards its axis.

(3) Jusepin Group

These accumulations lie on a trend running northeast from the Oficina fields. They actually comprise a single trap, discovered in 1938, to different parts of which individual names (*Tacat, Mata Grande* and *Travieso*) have been applied. Entrapment of oil is controlled by the local wedging-out of the Miocene La Pica sands.

(4) Temblador-Tucupita Group

These are a number of accumulations controlled by east-west regional faults and cross-faulting in the southern part of the basin, combined with gentle folds in a basinward-dipping homocline. The fields include *Morichal, Jobo, Temblador, Pilón, Uracoa* and *Isleno;* the reservoir beds are upper Oficina sandstones, and the crude produced is usually heavy. *Tucupita* lies on the same structural trend as *Temblador,* but has a thinner reservoir bed.

A belt of "tar sands"—sandstones impregnated with very heavy oil over an area more than 350 miles long and 40 miles wide—lies south of the Oficina and Temblador fields. An enormous volume of oil (of the order of 200 B brl) has been accumulated in lower Oficina sandstones lapping up against the "Shield" at relatively shallow depths, and awaits possible future development (p. 355). It is estimated that 10–50 B brl of oil may eventually be recoverable from these deposits.

(5) Orocual-Quiriquire-Guanoco

These three fields lie along a prominent trend line which extends northeastwards from Jusepin. *Orocual,* found in 1932, is a stratigraphic trap with Pliocene heavy oil; *Guanoco* has also produced relatively small volumes of asphaltic heavy oil for many years. In the same area, the *Bermudez Asphalt Lake* contains a large volume of solid and semi-solid asphalt derived from the inspissation of heavy crude oil at the unconformity between the Cretaceous Guayatu formation and the overlapping Miocene beds. The deposit covers some 1,000 acres and the average depth of bitumen is about 6ft, although considerably greater thicknesses occur locally. Active heavy oil seepages gradually refilled the Lake during the period (1910-1934) when more than a million barrels of asphalt were removed; these operations were terminated due to adverse economic factors.

Quiriquire, discovered in 1928, is the largest oilfield in East Venezuela, and is of special geological interest in that the crude here occurs in sandstones and conglomerates of the Pliocene continental (fresh-water)

Las Piedras formation, with some additional deeper Oligo-Miocene marine sandstone production. A section of at least 28,000ft of sediments is present here, ranging from Middle Cretaceous to Recent in age. The dominant structural feature is a homocline, dipping gently southeastwards, with no recognisable folding. Accumulation of oil is controlled locally by faulting, variations of permeability, and asphalt seals.

The origin of the oil at Quiriquire is thought[19] to be two-fold: from the truncated Miocene to Cretaceous marine shales that outcrop at the Pre-Pliocene unconformity, where they are in contact with the overlying reservoir sandstones; and also from shallow marine to brackish-water downdip equivalents of the continental-type Pliocene reservoir sediments. The thick gilsonite beds at the base of the Pliocene and the active seepages in Cretaceous outcrops north of the field are factors which favour a possible Cretaceous origin.

(6) Pedernales-Posa-112

Another northeast-southwest structural trend runs *en échelon* and to the south of the Orocual-Quiriquire-Guanoco uplift. It is associated with the extensive seepages in the Orinoco delta which encouraged drilling and the discovery of an accumulation at *Pedernales* in 1939. This appears to be a complex diapiric type of structure, with oil in sandstones of the Miocene La Pica formation, and also in shallower Plio-Pleistocene sands. At Pedernales and Guaripa the seepages are found in nearly vertical Miocene beds, and although commercial development of these deposits has been attempted, the workings have now been abandoned.

Some 12 miles to the northeast of *Pedernales,* the offshore field entitled *Posa-112* was discovered in the Gulf of Paria in 1958, lying on the northern flank of the syncline which parallels the Pedernales anticline. Accumulation here is due to stratigraphic trapping in La Pica sands. Further to the east, on the same structural trend but across the international boundary, is *Soldado,* the important Trinidad offshore oilfield (p. 313).

(7) In the northern part of the basin there are a number of small fields—*Bruzual, Cagigal, Quiamare, La Vieja, Maureasa.*

(8) Las Mercedes Group

These fields, in the westernmost part of the Maturin basin, were discovered in 1941. They comprise *Las Mercedes, Palacio, Guavinita, Piragua, Punzón, Valle* and *Dakoa* and have oil accumulations in the La Pascua-Roblecito (Oligocene) and the La Cruz (Lower Temblador) Cretaceous sandstones. Trapping has been controlled

---

[19] H. Borger, *Bull.AAPG,* 36 (12), 2291, 1952.

by a complex system of roughly east-west faults, with some strati-graphic "pinch-outs."

(9) The Valle de Pascua Group

This group of oilfields, near the town of the same name, was dis-covered in 1946. They are mainly stratigraphic accumulations resulting from lateral variations (with some faulting) of the productive Oligocene sandstones (Chaguaramas and Roblecito formations). The most important accumulations are those at *Tucupido, Saban, Ruiz, Monal, Taman* and *Bella Vista.*

(10) The only major non-associated gasfields in Venezuela occur in a group lying northwest of Las Mercedes. They are anticlinal accumula-tions (*Lechoso, Barbacoas, Placer, Tamanaco*) from which natural gas is currently being piped for industrial and domestic consumption in the Caracas and Valencia areas.

TABLE 88

VENEZUELA—BASIN SEDIMENTS
(Expressed in thousands of cu km; sediments down to 25,000 ft and more than 1,000ft thick)

	(thousand cu km)	Estimated Oil Resources (MM brl)
Maracaibo	295	51,200
Falcon	275	400
Maturin (Orinoco)	333	15,100
Barinas (Apure)	167	2,400
Cariaco	21	100
Gulf of Venezuela	95	?
Total	1,186	69,200*

* not including oil that may be obtained from the "tar sands"
Source: A. Martinez "Our Gift, Our Oil", 1965.

TABLE 89

VENEZUELA—OIL PRODUCTION BY AREAS, 1971

Western Venezuela	MM brl
Lake Maracaibo fields	854·0
Other Zulia fields	197·0
Falcon basin	0·4
Barinas basin	20·6
	1,072·0
Eastern Venezuela (Maturin basin)	
Anzoategui (southern area)	149·8
Monagas/Delta (eastern area)	65·8
Guarico (western area)	7·8
	223·4
Total	1,295·4

The Cariaco Basin is an isolated and elongated basin in northern Venezuela in which Tertiary sediments occur but in which to date no appreciable hydrocarbon concentrations have been found.

TABLE 90

VENEZUELA—CRUDE OIL RESOURCES

(At end-year, in MM brl)

	1961	1971
Cumulative production	14,932.5	27,638.3
Proved reserves	16,882.4	13,737.4
Secondary reserves	8,441.2	6,183.1
Estimated reserves still to be discovered	29,185.6	21,882.9
Total ...	69,441.7	69,441.7

Source: A. Martinez: "Recursos de Hidrocarburos de Venezuela".

OTHER AREAS

Except for Trinidad, and now perhaps Cuba, the production of oil in Central America and the islands that lie between the North and South American continents is negligible.

In **Barbados** there is much manjak and many mud flows, but extensive drilling has failed to find commercial oil in the complex Tertiary beds, although there are better prospects from the Cretaceous, if that series can be reached. Many surveys have been carried out in this area and unsuccessful test wells have been drilled on Andros and Cay Sal Bank in the Bahamas[20].

Cuba has had a small oil industry for many years, producing a few hundred barrels a day from shallow wells. In this island, there are two lines of folding running parallel to the north coast, and there are Cretaceous beds exposed in the cores with Eocene, Oligocene and Miocene strata on the flanks[21]. Most production has been obtained from the Colón basin, in the west of the island, near *Bacuranao* and *Jarahueca.* Oil comes here from the vicinity of serpentine plugs which upon decay form locally porous reservoirs. The age of this oil is thought to be Cretaceous. A sharp increase in the Cuban oil output to a rate of about 1·4 MM brl/a took place in 1968 as the result of the discovery early in that year of what seems to be an important accumulation at *Guanabo,* some 18 miles east of Havana.

The only commercial oilfield so far found in Central America is the small *Tortugas* salt dome accumulation in Guatemala.

Geophysical surveys and some exploration drilling, both onshore and offshore, have been carried out in a number of other countries in the Caribbean area and South America, with no announced commercial success to date.

[20] M. Spencer, *Bull.AAPG,* 51, 1967, 263.
[21] M. Iturralde-Vincent, *Bull.AAPG,* 53, 1969, 1938.

CHAPTER 12

Reserves, Requirements & Future Supply

SOURCES OF PRESENT SUPPLY

1. Primary and Secondary Recovery

The complex hydrocarbon mixture held in the pores of a subsurface reservoir is usually under a pressure at least as great as the hydrostatic head that would be exerted by a column of salt water extending to the surface (about 55 psi per 1000 ft). However, many cases of both abnormally high and low bottom-hole well pressures have been recorded, caused by local geological factors.

When a well is drilled into such a subsurface reservoir, a complex series of phase and pressure changes is initiated. A pressure gradient develops between the top of the well, where the pressure is lowest, and the undisturbed portion of the reservoir, where the pressure is highest. Liquid oil containing dissolved gas begins to move out of the rock pores into the well and upwards towards the surface, displacing the drilling fluid and evolving a proportion of its gas in the course of its passage. In modern practice, the "bringing-in" of a well is a carefully controlled operation, the specific gravity of the drilling fluid that initially fills the hole being so adjusted as to encourage the flow of oil from the reservoir at an acceptable rate. In the past, when such precautions were not taken, occasional disastrous "blow-outs" occurred when exploratory wells penetrated high-pressure reservoirs, as a consequence of the semi-explosive evolution of fluid into and up the well, leading to the destruction of the drilling rig, and often also to the disastrous ignition of the escaping jet of high-pressure oil and gas.

On the other hand, the flow of oil into the well may be very slow, perhaps because the drilling fluid used has been too heavy and has tended to enter and "mud-off" the potentially productive rock section, or because the reservoir formation is "tight", i.e. insufficiently permeable. Various artificial stimuli can then be applied to improve the rate at which oil will enter the well. Thus, where the reservoir rock is a carbonate, *acidization* is usually employed, using hydrochloric acid solvents to which various corrosion inhibitors have been added. Nitrogen gas is sometimes used to increase the acid mobility.

It is also possible to create new fracture channels in a compact reservoir rock by mechanical means. *Hydraulic fracturing* processes can increase the rate of flow by a factor of 2 or 3; they involve the high-pressure pumping into the reservoir formation of "fracturing fluids", consisting essentially of oil and acid gels containing suspended "propping agents" such as steel balls, glass beads, broken walnut shells, etc. When the pressure is subsequently reduced, the liquid drains away, leaving the propping agents to hold open the induced fractures.

Explosive fracturing ("well-shooting") can result in greatly increased productivity by creating new fractures in the reservoir bed. Originally, nitroglycerine was used, but nowadays high-energy ammonium nitrate based blasting liquids are usually pumped into the formation and detonated after the hole has been tamped with gravel.

In general, the rate at which a well can produce oil depends on a number of factors—the initial reservoir pressure, the porosity and permeability of the reservoir rock, the thickness of the productive section, the oil viscosity, the degree of gas saturation, etc. In practice, however, wells are seldom produced at their maximum capacities during the early history of an oilfield, but usually only at some fraction of this—the Maximum Economic Rate (MER)—which is calculated to result in the maximum output of "primary" oil over the lifetime of the oilfield—normally assumed to be 20 years*.

"Primary" oil is the volume of oil that can be produced at the surface by the unaided natural energy of the reservoir. It will amount to only a fraction of the original volume of oil-in-place, due to the physical and chemical forces which tend to restrict flow, and the inefficiency of natural gas expansion as a production mechanism. Thus, the "recovery factor" for gas expansion is only 20–25%—i.e. between 75% and 80% of the original oil content of the reservoir will be left unproduced when the gas pressure has declined so much that it can no longer propel oil up the wells to the surface.

Considerably higher recovery rates are possible in reservoirs in which there is a "gas drive" or "water drive" in operation, i.e. where the

---

* *The statistics available about the actual lives of oil accumulations under production suggest that this is a conservative figure. However, all such statistics have to be viewed strictly in the light of the prevailing economic conditions.*

expansion energy of the gas cap or of the bottom water layer is effective. In such cases, it is often possible, by drilling rings of peripheral wells at the appropriate structural contour levels, to ensure that a larger proportion of the oil-in-place is swept out of the reservoir into the wells by the advancing gas/oil or oil/water interfaces.

When the reservoir energy has finally fallen to the point that oil no longer reaches the surface by natural means, extraneous energy must be supplied. This may be achieved by mechanical pumping or by the use of "gas lift" techniques, involving the introduction of high-pressure gas through the annular space between the casing and tubing of individual wells, so that the gas mixes with and lightens the oil entering the well, enabling it to flow up the tubing to the surface.

These operations can be considered as extensions of the primary production process. However, it has been found that in order to obtain the maximum recovery of oil from a reservoir over the longest period of time, some form of "repressuring" is needed to restore at least in part the original reservoir energy. This may be achieved by the re-injection of fluids into the reservoir—either gas or water, and even sometimes both—at the appropriate structural contour levels.

Originally, such operations were only initiated after the flow of primary oil from a reservoir had virtually terminated; they were then referred to as "secondary recovery" processes. Nowadays, it is generally understood that the efficiency of primary production will be optimized by maintaining the reservoir pressure at as high a level as possible for as long as possible. The advance of the gas/oil and oil/water interfaces should also be as smooth and uniform as can be arranged, so that the minimum of oil is by-passed. Besides restricting the productivity of individual wells and thus avoiding local "coning" effects, it is therefore usual to carry out "pressure maintenance" operations quite early in the producing history of a reservoir, and to continue them, by injecting high-pressure gas into the gas cap or water into the bottom-water layer, throughout the life of the oilfield, which as a result is usually considerably extended.

In modern practice, "pressure maintenance" operations sometimes grade into "secondary recovery", but for convenience the term "supplementary oil" may be used to describe all the additional oil that can be obtained as a result of the restoration of part at least of the reservoir energy by any extraneous means.

Secondary recovery by gas reinjection was first practised in the United States as long ago as 1891, and was common in Pennsylvania by 1916. It was subsequently installed in many fields in Oklahoma, Illinois and Texas, and the annual output of secondary oil was of growing significance in the years between 1930 and 1945. Since the end of the last war, there has been a great increase in the use of secondary recovery schemes in the United States and also in other countries where large volumes of associated gas have been available, as for example, in Venezuela, Kuwait and Saudi Arabia.

The injection gas—unless a high-pressure gas is available from a nearby or deeper reservoir (as in Bahrein)—usually comes from the gas separated from the oil produced, which is recompressed and pumped back through input wells. (Atmospheric nitrogen has also been used with success as an injection gas where the volume of field gas available has been insufficient; it is preferable to air, which has corrosive properties and always involves the danger of explosion. Mixtures of nitrogen and carbon dioxide or engine exhaust gases have also been used.)

Secondary recovery by water injection ("water flooding") is generally more difficult to arrange, since the very large volumes of water required may not be readily available. However, many water flooding projects, occasionally initiated by accident or to dispose of surplus salt-water produced with the oil, have been remarkably successful in rejuvenating old reservoirs and increasing the recovery from newer oilfields.

Although fresh water is preferred for reinjection purposes because of corrosion problems that might otherwise occur, sea water has been successfully used in a number of cases—most notably and extensively in the Gulf Coast (Louisiana) offshore oilfields. The sea water has first to be treated with inhibitors to minimize its potentially corrosive chemical and bacterial activity.

Where water from a stratigraphically higher formation is used for gravity injection into a lower oil reservoir, the process is termed "dump flooding". An example is the enormous secondary water injection operation being carried out in the *Ghawar* and *Abqaiq* fields of Saudi Arabia.

In the United States, water flooding was initiated in the *Bradford* field in Pennsylvania in 1907, but progress was slow, since it was at that time prohibited by law to introduce water into oil sands. However, in the late 1930's, oil production by water flooding reached the rate of 150 MM brl/a as a result of a court ruling that water could be—in fact should be—reinjected into the *East Texas* field to avoid pollution of the Angelina and Sabine rivers. By 1952, more than 98% of the produced water was being returned to this field and it is estimated that as much as 90% of its original oil-in-place may ultimately be recovered by water flooding.

In Western Canada, where production comes mainly from fissured limestone reservoirs, secondary recovery schemes using water flooding techniques have been particularly successful, since in this area great volumes of suitable water are available from lakes and rivers. It is believed that both the ultimate oil outputs and the lives of many Canadian oilfields can be doubled by water injection, and in one particularly favourable case more than 90% of the original oil-in-place has actually been recovered. Water injection has also been used extensively with notable success in many oilfields in the USSR, notably the huge *Roma-schkino* accumulation and other fields in the Ural-Volga area.

2. Tertiary Recovery

Processes which seek to recover additional hydrocarbons from a reservoir after the completion of normal primary and secondary production operations are termed "tertiary" processes. They involve the use of steam or various miscible solvents introduced into the reservoir, or the controlled combustion of some of the residual crude oil *in situ* to produce an advancing hot "front" or "thermal flood" to sweep out some of the residual crude still retained in the reservoir rock. The rapidly growing importance of such processes is shown by the fact that as much as 8% of current United States oil production comes from oilfields where one or other of the tertiary processes described below has been installed.

(i) Solvent Injection (Modified Waterfloods)

A number of non-polar solvents and soluble surfactants have been added to the water used in secondary water floor operations to improve the efficiency of the operation. Thus, the "Maraflood" process* uses "slugs" of water containing micellar surfactant-stabilised dispersions of water in hydrocarbon oils; the injection of the initial "slug" is followed by a buffer of a second, cheaper liquid, such as water-external emulsions or "thickened water" containing water-soluble polymers, e.g. partially hydrolysed polyacrylamides. The buffer "slug" has the function of stabilising the movement of the micellar "slug" through the reservoir. Theoretically, this technique makes it possible to displace most of the oil from a reservoir, even after a conventional water flood. Carbonated water has also been used for solvent injection operations, with less success, although the CO_2 "slug" process is a possible alternative to the LPG "slug" *(see below)*.

(ii) Hot Fluid Injection

The "hot-water drive" technique has been employed in various heavy-oil fields, notably *Schoonebeek* in Holland, where it was initiated in 1954. Here, the injection of very hot water into the reservoir through selected wells is calculated to have improved the ultimate recovery of oil quite considerably (perhaps by 50% more than if only cold water drive had been used), at an overall cost of only 0·05 $/brl[1]. Cold water subsequently injected picks up heat from the warm reservoir sand and continues the hot water drive process further downstream.

The "steam-drive" oil recovery process has had considerable success

---

* *Marathon Oil Co trademark: see* World Oil, *Feb 1, 1970, 41.*

[1] D. Dietz, 7 WPC, Mexico, 1967, P.D. 12 (1).

in California, the US Midcontinent, Alberta and Venezuela. The term covers the injection of two-phase mixtures of water and steam, or of superheated steam, using one of several different techniques, of which the cyclic method (sometimes called "huff-and-puff") has proved most successful. This uses a single injector-producer well, through which steam is injected for a period of up to four weeks, followed by a period of up to five days of "steam-soaking" and then the production of oil for up to 7 months before the cycle is repeated.

Eventual recoveries of 5%–40% of the oil remaining in place before the start of the operation are possible by the use of this process, the increased output being due to the reduced viscosity of the oil and the expansion of the reservoir fluids. In addition, there may be some steam distillation of residual crude. The initial oil production rate after the "steam-soaking" period is claimed to be as much as 20 times the rate before injection, declining gradually to perhaps 10 times the original rate before the inception of a new cycle.

(iii) "Miscible Slug" Processes

In the "miscible slug" or "miscible front" techniques, a "slug", "cushion" or "bank" of LPG hydrocarbons (mainly propane) extracted from "wet" natural gas is injected into the reservoir, followed by "dry" gas designed to drive the LPG through the reservoir and thus remove a substantial proportion of the oil that has been left after normal primary and secondary recovery processes have terminated. It has been found that the most efficient oil recovery results when the size of the propane "slug" is about 3%–5% of the reservoir pore volume up-dip of the injection wells; in such cases about two volumes of oil can be recovered for each volume of LPG injected.

Unfortunately, this process has proved somewhat disappointing because of the low sweep efficiency of the miscible hydrocarbons. However, some successes have been achieved, notably at *Golden Spike,* the pinnacle-reef type reservoir in Alberta (p. 285), where 95% of the oil-in-place is expected eventually to be recoverable, using a "dry" gas-driven "bank" of LPG derived from the crude oil itself.

(iv) Thermal Recovery Processes

Where the reservoir oil is too viscous for recovery by other techniques, underground *in situ* combustion of a part of the crude has been used as a method of raising the temperature of the remainder, thus reducing viscosity and increasing the ultimate recovery[2]. A moving combustion zone in the reservoir is set up, in front of which the oil is effectively swept.

Air consumption is a key factor, and this depends on the rate at which air can be introduced into the reservoir. Several types of downhole

---

[2] P. Crawford, *World Oil,* Aug. 1971, p.47.

burner have been developed which can be used to initiate subsurface combustion, and airborne infra-red surveys can be used to follow the underground movement of the "fireflood" front once combustion is under way.

"Electrothermic heating", using the continuous flow of single-phase AC current has also been used with some success to stimulate the flow of viscous oils out of the reservoirs.

3. Classification of Reserves

The term "reserves" is conveniently used to denote the producible oil which is contained in a subsurface accumulation or within the limits of a particular geographic area, but it has to be qualified by certain adjectives if it is not to be dangerously misleading.

The first group of qualifying adjectives are those which express the degree of confidence that can be attached to the presence of the reserves —i.e. whether they are in fact "proved" (less elegantly "proven"), "probable", "possible", or only "speculative". Unfortunately, these terms have been used rather loosely and are inevitably to some degree subjective and overlapping, although they express important differences.

It is normal practice for all operators to calculate the volume of reserves in any structure within their concession areas in which oil is discovered, and to improve and refine such estimates as appraisal and development of the field progresses. However, many of the factors which are required for such calculations can only be obtained as a result of considerable production history, so that the true extent of the oil reserves of a field may not be known until long after that field has been put into production. In general, initial estimates are progressively increased as time goes on, and the reserves which were originally only "probable" or "possible" are converted to the "proved" category as the result of appraisal wells drilled to define the extent of the productive rock volume and its parameters.

In US practice, the area of an oil reservoir that is accepted as "proved" is defined (by the American Petroleum Institute) as: "(1) that portion delineated by drilling and defined by gas-oil or oil-water contacts, if any; and (2) the immediately adjoining portions not yet drilled but which can be reasonably judged as economically productive on the basis of available geological and engineering data. In the absence of information on fluid contacts, the lowest known structural occurrence of hydrocarbons controls the lower proved limit of the reservoir".

The second group of qualifying adjectives applied to the term "reserves" refers to their *degree of recoverability*. We have seen above that "primary" reserves can be defined as that portion of the original oil-in-place in a reservoir that can be expected to be recovered using the reservoir energy alone. Similarly, "supplementary" reserves denotes the additional oil that could be obtained by known secondary and tertiary recovery processes. Obviously, the extent to which any such processes can be applied will be controlled by economic as well as technical factors; thus, the reserves which are recoverable "under current economic conditions" may be only a fraction of the reserves which might be physically recoverable without such restrictions. In fact, whether supplementary reserves become available at all must depend on the balance existing between the price to be obtained for the additional oil and the capital and operating costs of the installations required. In American Petroleum Institute practice, the reserves which can be "produced economically through application of improved recovery techniques (such as fluid injection) are included in the "proved" classification when successful testing by a pilot project, or the operation of an installed programme in the reservoir, provides support for the engineering analysis on which the project or programme was based".

When the attempt is made to extend estimates of reserves from single fields to large geographic areas, many difficulties intervene. For instance, only in the United States is there any agreed basis for making such estimates. There, numerous committees set up by the API regularly examine the reports submitted by all companies engaged in exploration, and after due consideration publish consolidated estimates of the proved reserves held at the end of every year in each state and in the United States as a whole. Of course, such figures, however carefully compiled, provide nothing more than a kind of working inventory of the oil considered to be recoverable at a given time under the economic and operating conditions prevailing at that time. Such estimates refer to the primary oil plus the supplementary oil available as a result of fluid injection operations which are "known to be both actually in operation and economic".

An additional class of "indicated reserves" is also used in the United States to categorize reserves not "fully proved", and those also likely to be obtained from known reservoirs by the *future* application of supplementary recovery techniques. (Indicated additional reserves amounted to about 17% of the US proved reserves at end–1971.)

In the USSR, the systems of classification used are somewhat different. There, the "industrial reserves" of an economic mineral such as oil or gas are classified as the sum of three totals A_1, A_2 and B, where these classes are defined as: A_1—"reserves being currently exploited"; A_2—"reserves tested in detail by boreholes"; and B—"reserves established quantitatively, sufficiently accurately by prospecting. The form of the body or the distribution of the natural type of useful mineral or the technology of processing are insufficiently known."

In addition, the USSR classification includes groups C_1 and C_2, which are rather vaguely defined respectively as reserves adjacent to explored structures of a higher category, and reserves "determined by geological premises"; and D_1 and D_2, which are sub-groups of "predicted" reserves, believed to exist in relatively unprospected territory.

The 1968 OPEC* definition divides proved reserves into two categories, and adds classes of "semi-proved" and "possible" reserves:

1.A. *Developed Proved Reserves:* Proved reserves currently subjected to exploitation through existing wells, reservoir outlets and production facilities.

1.B. *Undeveloped Proved Reserves:* Proved reserves that are not currently subjected to exploitation due to lack of wells, or lack of reservoir outlets to already existing wells, or lack of production facilities.

2. *Semi-Proved Reserves:*

The volume of hydrocarbons which geological and engineering information indicate, with a fair degree of certainty, to be recoverable from a tested reservoir, or those volumes of hydrocarbons technically recoverable, to a high degree of certainty, but the exploitation of which is deemed uneconomic†.

3. *Possible Reserves:*

The volume of hydrocarbons expected to be recovered from untested reservoirs that have been penetrated by wells, or from the application of known supplementary recovery methods.

4. World Oil Reserves

Since estimates of reserves generally decline in reliability in proportion to the size of the geographical area concerned, at least outside the United States, and because there are few equivalent official bodies issuing the same sort of carefully considered data as does the American Petroleum Institute, it follows that compilations of national oil reserves must be viewed with great caution. Nevertheless, attempts are regularly made to produce such data, since information about a country's oil reserves is of obvious interest from many points of view. The information used inevitably varies in its reliability from country to country and from year to year; and in some cases it is simply only an arbitrary multiple—say 20 times—the last year's crude oil production figure. Furthermore, for various commercial, political or military reasons, published reserves estimates, where they are available, may differ considerably from the information held in confidential files by companies and governmental bodies.

A further difficulty is essentially one of terminology: it is seldom clear whether the reserves quoted are in fact "proved", or whether they contain a proportion of "probable" or even "possible" oil—added-in for prestige purposes or simply for lack of an agreed basis of definition.

---

* Organisation of Oil Producing and Exporting Countries—see footnote 17, p. 361.
† Presumably "at present".

Similarly, there is always some confusion as to the degree of recoverability that is implied in the data from different sources.

In general, the compilations of reserves which are published refer to "recoverable primary reserves"; but sometimes they also include an estimate of the "supplementary reserves" that could be recovered by known secondary techniques. Sometimes also, reserves of natural gas liquids are added to the crude oil statistics.

Table 93 lists the "published proved reserves" of all the countries of the world as at the end of 1971, based on one such set of statistics, which results in a total of 569·9 B brl of proved recoverable oil. Another estimate* is higher, amounting to 641·8 B brl at the same date, but including in this total, oil obtained from natural gas liquids in North America and a proportion of "supplementary" reserves.

It is remarkable to note how estimates of recoverable oil reserves have increased with time, multiplying by a factor of nearly ten over the period since 1945, when they were only 58·9 B brl, in spite of the steadily increasing annual withdrawals. Clearly, therefore, crude oil reserves have been developed during the last quarter-century or so in relation to the demand for oil, rather than with regard to their absolute availabililty. That this process may no longer be continuing to the same extent might be deduced from an examination of the ratios between world "proved reserves" and world production, calculated as shown in Table 91 for the period since 1930. It will be seen from this that while the ratio greatly increased from 1930 to 1960, it began to decline after 1960 and this decline appears to be continuing, since at end-1971 the world reserves/production ratio was less than 33.

TABLE 91
Ratio between World "published proved" recoverable oil reserves
and production in the same year

1930	18
1940	16
1950	21
1960	39
1970	36

The remarkable changes that have come about over the last quarter-century in the proportions of world overall reserves held in the different producing regions are shown in Table 92. These are mainly due to the high rates of consumption in areas such as the United States, Western Europe and Japan, combined with the large new discoveries made and developed during this period in the Middle East and Africa. Most remarkably, the United States proportion of world "proved reserves" declined from more than 35% at the end of 1945 to less than 7% at the end of 1971, while that of the Middle East almost doubled over the same

* BP Statistical Review, 1971.

period and now exceeds 60%—nearly all of which lies in only six areas (Kuwait, Iran, Iraq, Saudi Arabia, Neutral Zone, Abu Dhabi).

TABLE 92

WORLD PROVED CRUDE OIL RESERVES 1945–1971

	end-1945[1]		end-1971[2]	
	B brl	% of total	B brl	% of total
North America	21·9	37·1	49·2	8·7
South America and West Indies	8·3	14·2	26·2	4·6
Western Europe	0·8	1·4	6·8	1·2
USSR and Eastern Europe	8·0	13·6	62·8	11·0
Africa	0·1	0·1	50·9	8·9
Middle East	18·5	31·4	347·1	60·9
East Asia	1·3	2·2	24·9	4·4
Australasia	—	—	2·0	0·3
	58·9	100·0	569·9	100·0

Source of data: World Oil, Oct. 1972.

1. 1945 production 2·8 B brl.   2. 1971 production 17·5 B brl.

TABLE 93

WORLD CRUDE OIL PRODUCTION AND RESERVES*

	Production MM brl		Reserves MM brl	Ratio
	1971 (1)	1972 (2)	end-1971 (3)	col.3/col.1
USA	3,478·1	3,468·8	38,062·9	10·9
Canada	482·8	544·6	8,334·4	17·3
Mexico	155·9	161·4	2,837·1	18·2
North America	4,116·8	4,174·8	49,234·4	12·0
Trinidad	47·1	51·4	1,053·0	12·8
Others	1·4	1·5	8·0	5·7
West Indies	48·5	52·9	1,061·0	12·6
Argentina	154·5	159·1	2,000·0	12·9
Bolivia	13·2	12·8	262·0	19·8
Brazil	62·2	61·1	855·0	62·9
Chile	12·9	12·5	120·0	9·3
Colombia	78·2	73·2	1,626·0	21·8
Ecuador	1·6	9·9	6,070·5	—†
Peru	22·6	23·1	510·0	22·6
Venezuela	1,295·4	1,172·4	13,740·4	10·6
South America	1,640·6	1,524·1	25,183·9	15·4
Austria	17·1	17·6	186·5	10·9
Denmark	—	—	250·0	—
France	13·6	10·7	98·9	7·3
Germany, West	54·2	51·3	560·0	10·4
Italy	11·7	8·2	271·1	23·2
Netherlands	9·4	11·1	261·2	28·4
Norway	—	—	2,000·0	—†
Spain	0·9	1·0	162·0	—†
United Kingdom	0·6	0·6	3,018·0	—†
Western Europe	107·5	100·5	6,807·7	16·0
				(onshore only)

	Production MM brl 1971 (1)	1972 (2)	Reserves MM brl end-1971 (3)	Ratio col 3/col 1
Albania	9·7	11·2	630·0	—†
Bulgaria	2·2	1·8	245·0	—†
Czechoslovakia	1·4	1·3	245·0	—†
Germany, East	0·4	1·8	?	—†
Hungary	14·1	13·8	267·0	18·9
Poland	2·9	2·6	37·0	12·8
Romania	99·4	99·3	867·0	8·7
USSR	2,701·0	2,836·8	60,000·0	22·2
Yugoslavia	21·3	22·0	460·0	21·6
Eastern Europe	2,852·4	2,990·6	62,751·0	22·1
Algeria	289·6	290·7	9,839·6	34·0
Egypt	107·3	82·0	1,000·0	9·3
Libya	996·4	820·0	28,000·0	28·1
Morocco	0·3	0·2	7·8	39·0
Tunisia	31·9	29·1	410·0	12·9
Africa, Northern	1,425·5	1,222·0	39,257·4	27·3
Angola & Cabinda	41·3	50·9	1,000·0	24·2
Congo (Brazzaville)	0·3	0·3	8·0	2·2
Gabon	41·8	43·8	602·0	14·4
Nigeria	556·1	651·7	10,000·0	18·0
Africa, Central & Southern	639·5	746·7	11,610·0	18·1
Abu Dhabi	341·3	355·9	15,100·0	44·2
Bahrein	27·3	26·1	541·3	19·8
Dubai	45·7	51·2	1,500·0	32·8
Iran	1,657·0	1,838·5	60,450·0	36·5
Iraq	621·8	501·7	33,100·0	53·2
Israel	0·4	0·3	2·0	5·0
Kuwait	1,067·8	1,099·8	74,974·0	70·2
Neutral Zone	198·0	214·8	13,000·0	65·7
Oman	107·4	103·6	4,750·0	44·2
Qatar	156·9	163·3	4,800·0	30·6
Saudi Arabia	1,641·6	2,036·9	137,400·0	83·7
Syria	40·1	37·2	1,320·0	32·9
Turkey	24·0	24·1	200·0	8·3
Asia, Middle East	5,929·3	6,453·4	347,137·3	58·5
Brunei-Malaysia	70·1	99·9	820·0	11·7
Burma	6·9	6·5	39·5	5·7
China	145·0	216·9	12,500·0	67·2
India	50·6	54·9	739·6	24·0
Indonesia	325·7	390·1	10,673·4	32·8
Japan	5·5	5·3	24·9	4·5
Pakistan	3·0	4·0	35·5	11·8
Taiwan	0·2	0·2	20·9	104·5
Asia, Far East	607·0	777·8	24,853·8	40·9
Australia	113·5	119·0	1,958·2	17·3
New Zealand	0·8	0·8	30·0	37·5
Australasia	114·3	119·8	1,988·2	16·5
TOTAL WORLD	17,481·4	18,102·6	569,884·7	32·6

* *Data based on "World Oil", AAPG and PPS statistics, with some amendments by author. The 1972 production figures are provisional and subject to revision.*
† *No "reserves/production" ratio is meaningful in these cases, where production is in its initial stages only, or where information is particularly scanty.*

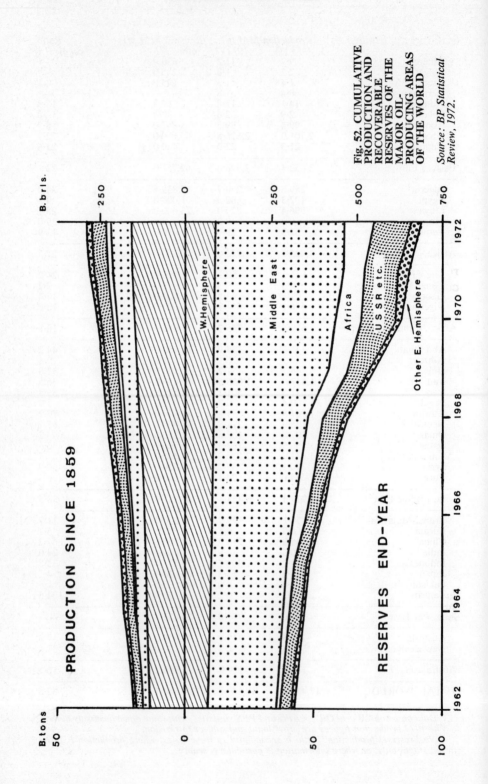

Fig. 52. CUMULATIVE
PRODUCTION AND
RECOVERABLE
RESERVES OF THE
MAJOR OIL-
PRODUCING AREAS
OF THE WORLD

Source: BP Statistical
Review, 1972.

5. Ultimate Resources

The term "resource base" is used to describe the total volume of petroleum that lies beneath the surface of a particular geographic area. The proportion of this amount that will ultimately be producible under a specified set of economic and technical conditions comprises the "ultimate resources" of the area. This is equivalent to the "ultimate production" calculated at a given date (i.e. the past production plus proved reserves as at that date) together with the additional volume of oil which it is thought may be found and recovered in the future.

In spite of the obvious uncertainties involved, a number of estimates have been made of the "ultimate resources" of the world's major petroleum provinces, and attempts have also been made to extend these estimates to the whole world. The area of suitable sediments has been estimated for each basin, average depths and proportions of reservoir rocks calculated, and factors applied for the likely oil concentration per cubic mile of reservoir rocks.

The hazards and guesswork involved in such operations are obvious, but for want of a better technique they have been thought worth trying. Unfortunately, since such a large proportion of the world's oil has been found in a relatively few "giant" fields whose occurrence is statistically unpredictable, any overall "basin richness" compilations can only be regarded with extreme caution.

TABLE 94

ESTIMATED QUANTITIES OF CRUDE OIL IN WORLD AREAS AND BORDERING CONTINENTAL SHELVES

	Crude oil originally in place B brl	Crude oil ultimately discoverable B brl	Crude oil ultimately recoverable B brl*
USSR, China and Mongolia	2,900	1,800	720
Africa	1,800	1,100	440
USA	1,600	1,000	400
Middle East	1,400	900	360†
South America	800	500	200
Rest of North and Central America and West Indies	500	300	120
Europe	500	300	120
Australasia and East Indies	300	200	80
South Asia	200	100	40
	10,000	6,200	2,480

* Based on 40% recovery factor. †see p. 342. Source of data: G. Hendricks, loc. cit., 1965

One of the most detailed of these exercises has been carried out by the US Geological Survey[3], who have estimated that the amount of oil

[3] T. Hendricks, *US Geological Survey Circular 522*, 1965.

"originally in place" in the sediments of the world was of the order of magnitude of 10,000 B brl, of which some 6,200 B brl "may eventually be discovered".

Applying an overall average recovery factor of 40% would give an ultimately recoverable total of 400 B brl in the USA and 2,080 B brl in the rest of the world—a total of 2,480 B brl (Table 94). Of this, about 240 B brl had been produced by the end of 1970 and about a further 544 B brl had been proved at that time—i.e. 784 B brl of recoverable oil had been found, so that, according to this estimate, a further 1,700 B brl of unknown but theoretically recoverable oil still remained to be discovered in the world's sediments†.

Other estimates of world oil resources generally arrive at rather lower totals—nowadays around 2,000 B brl, but in the 1950's only about 500–600 B brl. These estimates have obviously been expanded in proportion to the growth of proved reserves over this period, perhaps with little more than theoretical or mathematical justifications. Certainly, much of the world's undiscovered remaining oil may lie in very remote areas, deep waters or held in only small individual accumulations; there can be little more than guesswork to support calculations of what might be the economic costs and limits of its recovery in the distant future.

6. Future Oil Requirements

There have been three phases in the growth of world primary energy† demand since the beginning of this century. Between 1900 and 1960, demand grew at the average rate of about 2·1% p.a.; between 1960 and 1965 the rate of increase was 3·5% p.a.; and since 1965 it has accelerated to 5·3% p.a. (Table 95). The growth in energy requirements has of course been greatest among the major industrial nations, whose total requirements increased by 223% between 1950 and 1970.

TABLE 95
GROWTH OF WORLD PRIMARY ENERGY† OUTPUT

Year	Overall output of primary energy (1860 = 100)	Annual average growth rate during the previous period %
1900	240	2·1
1920	385	2·2
1940	580	2·0
1960	940	2·2
1965	1,120	3·5
1970	1,460	5·3

Source: N. Melnikov & M. Albegov, 7 WPC, Moscow, 1968
1970 figures estimated by author.

* However, the fact that by end-1971 the remaining "published proved" recoverable reserves of the Middle East were 347 B brl—compared with the forecast total of only 360 B brl "ultimately recoverable"—illustrates the uncertainty of all such estimates.

† Primary energy sources are defined as those found in the "natural" state—i.e. the solid fuels—coal and lignite; the liquid fuels—oils and oil products; natural gas; and "primary" electrical energy generated from hydroelectric or nuclear plants. It is usual to measure energy only when it first enters the energy economy; thus, the electricity produced by the conversion of primary fuels in thermal plants is considered as secondary.

Looking ahead to the future, it is estimated that, if the present rate of growth continues, as is probable in view of the almost explosive increase likely in the world's population (Table 96), coupled with the universal desire for higher living standards, then the world's primary energy needs which totalled about 6,800 MM tons c.e. in 1970 may reach 16,000 MM tons c.e. in 1985, and perhaps as much as 35,000 MM tons c.e. by the end of the century.

TABLE 96
WORLD—POPULATION GROWTH BY AREAS

	Population in 1967 MM	Rate of increase %p.a.	Population in 2000 at present rate of increase MM
Africa	328	2·5	741
America	479	2·1	951
Asia	1,907	2·0	3,665
Europe	452	0·8	592
Oceania	18	1·9	34
USSR	236	1·2	350
World	3,420	1·9	6,333

Sources: Data for 1967 from UN Statistical Yearbook 1968; estimates for 2000 after D. King-Hele (1969) "The End of the 20th Century?" p. 39 and O. Freeman "World Without Hunger" (1968) p. 7.

What proportion of these enormous requirements will be provided by crude oil?

If the historic rate of increase in crude oil consumption were to continue in the years between 1970 and 2000, at least 33 B brl of oil would be needed in 1980, over 66 B brl in 1990 and some 133 B brl in 2000. There are, however, certain factors which will reduce these potential demands.

The first is the development of nuclear-generated electric power, which, after a disappointingly slow start is now gathering momentum in several countries. It is believed that nuclear electricity will provide about 12% of the world's primary energy needs in 1985, and perhaps double this proportion in the year 2000. Much will depend on the economic success of the new generation of "breeder" reactors. (Of course, if fusion—as opposed to fission—energy becomes feasible before the end of the century, the whole balance of energy supply would alter.)

The proportion of world primary energy requirements provided by hydroelectricity is about 2%, and this proportion is likely to remain fairly constant in the future. Other sources of primary electricity, such as geothermal energy or solar-derived power may be developed before the end of the century, but to what degree it is still impossible to discern.

There is no doubt however about the increasing share of natural gas in the overall energy supply picture: it is likely to provide 22% of world

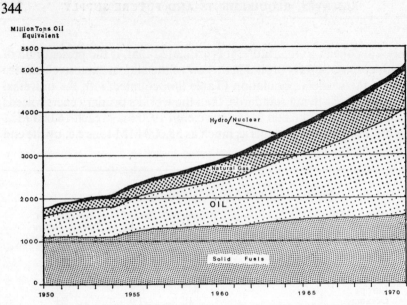

Fig. 53. GROWTH OF WORLD ENERGY DEMAND 1950-1971

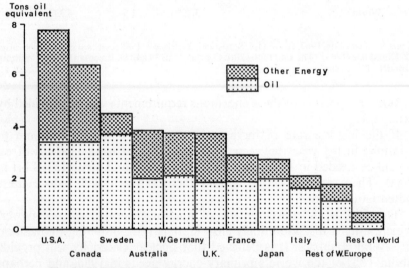

Fig. 54. ENERGY AND OIL CONSUMPTION PER HEAD, 1971

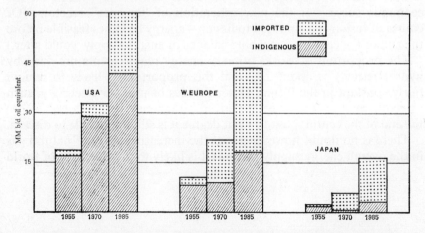

Fig. 55. INCREASING DEPENDENCE ON ENERGY IMPORTS 1955-1985

Source: BP Statistical Review.

primary energy requirements in 1985, rising to about 27% of the greatly increased total needed in 2000. Most of this increase will be at the expense of coal, but some of it will also be in replacement of oil. Natural gas has only become widely available relatively recently, and many of its applications are still being developed.

If as a result of these considerations the rate of increase in world crude oil consumption goes up at an average of only 5% in future, as against the historic average growth of 7·1% p.a., consumption in 1990 would be about 44 B brl (compared with 16·5 B brl in 1970) equivalent to perhaps 46% of world primary energy requirements in that year. But even at this modest rate of increase, 600 B brl of oil would be consumed in the two decades 1970-1990—appreciably more than the total proved reserves at the start of the period. Therefore, if in 1990 a modest world reserves/production ratio of no more than 15 is considered desirable, at least 650 B brl of "new" recoverable oil will have had to be discovered in the intervening years.

This is more than $2\frac{1}{2}$ times the world's total output between 1859 and 1970, and the difficulties and costs likely to be involved can be judged when it is recalled that the recoverable reserves of the world's two newest major petroleum provinces in Alaska and the North Sea, probably do not together exceed 60 B brl, i.e. less than one-tenth of the "new oil" requirement up to 1990, and both of these require the investment of huge sums of capital before production can begin.

Where will all the oil that will be needed during the remainder of this century come from? Some of the possible sources of supply are discussed on the following pages.

TABLE 97
GROWTH OF WORLD PRIMARY ENERGY DEMAND BY AREAS, 1960–1980

	1960		1970*		1980*	
	MM tons c.e.	% of total	MM tons c.e.	% of total	MM tons c.e.	approx %of total
Western Europe	845	19	1,300	19	1,920– 2,050	18
North America	1,580	36	2,270	33	3,170– 3,430	29
Japan	115	3	275	4	540– 665	5
Latin America	155	4	295	4	555 ⎤	5
Africa	110	2	110	2	180 ⎥	2
Middle East	50	1	95	2	190 ⎬ 1,555	2
South and Southwest Asia	115	3	210	3	380 ⎥	3
Polynesia	50	1	80	1	130 ⎦	1
Communist countries	1,365	31	2,175	32	3,835– 4,000	35
	4,385	100	6,810	100	10,900–11,700	100

* Estimated

Sources: 1 OECD Report "Energy Policy", 1966.
2 Commission of European Communities Report, 1966
3 W. Jensen "Energy and the Economy of Nations", 1970, IV.

TABLE 98
ENERGY DEMAND IN MAJOR REGIONS, 1970–1985
MM m. tons coal equivalent

		Liquid Fuels	Natural Gas	Solid Fuels	Primary Electricity	Total
EEC*	1970	685	88	363	83	1,219
	1985	1,410	345	255	345	2,355
	%change	+106	+292	−30	+316	+93
USA	1970	1,050	680	480	80	2,290
	1985	1,700]	1,130	670	700	4,200
	%change	+62	+66	+40	+775	+83
Japan	1970	290	5	90	25	410
	1985	760	10	105	175	1,050
	%change	+162	+100	+17	+600	+156
World	1970	2,685	1,215	2,415	385	6,700
	1985	7,050	3,450	3,200	2,250	15,950
	%change	+163	+184	+33	+484	+138

* Enlarged Community, also including figures for Norway.
Source: Commission of the European Community.

SOURCES OF FUTURE OIL SUPPLY

1. Increased production from known oilfields

(a) By the full development of known accumulations

Oil accumulations are only developed to their maximum productive capacity when there is sufficient economic incentive to justify the investment of the necessary capital. Production may sometimes also be restricted to a rate below MER by political intervention or as a result of protectionist legislation designed to maintain crude oil prices in a particular area. ("Prorationing" restrictions of this type have traditionally been practised in Texas, Oklahoma and California in the United States, and also in Alberta in Canada.)

It is difficult to quantify this type of excess potential capacity, part at least of which would be immediately available from existing wells and facilities, while another part would require the drilling of further wells and the construction of additional pipelines, storage tanks and terminals, and therefore would need considerable time and capital investment to develop. In any case, the development of known oilfields would not actually increase the *total* oil available, but would only ensure that it was produced sooner rather than later.

The definition of "productive capacity" which is used by the American Petroleum Institute is the maximum daily rate of production of crude oil attainable under certain specified conditions on March 31 of any year— i.e. 90 days after the end of the previous year, with "existing wells, well equipment and surface facilities, plus work and changes that can be reasonably accomplished within the period, using present service capabilities and personnel and with production declining as it would under capacity operation". On this basis, the US daily productive capacity attainable on March 31, 1972, was 10,649,000 brl, compared with an average actual production rate of 8,686,000 brl in December, 1971, i.e. an excess productive capacity at that time of 1,963,000 b/d or 22%, of which 1,383,000 b/d would have come from Texas. But it is doubtful whether the Texas oilfields can any longer be regarded as potential sources of immediate increased supply, since—for the first time since the last war—they were allowed to produce at 100% capacity during 1972*.

Obviously, excess crude availability will tend to vary from year to year and country to country according to the prevailing price of crude oil and the current level of demand. It was calculated at the end of 1964, for example, which was a time of plentiful, relatively cheap oil supply, that the overall non-Communist world's excess crude availability amounted to nearly 46% of the actual production in that year. Presumably today, at the start of an era of high-cost oil, world excess capacity is considerably less.

(b) By improvements in the techniques of oil recovery

At the present time, primary recovery factors of only about 25% are still common, and while the installation of pressure maintenance or secondary recovery facilities results in the extraction of a higher proportion of the oil-in-place, this must be paid for by capital and operating expenditures which raise the overall cost price of the crude. Clearly, a balance has to be struck at any given time between the cost of installing and running supplementary recovery operations, and the revenue that can be expected from the additional oil to be obtained from them. Many factors enter the balance, among them the rate of interest to be paid on the capital required and the marketability and likely selling price of the additional crude obtained, relative to the cost of other crude oils that might be available.

In the United States, the development of supplementary methods of oil recovery is more advanced than in most other countries of the world, but even there the overall recovery factor probably does not exceed about 32% of oil-in-place at the present time. Although this is a notable improvement upon the factor of 28% which prevailed in 1954, it is still obviously unacceptable in principle that two out of every three barrels of oil discovered should not be economically and technically recoverable. It is possible that improvements in the techniques available may raise the average recovery efficiency in the United States to 40% before the end of the 1980's, and a factor of 45% may eventually be reached. In the rest of the world, lower recovery rates must be expected

---

* The comment of the Chairman of the Texas Railroad Commission, the prorationing authority, was: "We feel this to be a historic occasion. . . and a sad one. Texas oilfields have been like a reliable old warrior that could rise to the task when needed. That old warrior can't rise any more".

to prevail, but the importance of achieving improvements is shown by the calculation that only a 1% increase in average recovery from the world's known oilfields would be equivalent to adding as much as a year's current production to the ultimate recovery, i.e. some 20 B brl of oil, equivalent perhaps to the ultimately recoverable reserves of a major new petroleum province.

2. The Discovery of New Oilfields

(a) By horizontal exploration

There are undoubtedly many as yet unknown hydrocarbon accumulations awaiting discovery in various parts of the world, especially in areas which have been either geographically or politically inaccessible in the past. In particular, much more oil will certainly be found in the Arctic, in Africa and Asia, regions where exploration has so far been relatively meagre in relation to the huge sedimentary areas available.

However, unless a new technique is developed whereby the existence of an oil accumulation in the subsurface can be located without the necessity for exploratory drilling, which seems improbable, the rate of future discovery in any part of the world must be related to the drilling effort deployed. This has traditionally been the case in the United States in the past. There, in some oil-producing areas, the density of exploration drilling has exceeded one well per two square miles. It is interesting to consider now whether if this degree of drilling intensity were to be extended over the whole sedimentary area of the United States, it would result in proportionately increased new reserves being discovered.

In 1972, 28,015 oil and gas wells were drilled in the USA—77% of the total drilled anywhere in the non-Communist world. Of this number, about 21% were true exploratory wells—i.e. "new-field wildcats": and only 9·7% of this number were successful in finding "some" oil or gas. The proportion of limited success has been slowly decreasing every year since 1945/6, when it was 11-12%; but probably more important is another trend—the decreasing proportion of "new-field wildcats" which after the elapse of six years could be shown to have found *significant* new oil or gas accumulations. This ratio has been falling steadily for many years—from 2·07% for the 1945 discoveries to 0·73% for those of 1965. Exploration for oil therefore seems to be subject to a law of diminishing returns in the United States, inasmuch as a decreasing volume of significant new oil is being discovered for each year's exploratory effort. More intense drilling would clearly therefore *not* result in a proportionate increase in new discoveries.

The conclusion that "undiscovered oil is becoming scarcer" in the USA has been borne out by the detailed analysis[4] of the rate of discovery

---

[4] M. K. Hubbert, Chapter 8, "Resources and Man" (Freeman), 1969.

of oil in that country in terms of barrels per foot of exploratory drilling. Thus, between 1859 and 1920 the average discovery rate was 194 brl/ft; between 1920 and 1928 it declined to 167 brl/ft; between 1928 and 1937 it rose again to 276 brl/ft, largely as the result of the discovery of the huge *East Texas* field; but, since 1937 it steadily declined to the 1967 level of only 35 brl/ft. It is particularly significant that the decline since 1937 should have coincided with the period of the most intensive research in and development of exploration techniques.

What of the rest of the world? In the non-Communist areas outside the USA, 8,592 wells were drilled in 1972, of which probably some 2,000 at most could be considered as exploratory—barely one-third of the US total. In fact, because of its long history of intensive development, the US industry has drilled very many more exploration wells than all the other countries of the world taken together. It is true, on the other hand, that in some parts of the world, notably the Middle East, the unitary concession system which was traditional for many years and the sheer size of the individual fields has resulted in the discovery of much more oil per unit of exploration effort than has generally been the case in the United States, where drilling has always been competitive and the average oilfield covers an area of considerably less than a square mile.

An enormous drilling programme would certainly be needed to bring the general intensity of exploration in other parts of the world up to anything like the present American pattern. But even if this were physically and economically possible, there must be considerable doubt as to whether it would result in anything like a *proportionate* increase in the new oil reserves which would be discovered.

TABLE 99

"NON-COMMUNIST WORLD" DRILLING EFFORT, 1972*

	No. of wells	
United States	28,015	
Canada	3,390	
Mexico	513	
North America		31,918
Central America and West Indies		228
South America		1,801
Western Europe		349
North Africa		265
Rest of Africa		465
Middle East		520
East Asia		815
Australasia		146
		36,507

* *Forecast on basis of "World Oil" data.*

The reason for this is related to the overwhelming contribution that "giant" fields make to world oil reserves, their unpredictable distribution, and their relative rarity. Thus, it has been estimated that about 75% of the oil ever found in the world has been found in "giant" accumulations holding at least 500 MM brl of oil apiece. Even more remarkably, the 187 "giant" oilfields discovered up to the beginning of 1970—amounting in number to only about 0·6% of the world's known oilfields—probably contained between them about 84% of the world's remaining recoverable reserves at that time.[5] According to another estimate, only six oilfields (five in the Middle East and one in Venezuela) between them contained almost 40% of the non-Communist world's remaining reserves in 1969[6].

Data such as these lead to the conclusion that it is the rate of discovery of these very large accumulations that is critical, and a recent analysis of the average size of fields discovered against time is significant in this respect[7]. After dating back the total reserves now known in major fields to the year in which the discovery well was drilled, it was concluded that there has in fact been a discernible downward trend since 1960, both in the rate of discovery of large accumulations and in the average size of the fields being discovered each year, in spite of developments in geophysical techniques, and the opening-up of new, previously inaccessible, offshore areas.

The decline in the discovery rate of large accumulations is most noticeable in the United States, as is shown by the statistics in Table 100.

TABLE 100
USA—RATE OF DISCOVERY OF "GIANT"* OILFIELDS

	discoveries per year
1920–30	7·8
1930–40	5·0
1950–60	3·3
1960–70	1·7 (estimated)

* Source of data: M. Halbouty, Bull. AAPG, 52 (7) 1115, July, 1968.
(In this Table, the term "giant" is applied to oilfields holding at least 100 MM brl of recoverable oil.)

It is not yet clear whether this decline is irreversible, but if the deduction is correct that the average size of new fields is falling and that most of the very large oilfields of the world have indeed already been found, then the conclusion must be that the average *cost* of future oil discoveries in terms of the exploration effort required per barrel of oil discovered must continue to rise in the rest of the world, just as in the United States, although probably at a slower rate. (In this context, it is of interest to note that the per-barrel cost of petroleum discovery

---

5 M. Halbouty et al., AAPG "Geology of Giant Petroleum Fields" 1970, 502.
6 R. Burke and F. Gardner, Oil and Gas J., Jan 13, 1969.
7 H. Warman, Petroleum Review, March 1971, 96.

in the United States increased by 2,200% over the period 1938-1970)[8].

In future also, a large and increasing proportion of exploratory drilling will inevitably be concentrated in offshore areas and in deeper and rougher waters—perhaps far out on the Continental slopes and rises. But the cost of drilling an offshore well even in shallow waters is already several times higher than the cost of an onshore well, and these costs increase sharply with the depth of water involved.

Furthermore, development costs for deep-water oilfields are likely to be very high, so that the cost of discovering and developing future oil reserves, of which an ever-increasing proportion will be found in offshore fields, will for this reason also inevitably increase.

In the past, most of the world's oil has been found in structural traps, due to the fact that accumulations of this type are most susceptible of discovery by geological and geophysical techniques. However, it seems likely that in the future more of the "subtle" or stratigraphic traps will need to be discovered to balance the declining number of structural traps remaining to be found. Of the "giant" fields known today, less than 10% are stratigraphic traps, and these were nearly all discovered more or less by accident in the course of drilling for structure. However, since no geophysical technique so far known can reliably detect stratigraphic traps in the subsurface—although seismic techniques have sometimes been used with limited success for this purpose—future stratigraphic traps will only be discovered by drilling and, from their nature, by drilling considerably more exploratory holes per trap discovered than has historically been the case in the search for structural accumulations. It seems likely, therefore, that this also will be an important factor making the crude oil reserves of the future considerably more expensive to discover than those of the past.

(b) By vertical exploration

New reserves of crude oil undoubtedly await discovery by future exploration wells drilled in the deeper, central parts of many sedimentary basins which have so far been inadequately explored in their vertical dimension.

The deepest hydrocarbon-producing well ever drilled anywhere in the world up to the end of 1972 was a gas producer in the central Anadarko Basin in Oklahoma. The reservoir here is an Ordovician limestone section lying between 24,065 and 24,584ft. (Not far away, the deepest well ever drilled reached 30,050ft in the same basin without, however, discovering commercial hydrocarbons.)

Of course, even in the United States such deep wells are still relatively rare; thus, in 1971 the average depth of the 4,462 "new-field wildcat" wells drilled was 5,924ft, and that of the 18,929 development wells was

---

[8] Calculated on a "value basis" for both oil and gas: F. Howell and H. Merklein, *World Oil*, Oct. 1972, 91.

4,429ft. There has been a steady increase in the average depth of exploration drilling in recent years (from 4,007ft in 1946 to 4,846ft in 1956 and 5,095ft in 1966), but the overall increase in the average depth reached in the last quarter century is only about 75ft p.a.

Outside the USA, very few wells have been drilled to depths greater than 15,000ft. The deepest wells in Europe are the Nassiet well in France (21,945ft), the offshore Italian Ernesto Nord well (20,254ft) in the Adriatic, and a well in East Germany said to have reached 23,049ft. These are exceptions to the generally rather shallow wells that have been traditionally drilled into the European Tertiary and Mesozoic reservoirs. In the Middle East, wells shallower than 10,000ft account for more than 90% of the oil produced; only in a few pools, such as *Minagish* (Kuwait), *Rumaila* and *Zubair* (Iraq), *Marun* (Iran), and *Darius* (offshore Iran) is there important deeper production. The deepest well drilled to date in the Middle East is a 19,119ft test drilled in 1970 (without announced success) in the Gorgan area of northern Iran (*Qezel Tappeh 2*). In Algeria, the *El Gassi el Agreb* and *Hassi Messaoud* reservoirs lie at depths of about 11,000ft, and account between them for about half the total Algerian oil production. In Nigeria, some of the producing formations are as deep as 12,000ft. In Venezuela, there are wells as deep as 13,000ft in the Lake Maracaibo area, but the bulk of the Bolivar coastal production comes from shallow fields at 2,000-5,000ft. In Indonesia the largest producing field (*Minas*) has a reservoir at only 2,400ft, and some of the oilfields of Oman produce oil from even shallower depths.

Improvements in drilling technology and equipment will certainly make it possible to drill progressively deeper wells in the future, although changes are likely to be relatively slow and will probably take the form of a steady, general increase in the average depths of exploration wells. More ultra-deep wells will be drilled in countries outside the USA, and as a result, deeper reservoirs will be found underlying many known accumulations. As a corollary of deeper drilling, an increasing proportion of new reserves will be discovered in Palaeozoic reservoirs.

However, the deeper a well is drilled the more difficult are the technical problems involved and the greater the resultant costs. This is illustrated by statistics about the sharp rise of cost with depth for wells drilled in the United States. There, it costs on average 5-6 times as much to drill to 12,000ft, and *nine times* as much to drill to 15,000ft, as it does to drill to 7,000ft.

Furthermore, the qualities which are desirable in reservoir rocks— high porosity and permeability—are generally seriously reduced at increasing depth as a result of the prevailing pressures and temperatures. The likelihood of finding gas rather than oil also seems to increase with depth (p. 39). It is certainly a fact that the hydrocarbons discovered took the form of gas rather than oil in the great majority of the 371 deep wells which were drilled to below 15,000ft in the USA in 1971. (In fact, more than a third of the gas reserves of the USA have been found in very deep fields in the southwestern states).

It may well be, therefore, that there are limitations on the occurrence of major new oil accumulations at depth which will result in rather less oil being discovered by deeper drilling in the future than has sometimes been thought likely.

3. The bringing into service of sources of petroleum not previously considered as economic

(a) Crude oil of qualities now non-commercial

Large accumulations of "non-commercial" heavy, viscous, and generally sulphurous crudes have been found in many parts of the world, as for example in the Qayarah area of Iraq, in Turkey, in Iran, and in parts of South America; some of these may become commercially acceptable as the result of future changes in technical and economic factors.*

In the United States, for example, up to 150 B brl of oils with API gravities of 25° or less are believed[9] to exist in more than 2,000 separate reservoir beds—many of which, however, are only a few feet thick. It has been estimated[10] that 5·2 B brl of these heavy oils could be recovered today by conventional methods, while a further 45·9 B brl would be recoverable by tertiary thermal techniques, given sufficient economic incentive. The balance of nearly 100 B brl which is at present classed as non-recoverable, may become worth development in the future as recovery techniques improve, provided there is a market for these heavy oils.

In Western Canada a number of large, shallow, viscous oil accumulations occur in a wide belt running from the Peace River Arch in Alberta southeastwards into Saskatchewan. These also await future development[11] when the qualities of these crudes becomes attractive.

(b) "Non-conventional" crude oil from "tar sands"

In North America it has been traditional to refer to sandstones impregnated with residual oils which are exposed near the surface of the ground as "tar sands", although a more accurate description would be "heavy oil sands". The largest occurrence of such "tar sands" known anywhere in the world occurs near Fort McMurray on the Athabaska River in Alberta, where an area of some 13,000 sq miles is underlain

---

* Of course, it should also be remembered that the development of more stringent anti-pollution legislation may tend to make some crude oils which are currently commercially acceptable eventually non-commercial, if the cost of extracting their contents of sulphur and other undesirable compounds is found to be excessive in the economic context.

9 US Bur. Mines Circ. Ic 8263, 1965.
10 S. Farouk, Proc. Penn. Oil Prod. June 1968, 32, 6, 10.
11 M. Westall, World Oil, May 1966, 119.

by sandstones impregnated with heavy oils. The sandstones here are of Early Cretaceous age and of mainly deltaic facies, resting unconformably on the tops of eroded Devonian beds. From east to west, they form three distinct stratigraphic units—Wabiskaw-McMurray, Bluesky-Gething and Grand Rapids. The impregnated sands are partially exposed along the river banks in the Mildred Lake area, but most of the deposits, which average about 150ft in thickness, are covered by an overburden of glacial drift and surface deposits which are often several hundred feet thick.

The tarry oil fills the pore spaces of the rocks, and forms a film around each grain, so that when it is removed by solution the residual sand grains are completely uncemented. Presumably these sands are the remains of deltas deposited by rivers which flowed northeastwards from the pre-Cambrian "Shield" into an elongated lagoon or lake, part of which may have been formed in a depression resulting from extensive evaporite solution.

A number of schemes have been considered for recovering some part of the enormous reserves of heavy oils that impregnate the Athabaska "tar sands". However, since about 75% of these sands lie under more than 250ft of non-productive overburden, open-cast mining is never likely to provide an economic process of recovery in view of the transport and handling problems involved. Even after the inert overburden is removed, at least one cubic yard of sand has to be removed for treatment for every barrel of oil to be recovered. Various *in situ* processes are more promising, including a technique of injecting a mixture of steam and sodium hydroxide into the sands through specially drilled shallow wells. In any case, the raw oil obtained must be mixed with lighter oil to make it fluid enough to be pumped off for full refining treatment.

The first plant designed to produce oil commercially from the Athabaska "tar sands" was put on stream in 1967, 260 miles north of Edmonton. It had an initial output of 45,000 b/d of oil, using a basically simple process in which the oil is washed from the sand grains by hot water, the alkalinity of the resulting mixture being adjusted until the oil floats to the surface, to be collected and treated by delayed coking and hydrogenation refining operations. Most of the high sulphur content of the oil is removed in the process and can be recovered and sold. The end-product is a "synthetic" or (more correctly) a "non-conventional" crude oil of 40° API gravity, with a low sulphur content, which yields on refining a higher-than-average proportion of light products.

The relationship between oil-in-place, recoverable oil and thickness of overburden for the Wabiskaw-McMurray sand deposit, which is by far the most important of these groups, is shown in Table 101, from which it will be seen that the Athabaska deposits may contain as much as 267 B brl of extractable "non-conventional" oil, a volume equivalent to nearly half the total proved "conventional" crude reserves of the whole world at the end of 1971. Some proportion of this huge quantity of oil will undoubtedly be recovered in the future as crude oil prices rise and the United States' requirement for liquid fuels increases. It is difficult to estimate, however, the likely time-scale of development. One prediction is that up to 1 MM b/d of "non-conventional" oils will be coming from the Canadian "tar sands" by 1985, and perhaps as much as 3·5 MM b/d by the end of the century—but such outputs would certainly require the investment of very large capital sums for 10-20 large-scale treatment plants. The amortization of this capital would inevitably add an element of costs to each barrel of the resultant oil that might render it uneconomic unless the operation was subsidised in some way (e.g. by tax concessions).

Another large "tar sand" deposit has been discovered in the Canadian Arctic on Melville Island, in Triassic sandstones. In the United States, the Bureau of Mines has estimated that 2·5-5·5 B brl of oil may eventually be recovered from various shallow deposits which are known in Utah, Kentucky and California. In East Venezuela, an extensive "tar belt" covers an area of some 9,000 sq miles in the Oficina-Temblador area, where there is a thickness of up to 100ft of oil-impregnated sandstone of Oligocene age, which is occasionally exposed but generally covered with overburden. It is believed that this deposit contains some 200 B brl of heavy oils, from which 10-50 B brl might one day be recoverable.

None of the world's other known "tar sands" are of the same order of magnitude as those of North America and Venezuela, nor are there any current prospects of their development. The largest deposit outside the Americas is at Bemolanga in Malagasy, where there is an area of 150 sq miles of Triassic "tar sands" which contains about 1·75 B brl of heavy oil; while smaller deposits are known in Albania (Selenizza), Romania (Derna), and in the USSR (at Cheildag, near Baku).

TABLE 101

ATHABASKA "TAR SANDS"

Saturation (oil % by wt.)	Reserves in place B brl	Recoverable "synthetic" crude oil (B brl)
Rich sands (>10%)	440·8	188·3
Intermediate sands (5–10%)	145·7	61·9
Lean sands (2–5%)	39·4	16·7
Total >2%	625·9	266·9

Source: P. Phizackerly & L. Scott, 7 WPC Mexico, 1967, PN 13(1).

4. The Use of Other Oils
(a) Shale Oil

Deposits of oil-shale are known on every continent and of many geological ages. They appear to have been formed from unicellular, gelatinous algae and resistant spores of terrestrial plants deposited in shallow seas, large lakes or coal-producing swamps. They are often associated with evidence of desiccation and of a very shallow-water environment of origin.

Oil-shales usually have a satin lustre and can vary in colour from reddish-brown to black; but even the richest in organic matter are not oily to the touch. The true oil-shales have been defined as "asphaltic pyrobituminous shales", because on being heated in a closed retort to $300°$-$400°C$ their organic content depolymerizes and produces a distillate which resembles in some respects a crude oil. The amount of oil produced varies between 10 and 150 gallons per ton, equivalent to between 4% and 50% of the weight of the rock retorted, and the calorific values of oil-shales range in consequence between 1,600 and 4,900 Btu/lb.

Shale oil (still sometimes called "shale tar") varies in its chemical constitution both with the temperature of distillation and the origin of the oil-shale processed. Thus, a single sample can yield oils containing proportions of 10-36% saturates, 36-46% olefins and 28-54% aromatics. Raising the temperature of pyrolysis increases the proportion of aromatics, so that a temperature of 600°C results in the production of aromatics only. In general, however, most shale oils contain aliphatic hydrocarbons, olefins, and dienes, together with various sulphur, oxygen and nitrogen compounds (cresols, phenols, etc.). Shale oil, like petroleum, is optically active; it is a dark-coloured liquid, with a greenish fluorescence, which can be refined to yield products varying from cleaning spirit to paraffin wax, including one type of motor fuel. Improved yield and product quality result from carrying out the distillation of the shale in a stream of hydrogen gas.

The by-products produced are ammonium sulphate, a range of organic dibasic acids, and residual "asphalt" or coke. Since the ash content of the shales is rarely less than 60% by weight, huge quantities rapidly accumulate and must be disposed of. Attempts have been made to use this ash as road material or building blocks. In Sweden, uranium has been commercially extracted from oil-shale ash, and the Ljunstrom in-situ process avoids the problems of transport and disposal of solid material but depends on the availability of abundant cheap electrical power.

Late in the seventeenth century, sufficient oil was produced by distilling oil-shale in the Modena district of Italy to provide the street lighting in several towns, while a British Patent was granted in 1694 to one Martin Eale who . . . "hath certainley found out a way to extract and make great quantities of pitch, tarr and oyle out of a sort of rock".

Although shale oil was produced in France as long ago as 1838, the Pumpherston area of Scotland was really the home of the oil-shale industry from 1848 until it became no longer economic after 1962. (The shales here occur in the upper part of the Calciferous Sandstone series, underlying the Carboniferous Limestone.) Shale oil production began in Germany in 1884, in Esthonia in 1921, in Spain in 1922, in Manchuria in 1929, and in Sweden in 1941. With the increasing availability of cheaper crude oil, shale oil became economically unattractive, so that operations in many areas were restricted or even terminated. The Esthonian industry has probably been the most economically successful, utilising the direct furnace combustion of pulverised shale as fuel for the Narva thermal electric plant, and also producing synthesis gas for Leningrad.

It has been estimated that in recent years the total world production of shale oil has been about 350 MM brl/a—of which China has produced perhaps 80% from the Fushun area of Manchuria, where the oil-shales directly overlie deposits of bituminous coal, thus providing unique conditions for economic shallow-mining operations. It is reported[12], however, that the adoption of improved techniques has resulted in the USSR shale oil output increasing to about 150 MM brl/a in recent years.

In the United States, oil-shale extraction is still only a semi-commercial process, although immense deposits are available in the Eocene Green River beds which underlie 8,000 sq miles of Colorado, Wyoming and Utah. The relative slowness of the progress that has been made in the establishment of an oil-shale industry in the United States is largely due to the taxation position and the relatively low depletion allowance it receives. (However, it was announced in 1973 that a consortium of 15 companies is to spend $7·5 MM in the next few years to develop the technology of oil-shale potentialities of the USA.)

Technical developments may improve the economics of the extraction processes, and experiments are being carried out ("Project Broncho") to test the feasibility of large-scale release of oil from shale by a subsurface nuclear explosion. A 50-kiloton atomic charge detonated at a depth of 5,000ft in a suitable oil-shale formation would, it is estimated, produce a rubble-filled "chimney" some 500ft tall, and from the debris as much as 480,000 brl of oil might then be distilled by driving a zone of controlled burning down through the rubble. The hot gases would leach oil from the fragmented shale, which would drain to the bottom of the chimney, whence it could be pumped to the surface.

In Brazil, a prototype plant was constructed in 1971 at São Mateus do Sul in Parana which has been designed to process 2,200 tons/day of Permian Irati oil-shales, from which 1,000 brl of oil, 17 tons of sulphur,

---

12 T. Volkov *et al.*, paper to Int. Gas. Conf., Moscow, 1970.

and 1·3 MM cf/d of fuel gas will be obtained. If its operations prove to be technically and economically viable, a much larger plant may be built to handle perhaps as much as 108,000 tons/day of the oil-shales of this province[13].

TABLE 102
PRODUCTIVITY OF OIL-SHALES ON DISTILLATION

Locality or Type	Yield of Shale Oil gallons/ton
Lothian Scotland	10–55
Kimmeridge (England)	10–40
Arcadian (Nova Scotia)	4–23
Eastern USA	4–45
Utah	6–10
Colorado and Wyoming	10–68
Kukkersite (Esthonia)	70–80

Because of the uncertainties that surround the technology and economics of the extraction processes, no reliable statistics are available about the amounts of oil that might realistically be obtained from the world's oil-shale deposits. In any case, the "reserves" of shale oil which are sometimes quoted are in quite a different category from any type of crude oil reserves; thus, *no oil at all* is recovered from the shale until it is retorted, but if completely retorted then all its oil is released. Hence, "reserves" only refer to the oil content of a particular shale deposit which *could* be released if all the shale were suitably treated.

On this basis of definition, it has been estimated[14] that there is the equivalent of some 2,500 B tons of oil-in-place in known accessible oil-shale deposits which are capable of yielding 10-40% of their weight of oil by distillation. In addition[15], there are much larger amounts—perhaps as much as 47,500 B tons—of oil tied up in shales which yield less than 10% by weight of oil but which could be used directly as pulverised furnace fuel for energy production.

No realistic estimates can yet be made of the volumes of oil likely to be derived in future by distillation of part at least of these enormous deposits. Up to the present time, only those countries which are poor in natural oil resources or unwilling or unable to import sufficient crude oil for their needs (e.g. China) have in general used shale-derived oil to provide a significant proportion of their fuel consumption.

Whether and when oil from shale will re-enter the world energy picture is speculative, since although numerous pilot-plant studies have been made, the cost of obtaining oil from shale as compared with the cost of conventional crude oil is still so high that it is unlikely to become an attractive large-scale source of fuel in the foreseeable future.

[13] C. Bruni et al., 8 WPC Moscow, PD 10) 2.
[14] D. Duncan, 7 WPC Mexico, PD (14) 1; D. Duncan and V. Swanson, USGS Circular 523, 1965.
[15] U.N. Dept. Econ. Soc. Affairs, "Utilization of Oil Shale", 1967.

(b) Oil from Coal

It has been known for years that a type of "synthetic" oil can be manufactured by the catalytic hydrogenation of coal under high pressures. This process has only so far been considered to be economically viable in special situations involving crude oil shortages. Thus, the German synthetic liquid fuels industry was successful in supplementing limited supplies of crude oil during the 1939-45 war, and more recently the Sasol plant near Johannesburg in South Africa has been successful in converting 14,000 tons/day of locally mined coal into liquid fuel at the rate of 66 gallons/ton.

Although the chemistry of converting coal into a liquid oil is basically simple enough, consisting essentially of increasing the proportion of hydrogen, the technology is more complex, involving as it does the removal of unwanted compounds of oxygen, sulphur and nitrogen, and requiring an economic supply of hydrogen as well as suitable catalysts, such as zinc chloride.

In the United States, the Office of Coal Research has sponsored several pilot-plant projects using different techniques for the production of coal-derived liquids capable of being "cracked" by present-day refining methods to produce acceptable gasolines[16]. However, because of the costs of handling and transport, oil from coal is not likely to compete with crude oil as long as the latter is available in abundance; however, when crude oil reserves tend towards exhaustion, improvements in the chemistry and technology of conversion may make these liquids of considerable importance, particularly in view of the immense reserves of coal (Table 103) that are likely to remain available long after the world's crude oil resources are finally exhausted.

TABLE 103

WORLD SOLID FUEL RESOURCES[1]
(B tons)

Fuel	Total Reserves[2]	Measured Reserves[3]	Economically Recoverable Reserves	Number of years' current consumption
Bituminous Coal	6,700	460	230	100
Brown Coal and Lignite	2,100	270	135	300
Peat	180?[4]	—	—	—

[1] World Power Conference "Survey of Energy Resources", 1968.
[2] Total reserves are made up of indicated and inferred reserves plus measured reserves.
[3] Measured reserves include seams not less than 30 cm thick, situated not more than 1,200 m below the surface in the case of bituminous coals and not more than 500 m in the case of brown coals and lignites.
[4] Little data is available about peat reserves. This figure is therefore probably a gross understatement.

[16] N. Cochran & G. Staber, J. Petrol. Technol., October 1967, 1345.

The production from coal (and from heavy oils) of high calorific value gas—"synthetic natural gas" or SNG—is also likely to provide an increasingly important source of fuel in the future.

It is possible also to produce "oily" liquids and hydrocarbon gases from the pyrolysis of other materials—for example municipal waste matter—usually by hydrogasification processes. However, such processes would obviously tend to be relatively small-scale and therefore expensive.

CONCLUSION

THIS work was compiled at a time when the world's petroleum industry was entering a new phase in its history as a result of the political and economic developments of the early 1970's. Therefore, although its main purpose is to present a readable synthesis of information about the geographical and geological occurrences of the world's major oil accumulations, some mention must be made here of other than technical matters.

Since the geology of petroleum bears no relationship to the accidents of political frontiers, the extent to which individual states are endowed with oil resources varies greatly. However, since the consumption of petroleum among the industrialised nations of the world has paralleled the growth of their Gross National Products, they have traditionally made up for the deficiencies of their domestic production by building up external sources of supply from less developed countries with large oil resources but small energy requirements. Of course, a number of complex factors affect the relationships between the oil-importing and oil-exporting countries, and all barrels of crude oil are by no means of the same value. Table 104, however, summarises the imbalances which existed between crude oil production and crude oil consumption in different parts of the world in 1971, and shows in particular how Western Europe, North America and Japan made up their oil deficiencies by imports from the Middle East, Africa and South America.

It will be noted that the position of Western Europe is particularly vulnerable, since, with only 10·3% of the world's population, it consumes 27%, but produces less than 1% of the total world oil output. It must therefore import more than 12·7 MM b/d of crude oil for its requirements. (Even when North Sea oil is produced at the forecast rate of 2 MM b/d in 1980, the need for imports will probably be no less, in view of the rapid rate of increase likely in consumption.) The United States, for long an oil-surplus area, but now experiencing the first intimations of incipient shortages, and Japan with negligible domestic oil production but a very rapidly growing consumption, are clearly also likely to become increasingly dependent on oil imports in the years ahead (Table 105).

The adverse economic consequences for these countries of this growing requirement for imported petroleum will have economic consequences far beyond their own frontiers. For example, the US deficit on trade in fuels will probably grow from the 1971 rate of $3 B p.a. to perhaps $25 B p.a. by the early 1980's, when its total oil imports may have increased to 12 MM b/d, or half its oil needs at that time. To support this huge deficit, the USA will require to increase its exports of goods and services to world markets by at least 40%. This will profoundly affect the pattern of world trade at a time when Europe will also be competing with the USA (and Japan) for oil imports to make up its own greatly increased requirements.

Most of these oil imports will inevitably have to come from the OPEC states[17] and will have to be paid for at rates which are even now rapidly rising. (For example, as a result of the 1971-72 negotiations between the producing states and the major international oil companies, the posted price of a typical Arabian light crude rose from $1·80 to $2·59 per barrel, not allowing for further increases following dollar devaluation; by 1975, allowing for the high "buy-back" prices written into the new "participation" agreements, North African and Middle East crude oils may be costing $4-$5 per barrel.)

As a consequence, it has been calculated that the OPEC states, most of which have very small populations (Table 107), will be receiving oil revenue of some $25 B p.a. by 1975, and possibly as much as $56 B p.a. by 1980[18].

Although there are thus obvious incentives for discovering and bringing into production new and alternative sources of future supply, it must be remembered that the oil industry is extremely capital-intensive, and very considerable sums have to be invested in exploration, production, transportation, refining, and distribution facilities before any financial return can be achieved. Thus, one recent estimate[19] is that about one *trillion* dollars of capital will have to be invested in the years between 1970 and 1985 if the oil requirements of the non-Communist world and its reasonable future reserves are to be made available during this period. Unfortunately, as the search for new accumulations moves into remoter areas, and more particularly into deeper waters, the expenditure required to establish new production inevitably increases. Thus, the cost of drilling a single exploratory offshore well in the North Sea may exceed £2 MM, and the expenditure needed to develop a major new discovery, such as the *Forties* field, is estimated to be as much as £400 MM.

---

[17] Organization of Oil Producing and Exporting Countries:—
Abu Dhabi, Algeria, Indonesia, Iran, Iraq, Kuwait, Libya, Nigeria, Qatar, Saudi Arabia, Venezuela. Of these, Algeria, Iraq and Libya have nationalised part or all of their oil industries.
[18] M. Adelman, *Foreign Policy*, 9, 1972/3.
[19] Chase Manhattan Bank, New York "Annual Financial Analysis of a Group of Petroleum Companies, 1971".

It seems clear therefore, that, the problem that the developed countries of the world now have to face is a not so much that of a forthcoming oil shortage, but rather the fact that the era of cheap oil on which our modern civilisation has largely been built has now come to an end.

TABLE 104

CRUDE OIL CONSUMPTION AND PRODUCTION, 1971

	Consumption[1]		Production[2]		Surplus/ deficit
	1000 b/d	% of total	1000 b/d	% of total	1000 b/d
North America*	16,910	34·7	13,325	25·6	−3,585
South America and West Indies	2,470	5·1	4,700	9·8	+2,230
Western Europe	13,180	27·0	440	0·9	−12,740
Middle East	1,040	2·1	16,165	32·4	+15,125
Africa	920	1·9	5,800	11·3	+4,880
South and Southeast Asia	1,840	3·8	{ 1,130	{ 2·3	{ −5,150
Japan	4,440	9·1			
Australasia	620	1·2	495	1·0	−125
USSR, East Europe and China	7,350	15·1	8,315	16·7	+965
	48,770	100·0	50,370	100·0	+1,600†

Sources of data: 1. BP Statistical Review, 1971.
                 2. World Oil, Aug, 15, 1972.

* *Includes natural gas liquids.*
† *Surplus overall due to internal stock movements and military uses.*

In general, world production of crude oil is about the same as world consumption in any year, since it is clearly cheaper to leave oil in the subsurface until required rather than to produce and store it at the surface. However, seasonal fluctuations must be accommodated, while some governments also insist on "stockpiles" of strategic fuels equivalent to several weeks' or months' normal consumption.

TABLE 105

DEPENDENCE ON IMPORTED ENERGY
% of total

	1960	1970
United Kingdom	25	45
EEC ("Six")	30	63
USA	6	10
Japan	44	84
Total of these countries	14	32

Source: "Coal and Energy Policy in Europe", 1972.

TABLE 106

SIGNIFICANT RESERVES/PRODUCTION RATIOS, 1971

	Years
Japan	4·5
USA	10·9
Western Europe *not including North Sea discoveries*	16·0
USSR	22·2
Middle East States	58·5

TABLE 107

SOME ESTIMATED OIL REVENUES AND POPULATIONS, 1980

	Revenue in 1980 in $MM	Pop. in M
Saudi Arabia	10,000	5,000
Iran	8,000	32,000
Libya	3,500	2,000
Kuwait	2,200	750
Iraq	1,500	10,000
Abu Dhabi	1,500	250

Source: "The Times", London, Feb. 2, 1973.

TABLE 108

COMPARISON OF OIL PRODUCED PER WELL IN DIFFERENT PARTS OF THE WORLD

	Approximate number of wells[1] × 1000 at end-1971	Average production b/d per well in 1971
North America[2]	533·1	21
South America and Trinidad	22·4	206
Western Europe	6·0	49
Eastern Europe	14·8	29
USSR	63·0	117
North Africa	2·1	1,874
Rest of Africa	0·8	2,313
Middle East[3]	3·0	5,512
East Asia	5·9	292
Australasia	0·4	778
Total	651·5	74
Total outside USA	139·1	279

Notes: [1] Flowing and artificial lift.
[2] Precise total for USA, Canada and Mexico was 533,072 wells, of which 96%, i.e. 512,471 were in the USA.
[3] Precise total: 3038

Source of data: World Oil, Aug. 15 1972, p.58, and author's estimates.

The shortages of the foreseeable future may therefore be essentially financial shortages, and the developed countries will have to find some means of controlling the inflationary effects on their economies of continually-increasing crude oil prices. Furthermore, the continuing availability of their oil supplies may not be as certain as the reserves/production ratios of some of the exporting states (Table 106) would seem to indicate. A time may come when the exporting countries may decide to retain a higher proportion of their petroleum in the subsurface for future use, rather than continue to convert increasing volumes of oil into surplus—even embarrassing—sums of money. Perhaps the first evidence of such a trend has already appeared: Libya decreed a cutback of output in 1972, and Kuwait has announced the stabilisation of production in future years at no higher than the 1972 rate[20]. It was,

[20] The Kuwait Minister of Finance and Oil, Mr. Abdel Raham al Attiqi, quoted in the "Financial Times", February 15th, 1973: "Why should we produce all this oil and then invest it in currencies that depreciate, when it will grow in value under the ground?".

after all, a Middle Eastern poet who pondered (according to Fitzgerald's translation)[21]:

> *"I often wonder what the Vintners buy*
> *One-half so precious as the ware they sell . . ."*

It is not yet possible to discern at what point it will become economically attractive to develop alternatives to crude oil in the form of oils derived from "tar sands", oil shales, or the hydrogenation of coal. It has been accepted, up to the present time, that any such "non-conventional" or "synthetic" oils would be many times more expensive than crude oils. How this balance of economy, availability and convenience will shift in the years to come is a fascinating subject for speculation. Certainly, the time cannot be far distant when some of the more profligate uses of the irreplaceable raw material which we call crude oil might well be curtailed, or replaced by other sources of energy. Perhaps indeed we should already be saying: "Oil is scarce—use it wisely".*

---

[21] Omar Khayyam, *Rubaiyat*, c.iii.

* *With acknowledgment to Mr. Thornton Bradshaw, President of Atlantic Richfield Co., in a speech in Los Angeles, November, 1972.*

TABLE 109 TIME-STRATIGRAPHICAL UNITS—EUROPE 365

ERA	SYSTEM	SUB-SYSTEM	SERIES	STAGE	SUB-STAGE
CENOZOIC	QUATERNARY		Holocene		
			Pleistocene	Calabrian	
	TERTIARY	Neogene	Pliocene	Piacenzian	Levantinian
					Dacian
				Pannonian	Pontian
					Meotian
			Miocene	Sarmatian	
				Vindobonian	Tortonian
					Helvetian
				Burdigalian	
				Aquitanian[3]	
		Paleogene	Oligocene	Chattian	
				Rupelian	
				Lattorfian (Sannoisian, Tongrian)	
			Eocene	Priabonian	Bartonian
					Ledian (Auversian)
				Lutetian	Bruxellian
				Ypresian	
			Paleocene	Landenian	Sparnacian
					Thanetian
				Montian	
				Danian[3]	
MESOZOIC	CRETACEOUS[2]	Upper	Senonian	Maestrichtian (Dordonian)	
				Campanian	
				Santonian	
				Coniacian	
			Turonian	Angoumian	
				Ligerian	
			Cenomanian	Vraconian	
		Lower	Albian		
			Aptian	Gargasian	
				Bedoulian	
			Barremian[4]		
			Hauterivian		
			Valanginian	Berriasian	
	JURASSIC	Upper (Malm)	Portlandian[5] (Tithonian)	Purbeckian	
				Bononian	
			Kimmeridg-ian	Virgulian	
				Pterocerian	
			Lusitanian (Corallian)	Sequanian (Astartian)	
				Argovian (Rauracian)	
			Oxfordian		
		Middle (Dogger)	Callovian		
			Bathonian		
			Bajocian		
			Aalenian		
		Lower (Lias)	Toarcian		
			Charmouth-ian	Domerian	
				Pliensbachian	
			Sinemurian		
			Hettangian		
	TRIASSIC	Upper	Rhaetian		
			Norian	Keuper	
			Carnian		
		Middle	Ladinian	Muschelkalk	
			Anisian		
		Lower	Skythian	Bunter	

ERA	SYSTEM	SUB-SYSTEM	SERIES	STAGE	SUB-STAGE
PALAEOZOIC	**PERMIAN**	Upper	Kazanian	Thuringian (Zechstein)	
		Middle	Kungurian	Saxonian (U. Rotliegendes)	
		Lower	Artinskian	Autunian (L. Rotliegendes)	
			Sakmarian		
	CARBONIFEROUS²	Upper	Stephanian (Uralian)		
			Westphalian (Moscovian)		
			Namurian		
		Lower	Dinantian	Visean	
			(Avonian)	Tournaisian	
	DEVONIAN	Upper	Famennian		
			Frasnian		
		Middle	Givetian		
			Couvinian (Eifelian)		
		Lower	Emsian		
			Siegenian		
			Gedinnian		
	SILURIAN	Gothlandian⁶	Ludlovian		
			Wenlockian		
			Llandoverian		
		Ordovician	Ashgillian		
			Caradocian		
			Llandillian		
			Llanvirnian		
			Arenigian		
			Tremadocian		
	CAMBRIAN	Upper			
		Middle			
		Lower			

Notes

1. Since the stratigraphy of Europe was the first to be worked out in detail, the names of the Systems and Series given here are generally used to describe strata of equivalent ages in other parts of the world. However, local names are often applied to the smaller stratigraphic sub-divisions (Stages and Sub-stages).

2. In the United States, the Cretaceous is divided into Upper (Gulfian) and Lower (Comanchean). The Carboniferous is divided into Upper (Pennsylvanian) and Lower (Comanchean).

3. Names given in brackets are synonyms. The Table represents a reasonable working synthesis, but there are a number of controversial questions unresolved—e.g. some stratigraphers place the Aquitanian in the Oligocene, the Danian in the Upper Cretaceous, etc.

4. "Neocomian" is generally taken to include the Barremian-Berriasian sequence of the Lower Cretaceous, but is imprecise.

5. "Wealden" is a rock Unit of Lower Cretaceous to Upper Jurassic age.

6. In Britain and North America, Gothlandian is a synonym of Silurian, and Ordovician is ranked as a System immediately below Silurian.

GENERAL INDEX

(Names of selected oilfields in italics)
Bold figures indicate first pages of major sections dealing with subjects.

LIST OF ILLUSTRATIONS